DATE   D0861612

RICHES TO RUST

A Guide to Mining in the Old West

By Eric Twitty

WESTERN REFLECTIONS PUBLISHING COMPANY®

Montrose, CO

First Edition
Printed in the United States of America

Library of Congress Catalog Card Number 2002101075

ISBN 1-890437-60-3

Cover photo: Nellie Mine, Telluride, CO by Laurie Goralka
Cover and text design by Laurie Goralka Design

Western Reflections Publishing Company®
P.O. Box 1647
Montrose, CO 81402
www.westernreflectionspub.com

Acknowledgments

This work would not have been possible were it not for the invaluable assistance of a handful of generous individuals. I wish to thank the notable historians Jay Fell, Ph.D, Tom Noel, Ph.D, Mark Foster, Ph.D, and Robert Spude, Ph.D for taking the time to review this work. I am grateful to J. Scott Altenbach, Ph.D for taking the time to guide me during my field research in New Mexico and for discussing his experiences building a headframe. I also offer thanks to consulting geologist Lane Griffin for providing guidance to important historic mine sites in the Goldfield, Echo, Schurz, and Garfield districts in Nevada. Last, I am grateful to Jim Watson, mayor of Victor, Colorado, and owner of the famous Strong Mine.

Several private and governmental entities proved to be absolutely crucial in facilitating my field research. I am deeply indebted to John Hardaway and the Cripple Creek & Victor Gold Mining Company for granting me access to numerous historic mine sites on company land in the famous Cripple Creek District. The extensive research I was able to conduct courtesy of Cripple Creek & Victor provided crucial. Matt Zietlow and the Homestake Mining Company's Ruby Hill Mine displayed similar hospitality and granted me access to several important mine sites in Eureka, Nevada which provided formative data for this work. I also wish to thank Homer Milford at New Mexico's Abandoned Mine Lands program for granting access to historic mine sites in the Lake Valley Mining District. Last, I am grateful to Paragon Archaeological Consultants, Incorporated in Denver for permitting the use of materials pertaining to the company's work in Cripple Creek.

I wholeheartedly dedicate this work to Lisa, who gave support during the years I conducted research, acquired data, and produced this work. Thank you for your patience, time, and sacrifice.

CONTENTS

CHAPTER 1

INTRODUCTION

Between the 1870s and 1930s precious metals mines in the West seemed to spring up from every hill, meadow, and mountain where prospectors searched for wealth. The same story is told, time and again, about how grizzled gold-seekers made lucky finds, staked their claims, and began producing gold or silver bullion. Missing from the popular accounts of mining history are the technical aspects of how a claim evolved from raw earth into a metals-producing mine. From the moment prospectors first pierced the soil with their picks, their claims underwent stages of improvement and upgrades, including both the underground workings and the support facilities located around the mine entrance. From the 1870s when our story begins, men involved in the mining industry, from financiers to miners, recognized those facilities at the mine collectively as the *surface plant*. The function of this fundamental facet of every hardrock mine was to administer to the needs of driving underground workings in a concert of men and machinery. By the 1930s, where our story ends, technology changed, but the basic role and form of surface plants remained the same.

The West hosted a hardrock mining industry on a scale greater than anything seen in the world up to that time. Mines abounded in districts that were scattered across the mountains and deserts, from British Columbia in the north to the Sierra Madre in Mexico to the south, and from the Coast Range overlooking the Pacific Ocean to the Rocky Mountains looming over the high plains. Hardrock mining began booming in the West immediately following the discovery of placer gold during the mid-nineteenth century, it reached a crescendo around the turn-of-the-century, and declined abruptly after World War I. Today, the gold and silver is largely gone, and many of the mining districts are empty, but the wealth seekers left behind a legacy of historic townsites and mines which modern culture celebrates as the physical remnants of this fascinating bygone era.

The ravages of time and human activities have taken a heavy toll on the West's historic hardrock mining sites. Today most buildings and other structures are gone, the machinery and equipment have been removed, and the once-bustling areas are overgrown. To the untrained eye, little apparently has been left of the mines themselves except for dark and dangerous holes in the ground and blasted piles of rock. Yet, in actuality, much remains of the long-forgotten mining operations and associated townsites. The evidence is there; it is merely subtle. Many historic mine sites still feature remains in the form of artifacts, structures, foundations, and topographic features. These material remains await examination by today's mining history researchers. This book seeks to provide the tools to examine a given hardrock mine site and use remaining material evidence to reconstruct what happened, and when. While the material presented here focuses on hardrock mining in the West, by the late nineteenth century mining technology had become fairly standardized throughout North America, and the reader may extrapolate the ideas presented here to hardrock mining regions elsewhere.

Hardrock mine sites in the West hold much importance to the field of American history. They constitute an untapped cultural resource that offers unique information regarding the rise of Gilded Age America and the Industrial Revolution in the far West. Hardrock mines drew skilled and unskilled laborers, capitalists, industrialists, and politicians who created cities and towns which boomed, flickered, and died. While historians and archaeologists traditionally focused their attention on the townsites and settlements, a survey of the surrounding landscape will often reveal an abundance of historic mines that speak of the very operations that supported life in the West.

These understudied and important cultural resources constitute the remains of an industrial work environment foreign to today's society. Miners spent their days drilling and blasting in dank underground workings while firemen, engineers, trammers, and pipemen attended to a variety of whirling, hissing, and clanging machines around the mine entrance. The work amid the surface facilities was difficult and dirty, the physical conditions ranged from intense heat, experienced while feeding fuel to boilers during the summer, to the arctic climate endured by workers outside of the buildings on windy winter days. The work amid the facilities that comprised the mine plant presented dangers that required vigilance of the work crew. Each machine presented unguarded limb-wrenching hazards. Dynamite was often thawed at the blacksmith's forge, miners rode open-topped ore buckets with no safety equipment down shafts, and men pumped forge bellows in feeble attempts to ventilate smoke-filled underground workings.

Historic mine sites also represent heavy industrial engineering adapted to the unique arena of the American wilderness. When looking over the remains of western hardrock mines, today's observer cannot help but marvel at how miners and engineers hewed mines out of raw mountainsides. The visitor cannot help but question who financed such operations, and how miners and engineers were able to carry out necessary work.

Research for this book proved interesting and complex. A strategy was implemented with the purpose of discovering how academic mining engineers recommended mines be constructed and organized from the Gilded Age through the Depression, how mines were actually developed, and what telltale remains visitors are likely to encounter today. In the first phase of work, models of surface plant technologies were assembled from primary sources such as mining engineers' texts, articles in trade journals, and trade catalogs. Then, the archaeological remains of approximately three-hundred mine complexes in over fifty historic mining districts in California, Colorado, Montana, Nevada, New Mexico, and Utah were analyzed. The extant remains were compared with mine-specific records to synthesize accurate histories of the operations. Last, the field findings were contrasted against the technologies and methods discussed by mining engineers in literature. The models constructed from literary materials and mine-specific documentation helped interpret the material culture encountered in the field. The material culture, in turn, demonstrated how mining engineers actually applied technology to solve the problems of mining. Employing an interdisciplinary research strategy helped determined how academic mining engineers, with access to sufficient capital, recommended mines be developed and equipped, and how mining companies and seasoned field engineers, working within the realities of scarce financing and a harsh western environment, actually established mines. Ultimately, this research strategy revealed patterns and relationships of capital, investor confidence, geographic location, time period, and geology, and how these factors influenced the building of a mine.

CHAPTER 2

BUILDING THE SURFACE PLANT

The Men of Mining

Mining in the West would have remained in a primitive labor-intensive state, and possibly suffered an early demise had it not been for the interaction of five central groups of people. *Prospectors* played an important role because they found the ore deposits and subsequently organized the mining districts where activity took place. Because they lacked capital, most prospectors were incapable of developing to significant depth the lodes they discovered, and as a result, they often sold their holdings to *promoters* or *investors*. In some cases a promoter or investor was the same individual, who played different roles in the opening and development of a mine. Promoters and investors relied on a fourth major group, *mining engineers*, who possessed the technical expertise necessary to develop a prospect into a profitable operation. Engineers relied on *miners* and *laborers*, the fifth important group of people, to carry out the grueling physical work of building and working the mine.

These five human keystones of mining contributed the elements necessary to develop a claim into a major operation. All of the groups shared a symbiotic relationship, one needing, but not necessarily liking, the other. In reality, their relationships were not as neat and well-defined as noted above, and often their capacities overlapped. Engineers also acted as promoters or miners. Prospectors labored for wages as miners until their coffers had been replenished for another search for wealth. Investors also acted as promoters. However, all five groups were necessary for the discovery, financing, and opening of mines in the West.

The prospector was the first individual to participate in the opening and development of a mine. After he struck an ore body and had samples of payrock assayed, he decided whether to sell the claim or establish his own mining company. Generally, when assay reports showed the ore to be of low to moderate value, prospectors attempted to sell their holdings because immense quantities of capital were required to profitably extract low-grade ores. Some prospectors formed small mining outfits when they knew their claim had high-grade ore, because such deposits were profitable to mine on a small scale. To fulfill either decision, the prospector was forced to seek out an investor who would purchase the property, or who would supply development capital. [1]

Through honest representation, and occasionally through deceit and chicanery such as claim salting, prospectors interested investors. When a claim proved to be barren, some crooked prospectors *salted* the property by strategically placing rich ore specimens in locations that made the claim seem promising. In general, claim-salting was rarely practiced. Hungry for profit and caught up in the romanticism of western mining, investors eagerly supplied capital to develop promising claims. Investors proved to be one of the underpinnings of western mining, and their money made possible drilling and blasting underground workings, and erecting the necessary surface plants. Joseph King, a noted western historian, accurately summarized this sentiment:

> *"It was the financier, not the romantic old prospector or quick-talking promoter, who ultimately built a hardrock industry in the rugged and remote mountains of Colorado. What the prospector discovered high in the Rockies and the promoter tried to peddle in towns and cities across the country would have certainly remained an undeveloped mineral resource without the millions of dollars invested by Easterners, Middle Westerners, and other distant and diverse capitalists, large and small alike."* [2]

Financing was absolutely necessary for all phases of discovering, opening, and developing mines. Capitalists funded not only the acquisition of supplies and the installation of machinery, but also the railroads and wagon roads that permitted the transportation of the crucial materials of mining. King aptly voiced the key role capital played in the installation of mine plants, and associated ore reduction mills:

> *"In capital lay the means of properly opening a mine, sinking the shaft, timbering it against sudden shifts of dirt and rock, and draining it dry. And capital made possible the hauling, milling, and treating of tons of ore, as well as a multitude of steps and processes required to produce bullion from raw material."* [3]

Between the 1870s and 1910s potential investors for western mines could be found across the United States. Old money and fortunes made from business and manufacturing awaited in the East; capital associated with business and commerce lay in the Midwest, and financiers with profits made in once over in mining were concentrated along the Pacific Coast. Large and small investors alike risked their dollars on western mines. Successful investors educated themselves on the economics of mining in the West, the realities of running a mine, and the properties and districts into which they sank money. In some cases investors felt compelled to travel west and personally inspect their mines.

David Marks Hyman was one financier, among the upper-middle class businessmen and professionals, who discovered that education in western mining was necessary to retain one's initial capital, let alone profit from it. Hyman was born in Germany in 1846, emigrated to the United States in 1864, settled in Chicago, and received a formal education in law and business.[4] After several successful years as a lawyer, Hyman had accumulated a tidy sum of investment capital, $5,000 of which he lent to a trusted business friend, Charles Hallam, who sank it into Colorado mines.

Hallam had no formal experience in the industry, but, by taking the time to become acquainted with Denver's mining financiers and promoters, he learned some of the basics of mining economics and investment, and found out which districts were truly promising. Hallam targeted the Aspen area, which was being opened in the early 1880s, and when he was offered the Smuggler and several other claims, he contacted Hyman about the $5,000 loan. Hyman, in the pattern of many distant investors, "paid the $5,000 and concluded to trust in luck to see what I could do."[5] Hyman grew anxious over Hallam's investments, and he felt compelled to travel into the field to ascertain exactly what his money had bought. Hyman met Hallam and together the two financiers traveled to the burgeoning Aspen area. Unimpressed with the primitive state of the properties, the partners realized that they needed more capital to build a surface plant, develop the ore bodies, and extract pay rock from their claim.

Shortly afterward, Adel D. Breed, who acquired a fortune making coffins and had gained considerable experience with Colorado mining investment at Caribou, located above Boulder, took an interest in the Smuggler. Breed offered Hyman $16,000 cash plus capital to develop the property in exchange for a one-third interest in the mine.[6] Hyman went for the deal, and Breed saw to it that the Smuggler was fully developed. Within a short period of time the mine began repaying its investment, and it ultimately became one of the richest operations in Colorado.

Hyman and Hallam represented a major constituency of mining investors. They came from middle and upper middle-class families which, while not directly providing them with capital, gave the support and empowerment necessary in the social climate of the nineteenth century to become upwardly mobile. Like many financiers in this group, Hyman and Hallam were working professionals earning enough income to allow for some savings, a portion of which they risked as investment capital. While they were certainly above the working-class, Hyman and Hallam were not part of the wealthy class, and in accordance they were capable of parting with only limited quantities of money for mining ventures. Because these people had to earn a living, they did not have much time to review economics and chart exactly which districts were most promising. Some individuals made lucky choices, as did Hallam, while most middle-class financiers sank their money into worthless holes. But the sum total of capital put up by modest financiers made possible the development of the numerous small and medium-sized operations considered by mining magnates to be trivial and not worth a second thought. In this way many prospect operations allocated funding.

Mining magnates and experienced investors tended to dominate the financing of large and highly productive mines. In addition to having access to the requisite pools of money, prominent investors also had the time and resources to become experts in the economics and geography of mining ventures, which gave them great advantage. Wealthy investors were willing to risk amounts of capital much larger than their professional working-class counterparts, and such well-financed capitalists considered small losses to be less critical, even inevitable. Their goal was to make large profits, often from an average of investments. The net result was the financing of the western mining industry, with a large proportion of capital going to promising mines in prominent mining districts.

One such investor, George Graham Rice, literally made a million dollars several times over from mining investment. Born Jacob Simon Herzig, he became involved in finance and business in New York City in the latter portion of the nineteenth century, and in the 1890s he was convicted of a number of white-collar crimes.[7] Ruined, Herzig literally reinvented himself as Rice and traveled west to Goldfield, Nevada, at the height of its boom, with the intent of applying his business acumen to profit from mining. Using capital borrowed from gaming house owner Larry Sullivan, Rice established the L.M. Sullivan Trust Company that used capital invested by individuals to finance the acquisition and development of central Nevada mines.[8] Through the Trust Company, Rice made a huge personal sum of money, both honestly and dishonestly. While many of Rice's ventures proved to be busts and small investors lost out, the portions of capital not directed into Rice's or Sullivan's pockets built many mines. While Rice attempted to invest in sound, productive mines when possible, he also put his money, as well as that of his clients, into small and unproven operations that showed promise.

George Graham Rice was the type of financier who also acted as a promoter, like many heavyweight investors attempting to profitably float their ventures. Mining promoters, usually men of some means, comprised an integral link in the chain of finance that built the mining industry of the West. Men in this profession were in the business of prospecting for investors to finance their mining ventures. To accomplish such a mission, promoters had to be socially outgoing, quick of wit, knowledgeable in mining economics and geography, and able salesmen. The most common means promoters used to obtain capital, either for fraudulent purposes or for claim development, was selling stock in mining companies.

When an honest and legitimate mining company formed, the directors decided how many *shares,* or units of stock, the company should possess and what price each share should fetch, in hopes of selling the majority to obtain capital for development. The promoter would then apply various methods to help management sell the stock. Sophisticated promoters like Rice launched elaborate public relations and advertising campaigns emphasizing the mine's virtues and exaggerating optimistic qualities. After fomenting demand, the owners of the stock, which often included the promoter, released it onto the market where investors hopefully snapped it up. Other promoters operating along simpler and more grounded lines tried physically selling stock certificates to reputed financiers, as well as to professionals, merchants, and government officials.

Many small mining companies made the task of selling stock difficult for themselves, because they often advertised huge volumes of shares for sale at low prices. Experienced investors came to learn that this was a tactic employed to leverage sales of large blocks of stock, and after sustaining losses many financiers no longer took this bait.[9] Shunned, these small mining companies with seemingly huge pools of stock actually became starved for capital, which was reflected in the primitive states of their surface plants and underground workings.

Generally, the wealthiest mines in the richest mining districts needed little promotion among educated investors. Rather, it was the medium and small-sized mines and prospect operations especially in poorly known mining districts that required the most promotion. Only with promotion could small mining outfits garner sufficient capital to build the necessary surface plant, drive underground workings, and where ore had been proven, begin extraction.

A few notorious promoters used the stock system of capitalization to represent fictitious mining ventures in hopes of pocketing investments. In the 1880s one such promoter, George D. Roberts, perpetrated one of the greatest mining stock scams. Roberts began his career in mining as a Kentucky farmboy turned "forty-niner" in the California goldfields, and after losing hard-earned money to merchants and card sharks, he promised he would learn to do to others the dishonest acts that had been done to him.[10] Roberts became a con man who made money from schemes ranging from salting prospect holes to mine and stock speculation. During the 1860s his activities became increasingly sophisticated, culminating in organizing the Great Diamond Hoax of 1872, and the opening of American Mining Stock Exchange, from which he conducted inside trading.[11] By 1880 Roberts, a master of deceit, readied himself for a grand scam involving the State Line Mine located on the northern edge of Death Valley, California.

Robert Shaw discovered the State Line in the 1860s, and working with several partners, extracted high-grade ore from a shallow vein that cropped out at ground-surface. Over the course of ten years Shaw and his partners worked the mine almost to exhaustion and sold it. Subsequent companies enlarged the shaft and explored the remainder of the vein, only to find it consisted of unprofitable low-grade quartz gold. Because the State Line was so remote the cost of operations were too high, and the mine stood idle until 1880, when Roberts purchased it for $100,000.[12] According to plan, Roberts used his own capital to install a new production-class surface plant, including a steam hoist, a boiler, an ore concentration mill, and a water pipeline, even through he knew the mine would not pay. In a cunning move, Roberts began his scheme by advertising that an "investor" had purchased the property for the grand sum of $100,000, and he then had an entourage of bribed engineers inspect the site and tout the wonders of the ore reserves, the surface plant, the mill, and the pipeline. After baiting the public with an announcement that the mine was privately owned and not public, Roberts began releasing stock in limited quantities in an effort to maintain a high price, and when the time seemed right, he dumped his remaining shares and reaped huge profits. When investors began questioning why the mine was not producing, Roberts, hiding behind the State Line Mining company, attempted to delay the coming protests by circulating stories about broken pipelines and other falsehoods. After several months of stalling tactics, Roberts abandoned the State Line company and slipped away from investors, who realized they had been cheated.

The majority of mine promoters were fairly honest, but the few exceptions such as Rice and Roberts earned the profession a bad name. Joel Parker Whitney, born in New England, raised in the California goldfields, and an emigrant to Colorado, represented the majority of honest promoters. Like many professional promoters, Whitney came from a business background where he learned the basics of finance. As with the case of Charles Hallam, fellow businessmen had given Whitney charge of $5,000 for investment in Colorado, and he traveled to Denver in 1865 to begin the search for a mine.[13]

Whitney stayed in Denver for a short while, then moved to Central City where he learned much about the economics and geography of mining investment. From his educated standpoint, Whitney used the capital in his possession to invest in several promising properties in Central City. Successful, Whitney remained in Colorado and continued to invest in mining, focusing his attention on the young districts in Summit County, such as the Robinson District northwest of Leadville. Realizing that unless he could interest other financiers in Summit County districts and townsites, the success of his operations there would remain stunted. As a result Whitney turned to promotion, and as part of his efforts, he published a booklet entitled *Silver Mining Regions of Colorado* which paved the way for similar promotional works. Due to his superior skills as a legitimate promoter, the state of Colorado appointed Whitney as a promotional commissioner.[14]

Promoters and financiers placed great trust in the opinions of mining engineers regarding the authenticity and potential profitability of a mine.[15] Trained in economics and practices of mining, engineers examined a mine and determined the quantity of ore by calculation as well as by estimation. Investors consulted with mining engineers to examine nascent prospect operations and to ascertain whether an adit or shaft showed promising characteristics, and if so, how much money would have been required to upgrade the surface plant to facilitate production. In the latter portion of the nineteenth century, barbed-wire baron Isaac L. Ellwood hired such an engineer, John B. Farish, to seek investments in Colorado. Farish, who experienced success with the wealthy Enterprise Mine in the Rico District in the San Juan Mountains, located a promising prospect operation near the mining town of Ouray. There, with Ellwood's money, he formed the Wedges Mining Company to further develop the claims. Through promotion and conservative management, Farish transformed the Wedges Mine from a simple prospect hole into a large and productive operation.[16]

Most mining engineers, however, did not act as promoters, and many did not even evaluate mines. Yet this group of gritty men was indispensable to mining in the West. The primary function of the mining engineer was to oversee the sequence of a claim's development, from systematic deep prospecting to the ever-hoped-for discovery of bonanza ore, and realization of the claim into a paying mine. Engineers charged with a property had serious responsibilities, including supervision of the driving of underground workings and the installation of the surface plant, all within physical, economic, and geological constraints. Performing such a duty was not easy, and successfully carrying out such work required men who were, as described by mining historian Clark C. Spence, "jacks of all trades" in the mining industry.[17]

The position of mining engineer began in Europe long before the opening of the American West, and mining engineers naturally became a part of the fabric of the American mining industry. As early as the 1860s mining company directors in the West realized the value of keeping mining engineers on staff to apply science and technology to maximize ore production.[18] From that time until the 1930s mining engineers entered the profession from two camps: formal technical training, or self-educated.

> *"The mining engineer it is said, achieved his status by doing the work of a mining engineer. This, of course, over-simplifies the matter. Once a young man had decided on the pursuit of such a career, he did not attain his goal merely by proclaiming his intent and going to work. He must enter by one of two doors — experience or technical training."* [19]

Many of the formally trained engineers who worked in the West during the first decades following the California Gold Rush received their education from British or German technical schools, and by the 1880s, from a handful of American universities. Self-taught engineers, however, began work as miners and ascended through the positions of shift boss, superintendent, and manager at small operations. The more they learned about mining, the applications of technology, of geology, and of economics, the greater their odds of ascendancy. According to Clark C. Spence, early in the Gilded Age the ability to become a mining engineer seemed to be a function of opportunity, intelligence, and attitude:

> *"During much of the nineteenth century, most mining engineers were of the practical variety — men who through circumstance, ability, hard work, and experience in the different aspects of mining and milling won their positions without special education."* [20]

In this fashion, mining in the American West was a true embodiment of how we today characterize the social and political climate of that distant time: freedom, autonomy, and mobility.

Between approximately 1860 and 1890, mine owners and companies favored self-taught engineers over those with formal education, in part because they felt their practical experience was superior for running a mine in the Wild West. As the decades marched onward, this preference changed, until by 1920 six out of seven mining engineers were college-trained.[21] In some ways this transition was a gain for mining, because the self-made engineers preferred to use well-understood, proven, conventional machines and methods, which had significant inherent limitations.[22] Progressive, technically trained engineers introduced and experimented with new technologies as well as fine-tuning the old methods in a calculated and planned manner, which ultimately reduced the costs of mining and milling.

Yet, there was no substitute for weathered and salty engineers because they possessed an uncanny ability to examine prospects deep in the backcountry, cobble together the machinery and materials necessary to open them with little wasted capital, and profitably extract ore. Winfield Scott Stratton and Frank A. Crampton were two such self-made mining engineers. In the self-made tradition, they began their careers working in mines, and perhaps most important, after experiencing successes and failures, they understood the realities of mining in the rugged West.

W.S. Stratton gained world-wide fame for discovering the fabulous Independence Mine in the Cripple Creek Mining District and selling it for a record $10,000,000, after he had already extracted millions of dollars in gold. Working the Independence was Stratton's high point, but before that he had learned mining and prospecting the hard way. Stratton was born in 1848 the son of a Midwest riverboat builder, and under his father he learned the delicate arts of both fine carpentry and heavy woodworking.[23] He came to Colorado in 1872 and went to work as a contract carpenter building houses in Colorado Springs, where he earned a reputation for fine work. The gold bug bit Stratton in 1874, and he departed in the spring to search the Rockies for wealth. Stratton subsequently fell into a pattern of prospecting in the late spring, summer, and early fall, and working as a carpenter in Colorado Springs and Breckenridge during the other months. Leaving no portion of Colorado's Mineral Belt unvisited, Stratton narrowed his search to areas he liked in the San Juan Mountains, South Park, Breckenridge, the Collegiate Mountains, and the Front Range. During his wanderings he acquired the characteristics typically attributed to a good prospector, such as the ability to live off the land, to find comfort in foul weather, to understand the landscape, and a knowledge of geology and mineralogy. The carpenter even found ore in the San Juans in 1874 and spent $2,800 developing the Ybretaba Silver Lode, but the promising lead did not pan out.[24]

After years of searching for riches, Stratton found them in his own backyard in 1891. During the 1880s local cowboy Bob Womack claimed that he had found telluride near his ramshackle ranch high above Colorado Springs, and that another time he encountered traces of placer gold in a nearby stream. Further, in 1890 Womack asserted that he had uncovered an ore vein at the bottom of a shallow prospect shaft at the head of Cripple Creek. Stratton was one of the few people in Colorado Springs who engaged Womack on a conversational level, and after much badgering, Womack convinced Stratton to apply his years of prospecting experience to verify the presence of ore. Skeptical, Stratton followed Womack to his ranch at Cripple Creek and after several days of searching, he chipped off fragments from a granite ledge south of the ranch and took them to Colorado Springs for assay results. The samples turned out to have just enough gold to raise Stratton's interest, and in accordance with classic prospecting methodology, he sank a small shaft along the ledge to track the mineral content as the ledge extended downward. The values increased, and Stratton realized he finally had struck a mother lode. The discovery helped touch off the Cripple Creek gold rush. Stratton christened the mine the Independence, and he wasted no time in erecting a surface plant and sinking a proper shaft.

During the seventeen years he spent combing Colorado's mountains, Stratton learned many practical skills pertaining to building and running a mine, which he applied to his Independence. Stratton designed the mine's surface plant, ordered and supervised the installation of the machinery and power plant, served as blacksmith, and even participated in construction of the buildings. The old prospector also did most of the drilling, blasting, timbering, and surveying in the underground workings. Between the skills he acquired as a professional carpenter and house builder, his years prospecting, and his experience developing the failed Ybretaba, Stratton typified the self-made mining engineer.

Frank A Crampton, another self-made mining engineer, came from a higher social class than W.S. Stratton. Crampton grew up in New York City in the 1890s as the son of an upper-middle-class family, and after a high school education in a military academy, he literally ran away from home with just the clothes on his back.[25] Crampton possessed the fundamental characteristics shared by many self-made mining engineers, including a basic education, fierce independence, a penchant for rustic travel, desire to learn, and experience in the mines that began from the bottom up:

> *"The prelude to my engineering career began in the jungles back of the railroad yards in Chicago. My practical education started on a November day when, as a lad of sixteen, busted from an old Ivy League college, broke after running away from home and a family that thought I had disgraced it forever, I was taken in tow by two hardrock mining stiffs and shown the ropes."* [26]

After learning the basics of underground work in Cripple Creek beginning in 1904, under the tutelage of miners known simply as John T. and Sully, Crampton purchased a defunct assaying and surveying agency in Goldfield, Nevada, where he became acquainted with fundamental mineralogy and surface engineering. While Crampton was at an impressionable age, Sully and John T. emphasized that each job he took should be an opportunity to gain new skills and learn more about practical mining, and Crampton latched onto this philosophy like a bulldog. Mining in Cripple Creek and surveying in Goldfield opened the door for Crampton, and he entered the western mining industry with a penchant for learning. By the early 1910s Crampton had gained a reputation for being able to set up and supervise small western mines under severe economic and environmental conditions. However, like many self-made mining engineers, Crampton was not well briefed on cutting-edge mining technology and he snubbed his technically-trained colleagues, in part out of professional jealousy, and in part out of disdain for their upper-class moorings.

Many technically-trained professional mining engineers in the West had come from the upper middle classes and better. Professionally-educated engineers formed an interesting contrast to the self-made variety in that they approached the problems of building and running mines in the West from more of an academic perspective. Professionally-trained engineers applied science, sophisticated mathematics, economics, and mechanical engineering when building and operating mines, rather than sheer cumulative experience and imitation of other operations. The best technically-trained engineers relied on their professional education, but tempered with intuition and experience. As a result, trained engineers were more willing to experiment and accept new technologies to answer the problems of mining, provided the solutions seemed to be at least hypothetically effective. The status and training of formally-educated engineers fetched a relatively high wage that was out of reach for small mining companies and prospect operations. Instead, technically-trained engineers were usually employed by prominent capitalists, investment firms, and profitable mining companies. The small mining operations were forced to hire the less-expensive self-made engineers, or superintendents who had a round, cursory working knowledge of how to set up and run a mine.

Herbert C. Hoover was a pinnacle of the professionally-trained breed of mining engineer. Hoover was born in Iowa, raised in Oregon, and received his introduction to mining with the California State Geological Survey during the 1860s, followed by a stint with the United States Geological Survey. Hoover received his formal engineering education at the respected mining school at Stanford University and went on to oversee mines in both the American and Australian West in the 1890s. During this time Hoover gained a solid reputation and entered a tight social circle of mining magnates, which became his doorway to substantial sums of money. The prominent engineer was proficient at mine evaluation, characterizing underground workings and surface plants, and discoursing on economics of large and profitable operations. Hoover went on to become involved in mining ventures in China, Korea, and Burma, and he wrote a respected mining engineering text book in the 1900s.[27] But his greatest claim to fame in the public eye was as head of the Food Administration during World War I, and as president of the United States from 1928 to 1932.

Thomas A Rickard was another world-renowned mining engineer, but he started somewhat more simply than Hoover. The master engineer was well-grounded in western mining, recognizing that the industry consisted of large and small operations, both of which deserved attention. Rickard came from a Cornish mining family. His grandfather was one of the first Cornish hardrock miners to work in California, and other members of his family, including many of his brothers, were involved in mining.[28] Rickard launched his engineering career by graduating from the Royal School of Mines in 1885, after which he went to work under his uncle in Idaho Springs, Colorado, as an assayer. In 1886 Rickard moved upward, literally and figuratively, to manage the Kansas, Kent County, and California mines in Central City, north of and above Idaho Springs. In the 1890s Rickard also managed the fabulous Yankee Girl Mine at Red Mountain Pass, and the heavily-producing Enterprise Mine in the Rico Mining District in the southwest San Juan Mountains, Colorado.[29] Rickard was even asked to evaluate Stratton's Independence Mine when an English syndicate was considering purchase.

Rickard was one of the best mining engineers in the world during the Gilded Age, because he understood mining from both the academic perspective and from the applied, experiential side. Combining formal training with practical experience made Rickard adept at valuation, judging the potential of prospects, and understanding that technology, surface plants, and underground workings all had to work together under an economic and managerial umbrella. Rickard, who was a prolific writer, stated:

> "In sizing up the situation it is necessary that a man should know what are likely to be the costs of stoping, timbering, road-making, erection of machinery equipment, etc.,

> *and these things he can only know through actual under-*
> *ground experience and personal participation in the admin-*
> *istration of mines."* [30]

By his statement, Rickard, one of the world's greatest formally trained mining engineers, identified the importance of practical experience, emphasizing the core of what defined the self-made mining engineer.

$$\times$$

The Roots of Mining

M ining methods and equipment became uniform throughout the West during the Gilded Age. This was largely a result of the diffusion of practices and technologies brought over from both Cornwall and Germany to North America. The Cornish probably had the greatest direct influence on American hardrock mining, but it is noteworthy that German mining specialists introduced systematic methods and engineering to Cornwall in the 1600s. The Cornish adapted the basic techniques taught them by the Germans to the unique tin and copper deposits that descended steeply under the Cornish Coast. To work the deep, wet, vertical ore bodies, several generations of Cornish miners and engineers developed technologies that were well-suited for transplantation to North America. [31]

The Cornish heavily influenced the evolution of the systematic use of shafts, drifts, raises, and winzes to block out and explore ore bodies, and they developed overhand stoping methods. Such approaches to the development of underground workings permitted calculated, quantified, and predictable ore extraction, which saw heavy application in the West. The Cornish also are credited with devising a revolutionary hoisting system in the 1700s that served as a basic template that persists in the mining industry to this day. The system they developed included a horse-drawn hoist known as a *whim*, a *headframe* standing over the shaft, and an *ore bucket*. [32] The whim consisted of a rope reel over five feet in diameter turned via draft animals that had been tethered to harness beams. As the animals pulled the harness beams they turned the reel, which raised the bucket in the shaft. The hoist rope extended over a heavy pulley, known as a *sheave*, suspended from the top of the headframe, and down the shaft. This system was a significant improvement over the slow, cumbersome and dangerous hand windlasses used until then, permitting greater production in deeper shafts.

During the 1600s or early 1700s Cornish miners developed the horse whim, which was an inexpensive and highly versatile mechanical hoist that literally revolutionized hardrock mining. The whim consisted of a rope reel, a harness beam, a headframe, and the mine shaft. The contraption depicted in the illustration represents a variation of the original Cornish design in that the rope reel was located overhead, and most of the components were wooden. Early miners in the American West erected similar versions, but by the latter half of the nineteenth century other forms of horse whims proliferated.

International Textbook Company International Library of Technology: Hoisting, Haulage, Mine Drainage International Textbook Company, Scranton, PA, 1906 A50 p3.

One invention that revolutionized the Cornish mining industry during the 1700s was the steam-powered beam engine. Cornish miners first used the beam engine to solve one of their greatest problems—flooding. Cornish mines were located on or near the coast and they often extended under the ocean floor. The ubiquitous seeping water proved to be a perpetual and expensive problem that inhibited the pursuit of deep ore. The Cornish had tried various means of dewatering their mines, all with limited success. When the vertical steam engine made its appearance in Britain in the early 1700s, mining engineers realized the machine could be adapted to run dewatering pumps on an enormous scale. Beam engines typically consisted of a large-diameter vertical steam piston connected to a walking beam that was hinged at center. When the piston rose under steam pressure, it pushed one end of the beam up, and the other end of the beam dropped. While such engines were adapted to a multitude of industrial purposes over the course of the 1700s, one of the first applications was lifting and letting fall dewatering pump rods in Cornish mines. The pump rods consisted of heavy timbers, often having come from the American colonies, spliced together to form a solid shaft extending into the sump of the mine shaft where the pump mechanism lay.

By 1727 five Cornish mines were equipped with *Cornish pumps,* and they experienced a huge success in draining the underground workings. By 1800 all large mines in Cornwall and many others in Britain featured Cornish pumps. The Cornish subsequently made a number of improvements to the beam engine and associated steam boilers through the late 1700s, and began adapting them to other mine applications, such as hoisting and driving stamp mills. [33]

Ironically, as improvement and implementation of labor-saving technology permitted Cornish miners to work at depths greater than ever and extract more ore in less time, British *adventurers,* the English term for investors, grew recalcitrant about providing more money to finance the improvements. As a result, the known ore reserves showed signs of depletion after several centuries of steady mining, threatening Cornwall's very economy. To the horror of the clannish Cornish, the mines that were their livelihood began closing in the 1840s. The decline accelerated in the 1850s and 1860s, stimulating migration of Cousin Jacks and their Jennies to the United States.[34] The Cornish miners brought their expertise, skills, methods, and technology at first to the Wisconsin lead mines in the 1830s, to Michigan copper mines in the 1840s and 1850s, to California's nascent quartz gold mines and mercury mines in the 1860s, and to the famed Comstock lode and Colorado's Central City in the 1860s and 1870s. Mining companies in each district recognized the Cornish for their superior talents. Framed by positions of prominence, the Cornish introduced their practices, which American miners quickly integrated. Through the 1850s and 1860s both Cornish and American miners carried these methods and technologies to nearly every Anglo hardrock mining district in the West. The speed of diffusion was quick, in part due to the expanding transportation and communication systems, and in part because of the mobility of western miners. But the Cornish did not have the last word in mining in the New World, because American miners and engineers adapted Cornish methods to the unique physical and economic conditions of the West. The result was a truly American mining industry.

$$\asymp$$

Factors Influential to Mining

D espite the fact that the western mining industry drew upon a fairly standardized pool of machines, technologies, and underground methods, each mine and prospect operation faced unique challenges presented by economic factors, the physical environment, and geographic location. Solving these problems meant that the mining engineer had to tailor his plant designs, selection of machines and materials, and means of transportation and construction to suit the conditions in what we can term *applied technology.*

The environmental, economic, and geographic elements that influenced the establishment and operation of a mine or prospect can be broken down into six basic categories. The presence or absence of ore, geology, and climate heavily influenced the manner in which engineers set up a mine, and they fall under the umbrella of the physical environment. The other factors include available capital, geographic location, and operating time frame. The presence or absence of ore proved to be one of the most fundamental influential factors. Huge ore reserves contributed to a long-term and prosperous endeavor, they attracted wealthy investors, and resulted in at least several incarnations of a mine. Minor ore reserves may have stimulated substantial underground exploration for payrock, but a brief, short-term occupation. By contrast, a total absence of ore ensured a quick abandonment after limited underground exploration.

The geology associated with a potential ore deposit influenced the nature of the underground workings, and it affected how a mining engineer designed the surface plant. Vertical ore bodies and terrain ranging from low to moderate topographical relief dictated that the mining company sink a shaft, while gently pitching ore bodies and/or high topographical relief was conducive to driving an adit.

The climate was another factor that affected how an engineer planned a mine. If the engineer expected the mine to operate all year in high altitude environments, he had to enclose the vital surface plant components in heated buildings, and link the tunnel portal or shaft collar with the shop and other facilities. In less severe climates, such as the Southwest and the Pacific Seaboard, heated buildings were not so crucial. The foul weather endemic to high altitude settings also affected transportation from points of commerce to mines, necessitating that roads be planned well and carefully graded. In the Southwest and Great Basin, flash floods had the potential to wash away roads in drainages, which mining engineers there had to take into account.

The availability of capital, a function of investor confidence, lies within the category of economics, and it may have been the single most influential factor to which a mining engineer had to respond. Mining revolved around money; everything from buying the claim, setting up the surface plant, paying miners and laborers, to shipping ore. The size of a mine, the sophistication of the surface plant and underground workings, the types of machines installed, the quantity of supplies consumed, and quantity and quality of the miners employed all were a function of the amount of available capital. Often the amount of investment in a property was a function of the presence or absence of ore, the district in which a mine was located, and of promotion.

Geographic location was a factor that became a severe hindrance to many mining and prospecting operations. The costs of mining increased with distance from railheads and commercial centers. Prior to the 1920s heavy freight wagons were the principal conveyance between mines and railroads, and they presented significant drawbacks that affected mining companies. Wagons limited the amount of ore that a mine could ship, and they dictated the sizes and types of machines and other plant facilities that could be brought to the site.

The last major factor influencing how a mining company or engineer set up a mine was the time frame. Mining machinery available in the 1860s and 1870s was simple, crude, limited in duty, and inefficient. But through the 1880s and into the 1900s the price of machinery dropped while the availability of technology expanded. What was a rare and state-of-the-art machine or surface plant component in the 1880s became commonplace by the 1900s. An increase in available capital among mining companies, a decrease in prices, and the sheer number of secondhand machines for sale ensured that mine plants and the technology used underground would continue to become increasingly sophisticated and efficient.

One unforeseen result of improvements in mining technologies and increase in the availability of capital was the periodic rehabilitation of defunct, previously productive mines, especially those which had stood idle for some time due to exhaustion of high-grade ore. Improved milling technology and a decrease in the costs of mining rendered previously uneconomical ores profitable. Thus, many mines thought to be exhausted in fact experienced a succession of operations.

$$\times\!\!\!\times$$

Developing the Mine

Generally, in the West ore bodies tended to take one of two forms. Miners and engineers had termed the first formation a *vein*, and they knew the other as being a massive and globular body. Typically, miners encountered free gold, telluride gold, tungsten, and silver in veins, and they found industrial metals such as copper, iron, and silver compounds in masses. At the point where a tunnel or shaft penetrated the ore body, miners *developed* the geological feature with internal workings consisting of *drifts, crosscuts,* internal shafts known as *winzes* which dropped down from the tunnel floor, and internal shafts known as *raises* which extended up. Drifts and crosscuts explored the length and width of the ore, and raises and winzes explored its height and depth.

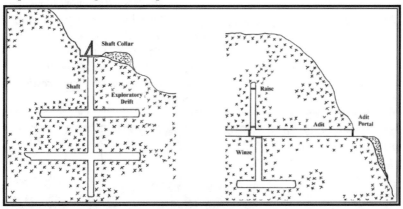

The cut-away views depict the nature of the underground workings typically associated with prospect shafts and prospect adits. The defining differences between prospect workings and the workings comprising paying mines primarily are a lack of extensive underground development and a lack of stopes where miners extracted ore. Prospect shafts typically feature limited underground workings, which is reflected on ground-surface by waste rock dumps with relatively little volume.

Author.

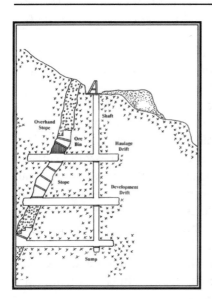

The cut-away view depicts the underground features typically associated with a mine shaft. The ore body has been developed from several drifts, from which miners drilled and blasted upward, using gravity to advantage. The shaft served as an umbilical cord linking the stopes underground with the support facilities on the surface. Where possible, well-capitalized mining companies attempted to connect the shaft sump to a tunnel driven horizontally through the hillside, which improved ventilation and decreased the costs of bringing the ore to daylight. Shafts were very popular in the West.

Author.

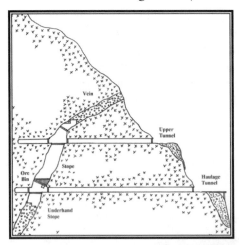

Where the topography was steep and the ore body's structure well-known, mining engineers attempted to develop it through a series of tunnels which afforded greater access and improved ventilation. The problem with mining through a tunnel was that in some cases the ore had to be drilled and blasted from the top down in a method known as underhand stoping, which proved costly.

Author.

Overhand stoping, as depicted in the profile view, involved drilling and blasting ore from the bottom up in a series of hanging stairsteps. When blasted, the ore usually dropped into a bin that miners tapped into ore cars for the trip up to the surface.

International Textbook Company
A Textbook on Metal Mining:
Preliminary Operations at Metal Mines, Metal Mining, Surface Arrangements at Metal Mines, Ore Dressing and Milling
International Textbook Company, Scranton, PA, 1899 A41 p22.

All of the underground exploration, as well as ore production, created a considerable demand for support from on-site facilities, which paralyzed the operation if ignored. These facilities, almost without exception, were located around the mouth of the adit or shaft collar, and engineers and miners alike knew them as the *surface plant*. It may be said that a mine was like an iceberg, and the surface plant formed the visible cap, hinting at what lay below the surface. Large, productive mines boasted sizable surface plants, while small prospect operations tended to have simple facilities. Regardless of whether the operation was a small underground prospect or large profitable mine, the surface plant had to meet five fundamental needs. First, it had to provide a stable and unobstructed entry into the underground workings. Second, the plant had to include a facility for tool and equipment maintenance and fabrication. Third, the plant had to allow for the transportation of materials and waste rock out of the underground workings and supplies in. Fourth, the workings had to be ventilated, and fifth, the plant had to facilitate the storage of up to tens of thousands of tons of waste rock generated during underground development, often within the boundaries of the mineral claim. Generally, productive mines, as well as complex and deep prospects, had needs in addition to the above basic five requirements, and their surface plants included the necessary associated components.

Miners and timbermen built stations underneath stopes that featured ore bins, space for moving equipment and supplies, junctions of air and ventilation pipes, and ore chutes for loading cars.

International Textbook Company A Textbook on Metal Mining: Preliminary Operations at Metal Mines, Metal Mining, Surface Arrangements at Metal Mines, Ore Dressing and Milling International Textbook Company, Scranton, PA, 1899 A41 p37.

The basic surface plant, whether haphazardly constructed by a party of inexperienced prospectors or designed by experienced mining engineers, consisted of a set of *components*. The entry underground usually consisted of a stabilized collar for either an inclined or vertical shaft, or a portal for an adit. While the exact differentiation between a tunnel and an adit is somewhat nebulous, mining engineers and self-made mining men have referred to narrow and low tunnels with limited space and length as *adits*. Passages wide enough to permit incoming miners to pass outgoing ore cars, high enough to accommodate air and water plumbing suspended from the ceiling, and extending into substantial workings have been loosely referred to as *tunnels*. Most surface plants featured transportation arteries permitting the free movement of men and materials into and out of the underground entry. Miners moved materials at adit operations in ore cars on baby-gauge mine rail lines, while shafts required an additional hoisting system to lift vehicles out of the workings. Materials and rock at shaft mines were usually transferred into an ore car for transportation on the surface. A blacksmith maintained and fabricated tools and equipment in a shop on site, and large mines often had additional machining and carpentry facilities. Most of these plant components were built on *cut-and-fill earthen platforms* made by excavating material from the hillslope and using the fill to extend the level surface. Once enough waste rock had been extracted from the underground workings and dumped around the mouth of the adit or shaft, additional facilities may have been built on the resultant level area. The physical size, degree of mechanization, and capital expenditure on these surface plants were relative to the constitution of the invisible portion of the iceberg that lay below ground.

Miners and prospectors consciously chose to sink a shaft or to drive an adit in response to fundamental criteria. Adits required surface plants different in composition and arrangement from those that served shafts. First, a shaft was easiest and less costly to keep open against fractured and weak ground. Second, a shaft permitted miners to stay in close contact with an ore body as they pursued it to depth, and they were able to sample the ore periodically. Third, in cases where miners sank a shaft on profitable ore, the payrock they extracted provided the company with almost instant income, which pleased stockholders and aided mine promotion. Last, a shaft lent itself well to driving a latticework of drifts, crosscuts, raises, and winzes to explore and block out the ore body.[35]

Mining engineers discerned between sinking vertical versus inclined shafts. One contingent of engineers, especially those working prior to the 1880s, preferred inclined shafts because, as they correctly pointed out, mineral bodies, especially veins, were rarely vertical, and instead descended at an angle. As a result vertical shafts were ineffectual for intimate tracking and immediate extraction of ore. In addition, inclined shafts needed smaller, less expensive hoists than those used for vertical shafts. The other camp of engineers, however, claimed that vertical shafts were in fact best because maintenance and upkeep on them cost less. Vertical shafts had to be timbered merely to resist swelling of the walls, while timbering in inclines had to also support the ceiling, which was more expensive, especially when the passage penetrated weak ground. Inclined shafts also required a weight-bearing track for the hoist vehicle, which, including maintenance such as replacing rotten timbers and corroded rails, consumed money.

Most engineers changed their perspective during the 1880s and favored vertical shafts instead because they reached depth in less distance, translating into lower sinking costs, and more rock could be extracted in less time with associated hoist vehicles, expediting production.[36] In light of the collective experience gained during five decades of mining in the West, by the 1900s mining engineers strongly recommended that vertical shafts be sunk in the footwalls of ore veins. Experience had taught the mining industry, often through expensive and dangerous lessons, that the hanging wall overlying the vein was likely to settle and shift after ore was extracted, throwing the shaft out of plumb.[37]

Often, the physical and economic variables that prospectors and miners grappled with clearly made the decision to drive an adit instead of sinking a shaft a wise one. The most fundamental consideration in deciding whether to drive an adit or sink a shaft lay with economics. Driving an adit was easier, faster, and required significantly less capital than sinking a shaft, despite the high costs of timbering in some areas. Mining engineers had determined that the cost of drilling and blasting a shaft was as much as three times more than excavating an adit.

In many cases geology proved to be a criterion. Steep hillsides, deep canyons, and gently-pitching ore bodies lent themselves well to exploration and extraction through adits. In many cases prospectors who had located an outcrop of ore high on a hillside elected to drive an adit from a point considerably downslope to intersect the formation at depth. If the ore body proved economical, then the mining company carried out extraction through the adit. One of the most problematic aspects of driving an adit to investigate an ore body was that miners had to labor at considerable dead work, drilling and blasting through barren ground, with no guarantee that they would locate the ore body where they had anticipated striking it. In many cases veins cropping out high on mountainsides disappeared at depth, or natural faulting broke them up and shifted the pieces around. In addition, adits were not as well suited as shafts for developing deep ore bodies, because interior hoisting and ore transfer stations had to be blasted out, which proved costly and created traffic congestion. One other problem, significant in districts where the rock was weak, lay in the enormous cost of timbering adits and tunnels against cave-in. However, much to the relief of mining companies, many western districts featured sound rock requiring little support.[38]

Prospectors and mining engineers understood that adits were self-draining, they required no hoisting equipment, and transporting rock out and materials into the mine was easier than it was through a shaft. Regardless, in many cases prospectors, those with the least access to capital, sank small shafts to explore ore bodies for the reasons cited on the previous pages, and for one additional significant factor.[39]

Historians of the West have aptly characterized mineral rushes to heavily promoted mining districts as a frenzy of prospectors who laid a quilt work of mineral claims. In most districts the recognized hardrock claim was restricted to being 1,500 feet long and 600 feet wide, which left limited work space both above and below ground.[40] In Colorado, prospectors were legally obligated to drive an adit or shaft, or sink a pit to a minimum depth of 10 feet to hold title to a hardrock claim, while they had to conduct $100 worth of labor in other states.[41] A small adit or pit was not adequate to fully explore the depths bounded by a 1,500 by 600 foot plot of ground, let alone to extract ore, forcing prospectors and mining companies to sink shafts.

The Cripple Creek Mining District serves as an excellent example of how crowded conditions forced mining companies to sink shafts to work at depth within their claim boundaries. The district was blanketed with individual claims during its heyday, and prospectors and mining companies were forced to sink shafts because they lacked the contiguous ground necessary to explore and develop ore bodies at depth through tunnels. In districts where competition for space was not as severe, mining companies had greater latitude to drive tunnels.

⚒

Equipping the Mine

In addition to differentiating between surface plants that served tunnels from those associated with shafts, mining engineers subdivided these mine facilities into two more classes. Engineers considered surface plants geared for shaft sinking, driving adits, and underground exploration to be different from those designed to facilitate ore production. Engineers referred to exploration facilities as *temporary plants,* and as *sinking plants* when associated with shafts. Such facilities were by nature small, labor-intensive, inefficient, portable, and most important, they required little capital. *Production plants,* on the other hand, usually represented long-term investment, and they were intended to maximize production while minimizing operating costs such as labor, maintenance, and energy consumption. Such facilities emphasized capital-intensive mechanization, engineering, planning, and scientific calculation.

Mines underwent an evolutionary process in which discovery, the driving of a prospect shaft or adit, installation of a temporary plant, upgrade to a production plant, and eventual abandonment of the property were all points along a spectrum. Depending on whether prospectors or a succeeding mining company found ore and how much, a mine could have been abandoned in any stage of evolution. Of course, the ultimate goal of most mining companies, capitalists, and engineers was to locate, prove, and develop fabulous ore reserves, and to install a surface plant large and efficient enough to earn accolades from the western mining industry. Most operations, however, did not succeed. Mining engineers and mining companies usually took a cautionary, pragmatic approach when upgrading a temporary plant to a production plant. Until significant ore reserves had been proven, most mining companies minimized their outlay of capital by installing inexpensive machines adequate only for meeting immediate needs. The circa 1890s mining engineering text series *A Textbook on Metal Mining* sums up the sentiment behind companies' approach toward temporary plants:

> *"The majority of mines are opened in a small way, or at least with little machinery, large plants being rarely seen until the mine has been proven by actual work of development to contain extensive bodies of ore. During the first stages of work at a mine, it is best to employ only such machinery as will perform the service most economically and at the same time safely"* [42]

It must be remembered that despite a showing of ore, many western mines never progressed beyond their temporary plants for want of capital and trained engineers.

Mining engineers and self-made mining men understood that temporary plants consisted of light-duty, inexpensive, and impermanent components. Many engineers classified the duty of these components, especially machines such as hoists, boilers, blowers, and air compressors by their size, energy efficiency, performance, and purchase price. Machine foundations, necessary to anchor and stabilize what were critical plant components, also fell under this scope of classification. Because of a low cost, ease of erection, and brief serviceable life, mining engineers considered timber and hewn log machine foundations to be strictly temporary, while production-class foundations consisted of concrete or masonry.[43] The structure of wooden foundations usually consisted of cribbing, a framed cube, or a frame fastened to a pallet, all of which were assembled with bolts and iron pins, and buried in waste rock ballast for stability.[44] Prior to around 1890 most production-class foundations consisted of rock or brick masonry. As concrete became an accepted building material by the early 1890s, it became universal for production-class machine foundations. The construction and classification of

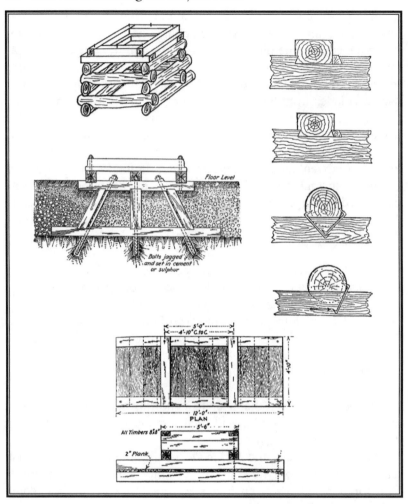

The line drawings depict temporary timber foundations built for machinery such as hoists, compressors, and steam engines. Clockwise from the top left, and in order of popularity: a log or timber cribbing cell, methods of fastening cribbing members, a timber pallet, and anchor bolts tied into bedrock. All of the foundations were usually buried up to the cap timbers with fill material for ballast. As the illustration shows, cribbing members were often assembled with a combination of saddle notches and timber spikes, which may still be encountered at historic mine sites today.

Clockwise from top left: Author.

Audel, Theo. & Co. New Chatechism of the Steam Engine Theo. Audel & Co. Publishers, New York, NY, 1900 p37.

Engineering & Mining Journal Details of Practical Mining McGraw-Hill Book Company, Inc., New York, NY, 1916 p9.

Engineering & Mining Journal Details of Practical Mining McGraw-Hill Book Company, Inc., New York, NY, 1916 p8.

machine foundations are of particular importance, because they often constitute the only remaining evidence at mine sites today capable of conveying the composition of the surface plant in terms of equipment.

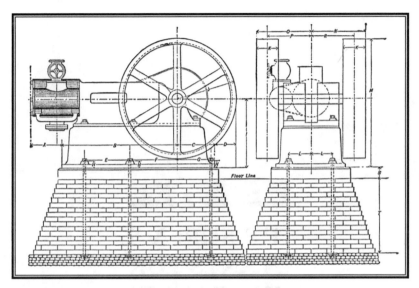

The profiles provide side and end views of a production-class brick compressor foundation. In most cases western mine crews built foundations with rock masonry or concrete, and despite the specific construction material, the foundations shared a similar form. Note that the masonry overlies a rock footing, and that the anchor bolts had to be put in place prior to erection of the masonry.

Audel, Theo. & Co. New Chatechism of the Steam Engine Theo. Audel & Co. Publishers, New York, NY, 1900 p36.

CHAPTER 3

THE SURFACE PLANTS FOR MINE TUNNELS

The photograph captures a mine crew at the Scotia Mine above Boulder, Colorado around 1900. The layout of the plant components and quality of the buildings depict a quintessential western adit operation. The structures, which consist of a blacksmith shop at rear left, a shed, and a utility building at right are clustered around the adit portal on cut-and-fill earthen platforms. The rough appearance of the structures and the lack of heavy machinery indicate that this operation suffered from limited funding. The fact that the miners used a horse to pull ore cars suggests that the adit is long. Well-capitalized, productive adit mines of this vintage may have had larger buildings housing a boiler, denoted by a tall smokestack, an air compressor, and a ventilation fan.

Preparing to drive underground workings on the scale necessary to explore and develop a hardrock mineral body was no easy task in the rugged West. Before miners could begin underground work, interested partnerships or mining companies had to make arrangements for financing, and they had to obtain the services of individuals capable of establishing and managing mines. The company often saddled a mining engineer or a supervisor capable of acting as such, with the onerous duty of obtaining equipment and supplies, hiring competent supervisors and labor, and seeing them delivered to the property. The engineer also had to carry out one of the most important aspects of prospecting and mining: setting up and running the physical infrastructure that was necessary for underground work. At the core of this infrastructure lay the *surface plant*, which this chapter seeks to examine in detail.

When a prospector staked a claim, his was an educated guess at what exactly lay below ground surface. The existence of ore was uncertain at best, requiring mining companies and prospecting outfits to undertake underground exploration to examine the geology at depth. To drive exploratory workings, they erected temporary plants, which consisted of small, inefficient, and inexpensive facilities. When the mining company finally had proven the existence of ore reserves in economical quantities, in many cases it sought to upgrade the small and inefficient temporary plant into a production-class plant. Depending on how much financing the mining company was willing to provide, in some cases the engineer could afford to supplant only the most vital temporary components with production-class equipment, leaving the remainder of the facilities in a primitive state.

How much a mine grew depended on the productivity, the depth at which miners had encountered ore, and how much underground exploration the company had initially undertaken. Mines with vast ore reserves found at great depths tended to be complex, reflected by large surface plants and voluminous waste rock dumps, while mines barely getting by on small ore reserves found close to the surface were small and simple.

Miners developed a few basic patterns of physically organizing the surface plant that they built around their adit, and through functionality and tradition, the patterns changed little through time. The most common pattern at small mines took shape when prospectors or miners dumped the waste rock immediately outside the adit portal and graded it flat. They located the shop and ventilation system adjacent to the adit portal on earthen platforms cut out of the hillslope. Usually a rail line served the miners' transportation needs, and it extended out of the adit, past the shop, and terminated at the dump's edge.

As miners labored away underground in search of ore, the mining company responded by upgrading the surface plant with additional facilities capable of supporting more intense activities. These expanded plants also adhered to common patterns of arrangement. The engineer had workers erect an ore bin on the shoulder of the waste rock dump, they completed a wagon road up to the adit, and they graded a spur down to the ore bin. The plant upgrade may have also included a ventilation system, enlarged shop facilities, and after the late 1880s, an air compressor and boiler. The miners and surface laborers lengthened the mine rail line by adding several spurs to the waste rock dump, the shop, and to the ore bin. If the engineer upgraded the surface plant with all of the above facilities simultaneously, he might have enclosed the shop and machinery in one or two large frame buildings partitioned into several rooms, or he may have enclosed them in a single tunnel house. The plant probably also included an outhouse, a water tank, an office, and bunkhouses on earthen platforms situated away from the central workings. If the mine was a heavy producer backed by substantial capital, it may have been serviced by a railroad line passing below the ore bins, or it may have had a custom ore concentration mill located downslope. Few mines in the West simultaneously possessed all of the above components, as most were not sufficiently profitable. Many western operations fell somewhere in the middle, but nearly all of them adhered to the basic spatial layout discussed above.

Surface Plants for Prospect Adits

We begin our detailed exploration of surface plants by examining the most fundamental facilities: those erected at prospect adits. The process of equipping a mine began when a mining company sent an engineer to inspect a poorly developed but promising claim. Usually the property merely featured a small shaft or adit driven by the original prospecting outfit to examine the mineral body below surface. The engineer examined the underground workings, which were often shallow and limited in extent, and attempted to characterize the mineral body. If he felt that the claim held potential, he made preparations for further development.

The first surface plants erected by prospect outfits almost always consisted of simple temporary-class facilities. Such plants had to be simple for several reasons. First, areas being prospected were usually remote and undeveloped, and hauling in heavy machinery was costly if not impossible because of a lack of roads. Second, small outfits did not have the capital necessary to purchase large surface plant components, nor could they afford qualified engineers. The fact that driving an adit cost up to one-third less than sinking a shaft attracted outfits with little capital, and by default they tended to erect simple and inexpensive plants. As a result of high costs and uncertain ore reserves, prospecting outfits built crude plants that incorporated little mechanization, and structures constructed of locally obtained materials. In the warm areas of the Southwest and Great Basin some prospecting plants did not even include buildings, leaving the facilities unprotected from the weather. Why worry about shelter when the goal of the prospector or partnership was to hastily prove the existence of ore and sell the claim?[1]

The adit was the primary component and nucleus of prospecting plants. Prospectors drove the adit with hammers and drill-steels, a simple and labor-intensive technology well suited for backcountry regions. Hand-drilling was not easy work and the progress proved slow, giving the miners great incentive to minimize the amount of rock they attempted to blast. As a result, prospect adits were usually low, narrow, and short, and while they served the prospectors' needs, they became bottlenecks when ore production began.

Professionally-trained mining engineers recognized a difference between these prospect adits and production-class tunnels. Height and width were the primary defining criteria. A production-class tunnel was wide enough to permit an outgoing ore car to pass an ingoing miner, and headroom had to be sufficient to house compressed air lines and ventilation tubing. Some mining engineers working in the twentieth century attempted to quantify the minimum size of a production-class tunnel as being at least three-and-one-half to four feet wide and six to six-and-one-half feet high. Anything smaller, they claimed, served merely as a prospect adit. This was after the acceptance of compressed air-powered rock drills, which reduced the costs of drilling and blasting.[2]

Miners generated tons of waste rock that they had to haul out of the underground workings, while they also had to send in tools, timbers, and explosives. As a result, prospect operations had to rely on some form of a transportation system, which had to be inexpensive, adaptable to tight workings, and capable of being carried overland into the backcountry. To meet these needs some prospect outfits found the old-fashioned wheelbarrow on a plank runway to be excellent. A wheelbarrow cost as little as twelve dollars during the Gilded Age, it was easily packed on a mule, and it fit into tight workings.[3] Mining engineers recognized the functionality of wheelbarrows, but classified them as strictly serving the needs of subsurface prospecting because of their limited load capacity, awkwardness of handling, and propensity for being crushed.[4]

Outfits driving substantial underground workings required a vehicle with a capacity greater than the few hundred pounds that a prospector could trundle in a wheelbarrow. The vehicle most mining outfits chose was the ore car—today the immortalized symbol of hardrock mining. The ore car commonly associated with metal mining consisted of a plate iron body mounted on a turntable that was riveted to a rail truck. Cars were approximately two feet high, four feet long and two-and-one-half feet wide, they held at least a ton of rock, and they had a swing gate at the front to facilitate dumping. Further, the body pivoted on the turntable to permit the operator to deposit a load of rock on either side of or at the end of the rail line.

While iron ore cars were extremely durable, often outlasting the mining companies that purchased them, the iron components were heavy. Even when disassembled, it was difficult to haul the ungainly parts into the backcountry. In response, prospect operations working in remote areas used a variety suited to the difficult economic and environmental conditions. The car's body consisted of heavy planks held together by an iron framework, and the truck chassis consisted of two heavy timbers fitted with wheel axles and a turntable. The benefit of this type of car was that prospecting outfits needed only to haul in lumber, iron stock, and parts, and the blacksmith could assemble it on-site.

Ore cars had to run on rails, and that created another problem for remote prospect operations. Iron manufacturers, such as Colorado Fuel & Iron Company and Bethlehem Steel, sold rail in a variety of standard sizes, the units of measure being weight-per-yard. Light-duty rail ranged from six to twelve pounds-per-yard, medium-duty weight rails included twelve, sixteen, eighteen, and twenty pounds-per-yard, heavy mine rail weighed from twenty-four to fifty pounds-per-yard, and anything heavier was used by railroads. Prospecting outfits installing temporary plants usually purchased light-duty rails because of their ease of transport and low cost. To illustrate the point, four-hundred feet of track required two-and-a-half tons of sixteen-pound rail, one-and-one-quarter tons of eight pound rail, or only one ton of six pound rail. Why purchase the heavier sixteen-pound rail when eight or six-pound rail would still permit the use of ore cars while costing half to purchase and haul to the site?[5]

Some prospect operations were so remote that transporting even as little as a ton of rail proved arduous and costly. In response, some prospecting outfits improvised a clever alternative known as *strap rail*. Well-suited for remote prospect adits especially in arid climates, strap rail consisted of flat strap iron or a half-round iron bar pinned onto the edges of two-by-four boards nailed on-edge to cross ties. In essence the two-by-four acted as the rail and the iron hardware served as an armored face for the car's wheels. While strap rail was strictly for temporary operations, having a light load capacity, being incapable of conforming to smooth curves, and decaying quickly in wet adits, prospecting outfits merely had to pack the iron and two-by-four boards to their claims, instead of heavy iron rails. Strap rail experienced mild popularity in the Great Basin and the high Rockies during the 1870s and 1880s, when these regions lacked well-developed transportation infrastructures. But when railroads and wagon roads arrived in mining districts, shipping prices fell and the preferred iron rails became affordable and replaced strap rail.[6]

Impoverished mining outfits working in the backcountry built strap rail out of 2x4 boards and iron strapping, as this example at Death Valley's Gunsight Mine illustrates. Strap rail was inexpensive to make and transport, but it was adequate only for light duty use. Most prospect operations and mining outfits preferred long-lasting steel rail.

Author.

At prospect adits the main purpose of a transportation system was to move waste rock out of the underground workings. Therefore, the size of the waste rock dump associated with an adit became a direct reflection of the extensiveness of the underground workings, as miners turned the earth inside out in their search for wealth. Whether miners moved the shattered rock in ore cars or in wheelbarrows, they dumped the material directly out of the adit portal, creating a semicircular pad much like a river delta. By nature prospect operations were small and their dumps never attained the substantial sizes that plagued large operations. Miners made an effort to maintain a smooth, flat surface upon which they could place surface plant components, cut mine timbers, and bend mine rails for curves in the track.

Every prospect adit required the services of a blacksmith who maintained and fabricated equipment, tools, and hardware. Therefore, surface plants dedicated to prospect adits almost always included a shop for the work. The common rate for driving a prospect adit with hand drills and dynamite in hard rock was approximately one to three feet per ten-hour shift. Over the course of such a day miners blunted drill-steels in substantial quantities, and for this reason, the blacksmith's primary duty was to sharpen the dulled drill-steels. Prospecting outfits almost invariably located the blacksmith shop adjacent to the adit portal to minimize handling heavy iron items. Due to a lack of money, the outfits often constructed their crude buildings with local materials, such as logs or stones.[7]

Hauling equipment and supplies to remote prospecting and mining operations was expensive, slow, and hard on man and beast. The photograph depicts a burro dragging the body of an ore car on a travois to a mine in Colorado's impenetrable San Juan Mountains around 1900. Heavier items might have been lashed onto sleds and pulled by teams, while mule skinners packed small items on the animal's back.
Courtesy of Colorado Historical Society, Denver, CO.

The blacksmith required few tools and much skill for his work. A typical basic field shop consisted of a forge, a bellows or blower, an anvil, an anvil block, a quenching tank, several hammers, tongs, a swage, a cutter, a chisel, a hacksaw, snips, a small drill, a workbench, iron stock, hardware, and basic woodworking tools. Prior to the 1910s some prospecting outfits working deep in the backcountry far from commercial centers dispensed with factory-made forges and used local building materials to make vernacular forges. The most popular type of custom-made forge consisted of a gravel-filled dry-laid rock enclosure usually three-by-three feet in area and two feet high. Miners working in forested regions substituted small hewn log walls for rock. A tuyere, often made of a two-foot length of pipe with a hole punched through the side, was carefully embedded in the gravel, and its function was to direct the air blast from the blower or bellows upward into the fire in the forge.[8]

Sharpening drill-steels was a delicate and exacting process that required an experienced mine blacksmith. Drill-steels, specialized tools that had to withstand the brutal work of mining, were of the utmost importance for driving underground workings. Miners used them to bore blast holes, which was the primary method of breaking ground in western mines. These unique tools were made of hardened hexagonal or octagonal three-quarter to one-and one-quarter inch-diameter bars of high-quality steel, and miners always used them in graduated sets. *Starter-steels*, also known as *bull-steels*, were often twelve-inches long, but numerous trips to the blacksmith's forge reduced them to as short as eight inches, and the rest of the steels followed in successive six to ten-inch increments. With each increase in length, a steel's blade decreased slightly in width, ensuring that it did not wedge tightly in the drill-hole. Generally, drill-steels for single jacking were no longer than three feet, and the longest steels used for double jacking were usually four to six feet long.

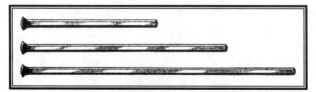

Miners used drill-steels in graduated sets for boring blast-holes in rock underground, blasting being the prime mover of rock at most prospect operations and mines.

International Textbook Company International Correspondence School Reference Library: Rock Boring, Rock Drilling, Explosives and Blasting, Coal-Cutting Machinery, Timbering, Timber Trees, Trackwork International Textbook Company, Scranton, PA, 1907 A35 p5.

Sharpening drill-steels began at the forge, where the blacksmith carefully arranged a layer of fuel over the gravel bed surrounding the tuyere. The choice of fuel for working iron was limited to a few sources that were clean-burning, fairly inexpensive, and easily sacked for transportation. Prior to the 1870s blacksmiths used mostly wood charcoal, but they substituted coke and metallurgical coal, also known as forge coal, by the 1880s. Metallurgical coal included anthracite, semi-anthracite, and unusually pure bituminous coals, all other grades having too much sulfur and other impurities. While metallurgical coal burned relatively cleanly, over time it left deposits of ash and clinker in the forge. Clinker is a residue which appears dark and vitreous with a scoria-like texture, and it forms in nodules up to three-quarters of an inch in diameter. The soot-smudged blacksmith had to periodically clean this nuisance out, and he either dumped it on the shop floor or threw it out of the building's doorway.[9]

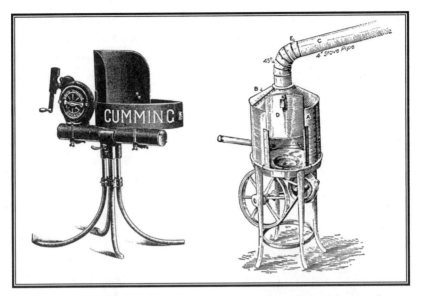

Blacksmiths at prospect operations relied on one of a variety of inexpensive forges for sharpening dulled drill-steels and for general mine work. The forges fall into two categories: portable iron pan units, and vernacular types constructed from local materials. At left is a portable pan forge with blower, and at right is a portable pan forge with blower and hood.

Left: Engineering & Mining Journal. Right: Engineering & Mining Journal Oct. 6, 1917.

After the blacksmith received a load of dull steels, he either pumped a bellows or slowly turned a hand blower connected to the tuyere, which fed oxygen to the fire. As the fire grew hot and began consuming fuel, he used a forge sprinkler to create a perimeter of wet coal to stop the fire from spreading. Blacksmiths often made forge sprinklers from food cans by perforating the bottom with small holes. The smith placed the ends of several drill-steels in the center of the fire until they grew almost white-hot. One by one he extracted them, hammered the blade against the step between the heel and top face of the anvil to reform the drill's sharp angle of attack, placed the steels back in the fire, and repeated the process using a special swage fitted into a socket in the anvil. The swage had a better-defined crevice, which gave the final steep profile to the sharpened blade. The steels went back into the fire yet again, and the denizen of the shop extracted them one-at-a-time for quenching in a small tank of cool water. Quick submersion hardened the steel so it would remain sharp. A second, slower immersion tempered the steel, adjusting the softness of the blade tip after hardening, which prevented fragments from spauling off in the drill-hole. In the event the miners managed to crack or damage the drill-steel blade, the blacksmith heated it white-hot and *upset* the steel before sharpening, meaning he used a cold chisel to cut off the damaged end. After upsetting the tip, the smith had to reform a fresh blade.

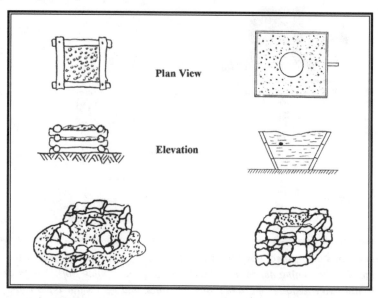

Both profitable mining outfits and prospect operations attempted to save capital by building vernacular forges with local materials. Clockwise from top left: profile and plan view of vernacular log cribbing forge, profile and plan view of vernacular wood box forge, quarter view of vernacular dry-laid rock forge, and quarter view of how a dry-laid rock forge appears following decades of abandonment. Note that the box forge has been capped with grout except for a center port that accommodated a tuyere. All vernacular forges featured a tuyere, which was a blower tube with a hole in the top to feed the fire with oxygen.

Author.

To temper a drill-steel, the blacksmith extracted it from the fire again while it was in a white-hot state and briefly immersed the blade in the quenching tank, quickly extracted it, and permitted the steel to cool in the open air. The incandescent colors of the steel changed as it cooled, and when it reached the desired temperature, as indicated by color, the blacksmith plunged it into the quenching tank to arrest the cooling. During the time the steel lay in the open, the skin cooled faster than the core and turned brown to gray, masking over the steel's true incandescent colors. To examine the colors of the inner steel, the blacksmith rubbed the blade on either a brick or whetstone, which scratched off the grayish scale. Experienced mine blacksmiths were able to complete the above processes proficiently and quickly. They were also able to subtly modify the angle of a drill-steel's blade to better suit it to different types of rock.

Blacksmiths could forge curved blades for bull-steels and judge the incremental widths of the blades by eye.

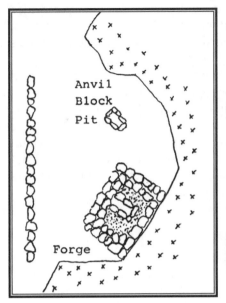

Plan view of the remnants of a crude blacksmith shop typical of impoverished prospect outfits operating in remote temperate areas, in this case at the Surprise Mine in Goodsprings, Nevada. Mine workers built the shop on a cut-and-fill earthen platform against a bedrock outcrop. They may have sheltered the shop with a plank wall or, more likely, a canvas tarp. The shop appliances included a vernacular rock forge, an anvil, a quenching tank, a workbench, and hand tools.

Author.

Plan view of a small temporary-class shop sheltered in a dry-laid rock cabin at a prospect adit in California's Clark Mountain District. The building featured either a plank or canvas roof. Prospectors often located their shops in stone cabins in the desert states between the 1860s and 1890s.

Author.

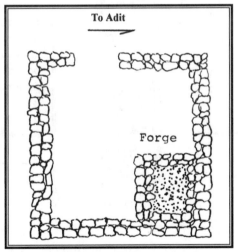

Because blacksmith shops associated with prospect adits were simple and temporary, consisting of small implements intended to be portable, they left scant evidence visible at prospect sites today. Telltale remains by which modern-day visitors can identify the former location of a shop often consist of a concentration of artifacts including a sparse scatter of anthracite coal or coke, forge clinker, forge-cut iron scraps, the blades of upset drill-steels, a brick or whet stone, and occasionally an anvil block. In some cases the remains of a vernacular forge, often reduced to a mound of gravel impregnated with coal and clinker and surrounded by collapsed cobble walls, may also denote the location where a shop stood.[10]

The surface plants erected by prospectors sometimes provided ventilation for the underground workings. The use of explosives for blasting, open flame lights, and the respiration of laboring miners turned the atmosphere of the underground workings intolerably stifling and even poisonous. Ventilating dead-end adits of foul gases was not an easy proposition for capital-poor operations, and as late as the 1910s many outfits completely ignored the problem until the workings attained significant length. *Passive ventilation systems* relied on natural air currents to remove foul air, but they proved marginal to ineffective in dead-end workings. *Mechanically assisted systems* were expensive and intended for production plants, and as a result they were rarely used at prospect adits.

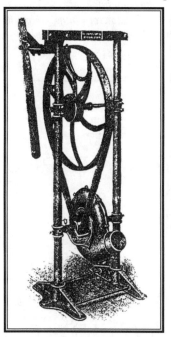

Miners' respiration, open flame lights, and especially the use of explosives for blasting rendered the atmosphere in underground workings lethal, necessitating ventilation. Mining machinery makers offered manual-powered blowers that were well-suited for prospect operations because they were simple, cheap, and easily packed into the backcountry, as exemplified by the illustrated model.

Engineering & Mining Journal June 2, 1904.

Necessity being the mother of invention, prospecting outfits employed several variations of ventilation systems that cleverly combined passive and mechanical means. One of the simplest semi-mechanical ventilation systems that saw extensive use consisted of a canvas wind sock fastened to a wooden pole. The wind sock collected natural breezes and directed the air through either canvas tubing or stovepipes into the underground workings. The obvious drawback to this system was poor performance on calm days, forcing miners to work in suffocating gases. Prospecting outfits employed another semi-mechanical system in which they linked the air intake

Some prospect operations used wood stoves as ventilation devices, as miners had done at the Small Hopes Mine above Boulder, Colorado around 1900. They installed a line of ventilation ducting deep into the adit, linked it to an air intake on a furnace, and used the convection of the fire to draw foul air out of the underground workings. Miners placed ventilation furnaces on small platforms adjacent to the adit portal, as shown. The remainder of the surface plant is typical of small western mining outfits with limited funding.

Courtesy of the Carnegie Branch Library for Local History, Boulder Historical Society Collection

on a stove or furnace, located at the adit portal, to tubing ducted into the workings. A surface worker stoked a fire in the stove, which drew foul air out of the underground through the ducts, combusted the gases, and released them out the exhaust chimney. Despite the simplicity of this ingenious system, most prospecting outfits declined to take the trouble of installing it.[11]

Some western prospecting operations made an earnest commitment to adequately ventilate their workings. From the 1870s into the 1910s, some of these outfits installed large forge bellows at the mouths of adits and used stovepipes or canvas tubing to duct the air into the workings. Bellows effectively ventilated shallow workings, but they lacked the pressure to clear gases out of relatively deep adits and shafts. By the 1880s mining machinery makers offered small hand-turned blowers, which cost more money than the above systems and took greater effort to pack to a prospect operation, but they forced foul air much more surely from workings.[12]

Of all the surface plant components associated with prospect operations, ventilation systems were particularly transitory and left little evidence obvious today after prospectors had abandoned a site. Usually, the only visible remains left at prospect adits today consist of stovepipe sections, duck canvas scraps, and baling wire. The adit interior may feature stovepipe sections still hanging from the ceiling, or placed along one corner of the floor. Such relics usually indicate that a prospecting outfit installed a ventilation system, but the remains are often inconclusive as to which type. Occasionally artifacts such as large forge bellows and blower parts, as well as standing wind sock poles may still be encountered, providing the visitor with a greater certainty about the system.

Surface Plants for Deep Prospect Adits and Mine Tunnels

Mining districts across the West are full of prospect adits abandoned in nascent stages of development. A few mining outfits, however, were graced with luck and a tantalizing showing of pay-rock, and in rare cases they encountered a bonanza. While the surface plants for small prospect adits usually permitted an outfit to drive shallow underground workings, such plants were inadequate for deep exploration, and ultimately ore extraction. Mining companies that possessed claims with ore had to erect large surface plants capable of supporting more intense activity.

The first problem that a mining engineer faced was upgrading the existing, simple surface plant into something capable of meeting the needs of a larger operation. Like the surface plants associated with shafts, the plants for adits met either temporary or production-class definitions. Money being one of the underpinnings of mining, the size of a plant was a function of how much capital the company, i.e. the investors, were willing to part with, and how much ore lay in the ground.

Once the mining engineer had finished evaluating a claim, he entered a planning phase to determine how best to upgrade the property in accordance with anticipated deep exploration and subsequent ore production. He defined which plant components and structures were needed, and where they should be placed. Upgrading a small prospecting plant required a variety of goods and supplies. The mining company had to furnish building materials, heavy tools and hardware, machinery, and other items far greater in size and weight than the carrying capacity of the pack animals used by prospect operations. Therefore, one of the first steps an engineer made toward developing a mine was to establish a transportation artery to the site, preferably with the capability of accommodating heavy wagons. When faced with such an expensive and labor-intensive proposition, mining engineers were able to take solace in the fact that their operations were rarely totally isolated, often being located in mining districts where other companies were engaged in similar development. Neighboring mining companies, ordinarily incapable of agreeing on basic matters such as possession of mineral and water rights, did concur on their common need for roads, and they often cooperated and shared the costs of road building.

Some prospects with excellent showings of ore were located in the most remote, rugged, and inaccessible regions, and road-grading remained an economic impossibility. Such mines were accessed only by mule train and as a result, most of these operations remained retarded because bringing in the necessary materials was physically impossible, rendering the costs of production high. Many prospects in the eastern Sierra Mountains in California, the San Juan and Collegiate Mountains in Colorado, and the Cascade Mountains in Washington not only were too remote for roads, but had the added difficulty of being snowbound during the winter. Mining companies attempting to develop such claims ironically found the winter to be the most conducive time of year to upgrade their surface plants. Rather than relying on mules during the warm months, they used the snowpack to advantage, sending out teams of miners to winch supplies and machinery on sleds across ordinarily impossible terrain. In cases where such claims proved highly profitable, engineers convinced the investors to part with enough capital to build an aerial tramway, which we will examine at the end of the chapter.

Once the road grading crew had completed a transportation artery to the site, all was ready to begin moving machinery and materials. The road acted, in essence, as an umbilical cord linking the mine to commercial centers, giving engineers a freer hand on what types of machines and other facilities they could install. Professionally trained engineers recommended installing production-class plant components from the outset, economic conditions permitting. But seasoned mining engineers, such as Frank Crampton and W.S. Stratton, who had experienced firsthand the vagaries of western mining, often thought otherwise, remaining hesitant until the mine truly proved itself capable of sustained production. Part of the reason for the seasoned engineers' hesitance was that ore reserves often pinched out relatively close to ground surface, which could have left the mining operation in debt. Another major reason for their hesitance was that they had slightly lower standards for the tonnages of ore produced per shift than did professionally trained engineers. Seasoned and self-taught engineers, with their skepticism, installed smaller, less-efficient plants, but in so doing they risked less capital than was typical of their professionally-educated counterparts. As deep exploration revealed the existence of ore, the seasoned and skeptical engineers then upgraded their surface plants in a piecemeal fashion, installing or improving the facilities only on demand.[13]

Whether the operation was under the charge of professionally-trained engineers or the self-made variety, improvements to the plant required several fundamental steps that both schools of engineers followed. The first stage was preparing the overall site for the installation of a larger and more complete surface plant. Between the 1870s and the 1890s, this meant at a minimum building a new shop, transportation system, storehouses, an office, and an ore bin. Improvements made after this time often included an air compressor, power appliances for shops, and a power source. According to plan, the mining engineer put a labor crew to work with pick and shovel to create cut-and-fill earthen platforms around the adit portal on which they sited the necessary facilities. With this accomplished, ore production could begin in earnest.

⛏

The Adit Portal

Engineers recognized that narrow, low-clearance adits driven for prospecting were wholly inadequate for deep exploration and ore production. When improving a shallow prospect for deep exploration and ore production, the engineer put miners to work widening the passageway with drill-steels and dynamite.

Improving the adit included giving due attention to the adit portal, which guarded against cave-in of loose rock and soil. Aridity in the Southwest and many areas of the Great Basin discouraged the development of deep soils, and, in accordance bedrock lay near ground surface, often permitting adit portals to remain unsupported. However, in the mountain states where rain and snow were more frequent, mining companies had to use timbering. Mining engineers recognized *cap-and-post timber sets* to be best suited for supporting both the adit portal and areas of fractured rock farther in. This ubiquitous means of support consisted of two upright posts and a cross-member, which miners fitted together with precision using measuring rules and carpentry tools. They cut square notches into the cap member, nailed it onto the tops of the posts, and raised the set into place. Afterward, the miners hammered wooden wedges between the cap and the adit ceiling, and between the posts and adit walls, to make the set weight bearing. Because the adit usually penetrated tons of loose soil and fractured rock, a series of numerous cap-and-post sets were required to resist the heavy forces, and they had to be lined with *lagging* to fend off loose rock and earth. In areas penetrating swelling ground, the bottoms of the posts had to be secured to a floor-level cross-timber or log footer to prevent them from being pushed inward. When adits penetrated great lengths of heavy ground, miners spaced the timber sets as close as two feet together and tied each set to the next with horizontal stringers.

When miners encountered bad ground while driving a tunnel, they erected an internal skeleton of cap and post support timbering. Desert operations like the Moorehouse Mine near Death Valley found dimension lumber to be as economical to use as hewn logs, which mountain outfits favored. In wet ground, mining companies had to endure the costs of periodically replacing the timbers when they rotted. The photo captures a typical small production-class tunnel with 18 inch gauge track and a compressed air line extending along the left wall.
Author.

Wood used for the purposes of supporting wet ground decayed quickly and had to be replaced as often as several times a year, but as infrequently as every few decades in dry mines. Professionally-trained mining engineers claimed that dimension lumber was best for timber sets because it decayed slowly and was easy to frame, but a relatively high purchase price and the cost of transportation discouraged its use where cheaper alternatives were available. Most down-to-earth miners and engineers favored using hewn logs for their timber sets and lagging because they cost less than milled lumber, and they were often ready at hand in the mountain states. In the desert states where logs were a rarity, mining companies found it economical to use dimension lumber.[14]

A few highly profitable and well-financed operations featured portals with permanent support. For example, the portals of Colorado's Bobtail Mine in the Central City Mining District and the Argo Tunnel in the neighboring Idaho Mining District have been supported with faced stone masonry and decorative poured concrete, respectively. In addition to the functionality of extravagant facades, such portals served as intentional statements of wealth, productivity, and permanence, which inspired confidence among investors, as well as in the community.

Constructing the Mine Shop

A fully equipped blacksmith shop was one of the most important of the new plant components an engineer included in the mine's upgrade. The physical composure of a shop reflected the financial state of a mining company. Small outfits with limited financing erected simple, labor-intensive facilities, while well-capitalized mining companies proximal to commercial centers often erected large, well-equipped facilities.

The size, complexity, and makeup of mine shops changed during the 1890s, paralleling an overall surge in affordable and practicable technology. In many cases professionally-trained mining engineers such as T.A. Rickard and Herbert C. Hoover formulated models and ideals of efficient and economical mine shops. The engineers recommended that woodwork and metalwork be conducted in individual buildings, and that the respective shops be equipped with modern, energy-efficient, and time-saving appliances. Some of these mining men even applied hypotheses of materials handling efficiencies to the layouts and makeup of shops at the largest mines. They asserted that the facilities comprising a shop be arranged according to materials flow, and more specifically to the steps required for sharpening drill-steels. However, these recommendations proved impractical for most western mines due to limited funding and space.[15]

The shops at small mines typically occupied a space approximately ten-by-fifteen feet in area. They featured a forge and blower in one corner, an anvil and quenching tank next to the forge, a work bench with a vice located along one of the walls, and a lathe and drill-press. Forges at mines in both the Great Basin and mountain states tended to be either free-standing portable iron pan types, or vernacular fieldstone forges. Rarely did shops at small mines include power appliances. Instead, most of these shops were equipped with manually operated machinery.[16]

The line drawing is representative of the austere interior of the blacksmith shops at small western mines and deep prospects. The shop consists of basic appliances, most of which are out of view. Center is an immobile tank forge fed air from a bellows at right. Few shops at western mines featured brick floors like the shop in the illustration. Note the anvil block, which today's visitor may encounter at historic mine sites.

Drew, J.M. Farm Blacksmithing: A Manual for Farmers and Agricultural Schools Webb Publishing Co., St. Paul, MN, 1910 p1.

Prior to the 1890s, production-class shops at medium-sized mines were between fifteen-by-fifteen feet to fifteen-by-twenty feet in area, and they featured more appliances than their small-mine counterparts. They usually included several workbenches equipped with vices, a large manual-powered lathe, a drill press, a full array of hand tools for forge and machine work, and pipe threaders. The blacksmith also nailed small parts bins to the walls to contain the mine's supply of basic hardware. In addition to the usual metalworking and mechanic's implements, the shops at medium-sized mines were large enough to accommodate a carpentry work area where workers manufactured small wood items. Heavy carpentry work such as dressing mine timbers, framing, and other types of fabrication presented space problems in such shops, driving workers out onto the flat surface of the mine dump. If the mine lay deep in the mountains where winters presented arctic conditions, the mining company financed construction of a *tunnel house,* which offered drafty shelter to miners, laborers, and shop workers engaged in critical support functions. Tunnel houses usually took the form of a gabled frame building, and they enclosed the adit portal, the shop, a work area where carpenters wrestled with heavy woodwork, space for limited materials storage, and possibly an office.

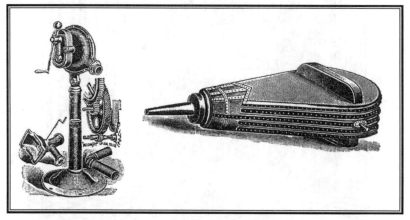

Blacksmith shops at small mines and deep prospects relied on either manual-powered mechanical blowers, like the model at left, or bellows, right, to feed oxygen to the fire. Bellows fell out of favor during the 1890s.
Mine & Smelter Supply Co. Catalog No. 22: Machinery and General Supplies J.D. Abraham Publishing Co., 1912 p724, 725.

Mining engineers working for large companies had the resources to erect well-equipped and spacious shops. To permit two blacksmiths to simultaneously sharpen drill-steels, manufacture hardware, and repair items, the commodious shops featured one large forge, and in some cases two separate forges. In addition, such shops featured belt-driven metal and woodworking appliances, such as trip hammers, lathes, drill-presses, and wood saws. A small upright steam engine drove the machinery by canvas or leather belts descending from an overhead driveshaft. Mining companies rarely installed a steam system solely to power shop appliances; they usually did so in concert with additional machinery such as a ventilating fan and an air compressor.[17]

Shops at medium-sized western mines built during and after the 1890s remained rough, sooty, and dark like their old counterparts, but they were larger and better equipped. A greater availability and affordability of steam engines, air compressors, and electricity during the 1890s brought power appliances within reach of more mining operations than in decades past. Still, available capital and geographic location heavily influenced the degree to which mining companies mechanized the shops at medium-sized mines. Companies working remote properties were less enthusiastic about paying the high transportation costs to haul power appliances to their sites.

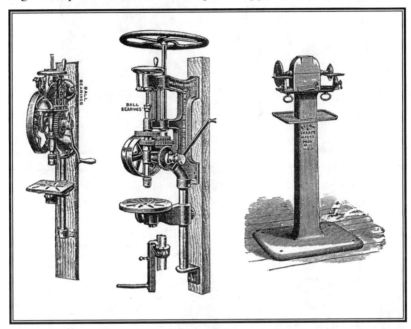

Mining companies with limited financing were forced to equip their shops with manual-powered metal and woodworking appliances, such as the drill-press at the left. They also installed manual-powered lathes, grinders, and used hand tools. Well-funded mining companies installed belt-driven drill-presses and grinders, as depicted by the images at center and right, which enhanced their materials-handling capabilities. The drill-presses were bolted onto the shop building's support frame.

Mine & Smelter Supply Co. Catalog No. 22: Machinery and General Supplies J.D. Abraham Publishing Co., 1912 p744.

Brown & Sharpe Mfg. Co. Catalog: Machinery and Tools Providence, RI, 1904 p111.

By the 1890s, medium-sized mines featured shops usually enclosed in frame buildings between fifteen-by-twenty and fifteen-by-thirty feet in area. These shops were equipped with at least one forge, an accompanying blower, an anvil, a quenching tank, two stout workbenches, a lathe, a drill-press, and an array of machine and carpentry tools. In addition to the above appliances, many such shops were also equipped with a mechanical saw, a grinder, and a pipe threader, which may have been power-driven.[18]

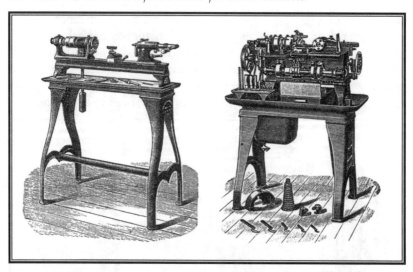

Well capitalized mining companies installed additional belt-driven appliances such as the lathe at left and threading machine at right. Belt-driven appliances usually were free-standing and they had to be anchored to a foundation to prevent shifting while running.

Brown & Sharpe Mfg. Co. Catalog: Machinery and Tools Providence, RI, 1904 p156, 196.

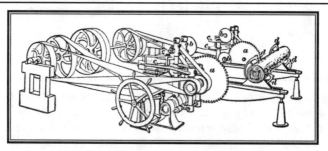

Large and heavily capitalized mining companies often erected a carpentry shop, in which they may have installed a complex belt-driven timber-framing machine, as illustrated. Such appliances could cut notches and joints in hewn log and dimension lumber mine timbers, which was a great aid to over-worked timbermen.

International Textbook Company A Textbook on Metal Mining: Preliminary Operations at Metal Mines, Metal Mining, Surface Arrangements at Metal Mines, Ore Dressing and Milling International Textbook Company, Scranton, PA, 1899 A41 p113.

Blacksmiths working for large mines were more sophisticated than their brethren at small operations in tempering and hardening their ironwork. Blacksmiths had recognized that using a tank of stagnant water for quenching became problematic as the water tended to absorb heat after the repeated immersion of searing-hot metal. The water's ever-changing temperature interfered with precise and exact hardening and tempering. Blacksmiths and mine machinists at larger mines avoided this problem by installing continuous-flow quenching tanks, which maintained an even water temperature. These innovative shop appliances were costly to purchase because they were heavy galvanized sheet iron troughs capable of holding over twenty-five gallons of liquid, they featured inflow and drain lines, and they required a source of water. Well-financed mine shops at large mines also used quenching tanks filled with oil for the express purpose of extreme hardening, and they also hardened steel in brine solutions, both of which were more efficient than plain water.[19]

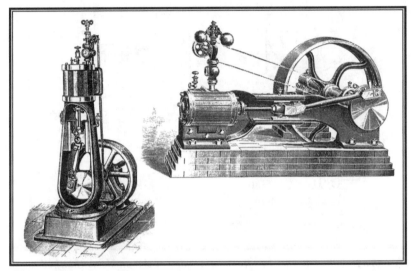

Belt-driven shop appliances required a power source. Prior to the adoption of electricity around 1900, well-capitalized mining companies installed steam engines that turned a central drive shaft suspended from bearings in the building's rafters. The illustration depicts the ubiquitous vertical engine, at left, and the less-common horizontal steam engine at right. Note the foundations, which can still be seen at the sites of large mines today. These machines required steam supplied from a boiler.

Ingersoll Rock Drill Company Catalog No.7: Rock Drills, Air Compressors and Air Receivers
Ingersoll Rock Drill Company, New York, NY, [1887] p53.

In the tradition of western mining, between the 1890s and the 1920s, shop laborers continued to focus their efforts on drill-steel sharpening. However, the widespread embrace of compressed air-powered rock drills required shop workers to adapt their skills and materials-handling processes. While the drills proved to be a mixed blessing for their operators, generating silicosis-causing rockdust and being backbreaking to handle, they were a boon for shop workers. The noisy and greasy machines produced volumes of dulled steels and broken fittings. Contrary to today's popular misconceptions, rock drills replaced hand-drilling wholesale in western mines by the late 1910s, and not earlier as supposed. The conversion evolved over the course of thirty years, progressing more rapidly among well-financed mining companies than at small operations. During the conversion period, blacksmiths became proficient in sharpening both hand-steels and machine drill-steels, each of which had specific requirements.[20]

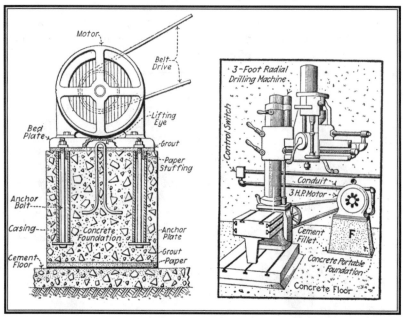

By around 1900 engineers had ironed out many of the problems that had rendered electric power impracticable for mine work, and well-capitalized mining companies employed motors to power shop appliances. The illustration details a typical motor mount at left, and the means of connecting it to a heavy drill-press at right. Such appliances were out of the economic reach of many modestly capitalized mining outfits.

Croft, Terrell Machinery Foundations and Erection McGraw-Hill Book Co., New York, NY, 1923.

The large volume of dull rock drill steels, machine repair work, and the manufacture of fittings constituted a heavy workload for shop workers. In an effort to facilitate the completion of projects in a timely manner, mining companies usually hired a blacksmith and a helper for metalwork, and a carpenter and another assistant for woodwork. In terms of metalwork, the blacksmith's helper proved to be particularly important. Blacksmiths traditionally sharpened hand-steels alone because the implements were relatively short, light, and easily managed. But this was not the case with machine drill-steels, which were made of heavy iron rods up to eight feet in length.[21]

Before discussing the specific processes blacksmiths employed for sharpening machine drill-steels, it is important to become familiar with the basic forms commonly used by western mines prior to World War II. Simon Ingersoll and the Rand brothers introduced the first commercial rock drilling machines in the early 1870s. Termed by mining machinery makers the *piston drill*, the early rock drills consisted of a compressed air-powered piston in a tubular body, with a drill-steel chuck cast as part of the piston. As the piston chugged back and forth at the rate of several hundred strokes per second, it repeatedly rammed the drill-steel against rock in a manner similar to a high-speed battering ram. When in operation the mechanical drill also imparted a spinning motion to the piston and drill-steel to keep the hole round and prevent the steel from wedging tight.

It may be apparent to the reader that drill-steels used in conjunction with the heavy machines were specialized implements that had to withstand tremendous forces. As early as the 1870s *machine runners*, also known as *machine men*, found that single-blade cutting bits like those used for hand-drilling dulled quickly, impeded progress, and interfered with the rotation imparted by the machine. The most effective bit proved to be a cruciform shape where two chisel blades crossed at dead center. This *star bit* better withstood punishment, it cut faster, and was conducive to rotation. The butt of rock drill steels was round to fit into the drill chuck, and the steel was usually made from one to one-and-one-half inch hexagonal steel rod.[22]

Many miners found that piston drills had severe limitations and inconveniences. For example, every time the *chuck tender*, the machine runner's assistant, changed a dull steel for a fresh one, he had to use a heavy wrench to unbolt the chuck shackle, trade steels, and refasten the nuts using tremendous strength. In addition, miners agreed that the monstrous piston drills were exceedingly heavy, often weighing between 200 and 350 pounds without accessories, and their drilling speeds were limited. George Leyner, Denver machinist and former Colorado hardrock miner, invented a superior rock drill in 1893 that was based on a mechanical simulation of double jacking. Instead of repeatedly ramming the rock as did piston drills, Leyner's drill employed a loose piston known as a *hammer* that cycled back and forth inside the drill and struck the butt-end of the drill steel, which rested loosely in the chuck. Like most rock drill makers, Leyner designed his drill for positive chuck rotation to make round holes and to keep the drill-steel from jamming. Leyner patented the first marketable *hammer drill* in 1897 and began producing an improved version in 1899.[23]

Between the 1900s and the 1910s, Leyner's drill began finding favor with the hardrock mining industry. Time and again miners demonstrated that hammer drills bored holes faster than piston drills, and miners found them easier to work with in terms of changing steels. All the chuck tender had to do was give the dull drill-steel a twist to unlock it, and twist in a fresh drill-steel; no longer did miners have to deal with clumsy shackle bolts. Leyner's steels were made of one-and-one-quarter inch round bar stock, and they featured star-shaped cutting bits like piston drill steels. A crew of two miners was necessary to handle Leyner's machine, and it too had the drawback of running dry like the old piston drills. To this regard Leyner devised a hollow drill-steel which jetted water into the drill hole while the drill was running, allaying rock dust. Leyner's technology gradually caught on throughout the mining industry until, by the mid-1910s, drill companies curtailed the manufacture of piston drills in favor of the hammer drill.

From 1897 to 1912, mechanical engineers introduced new types of rock drills using Leyner's hammer principle. The first new drill was the *stoper,* which was a lightweight machine designed to bore holes upward. Early stopers lacked a chuck rotation mechanism, and as a result the miners running them had to use a long handle that extended out of the machine's body to turn the unit side to side to keep the drill

A pair of miners operate piston drills to bore blast-holes in a shaft-sinking operation. As the drill deepened the hole the operator slowly fed the machine forward by advancing the crank. When he needed a longer drill-steel, he shut off the air valve, retracted the machine, unscrewed the nuts on the chuck, and replaced the short dull steel with a longer unit. These noisy machines required compressed air, they had to be affixed onto a heavy iron column, and they had to be moved by two miners. However, rock drills expedited the blasting process, and they were in some ways easier than drilling by hand.

Rand Drill Company Illustrated Catalog of the Rand Drill Company, New York, U.S.A. Rand Drill Company, New York, NY, 1886 p10.

hole round. Miners and stoper manufacturers found that the best type of drill-steel proved to be cruciform in shape, which prevented the steel from twisting and jamming in the machine.

A drilling and blasting team operates a hammer drill deep in the Colorado Consolidated Mine No.4, Eureka, Colorado during the late 1910s. Hammer drills were faster than piston drills and changing drill-steels was much easier. The logistics of setting up and running a hammer drill were the same as with the old piston units. The mustachioed miner at left clutches a wrench used to tighten the nuts clamping the drill's saddle mount.

Ingersoll-Rand Co. Leyner-Ingersoll Drifters Ingersoll-Rand, New York, NY, 1921
[Trade Catalog] p20.

In 1912 Ingersoll-Rand, formed by the 1906 merger of the Rand and Ingersoll companies, developed a revolutionary hammer drill for boring down-holes. Known among miners generically as a *plugger, shaft sinker,* and as simply a *sinker,* Ingersoll-Rand named its model the *Jackhammer,* which is the origin of the slang name used today. The machine consisted of a gracile hammer drill fitted with handles, and a mechanism for rotating the chuck. The relatively small hand-held machine required a drill-steel lighter than those used with the larger Leyner hammer drill, and Ingersoll-Rand and subsequent manufacturers found that seven-eighths inch hexagonal bar steel proved best. The butt of sinker steels was hexagonal and featured a collar that fit into a special hinged clamp. By the 1910s, hollow drill-steels manufactured from round and square stock became standard for hammer drills, and hexagonal steels became standard for sinker drills. During the 1920s drill makers ceased production of cruciform steel, and equipped their stopers for hexagonal and square steels.

A miner, spotted with candle wax drips, operates a stoper drill to bore an uphole during the first years of the twentieth century. Stopers usually required cruciform drill-steels which miners used in graduated sets. Instead of a hand-crank mechanism for feeding the drill into the rock, the stoper featured a compressed air-powered telescoping foot which provided constant pressure. These machines produced high volumes of silicosis-causing rock dust.

Cleveland Pneumatic Tool Co. Cleveland Air Hammer Drills Cleveland Pneumatic Tool Co., Cleveland, OH, ca. 1905 [Trade Catalog] p3.

The illustration depicts a graduated set of drill-steels for piston rockdrills. Note that one end of the steels feature star cutting bits and the other ends are round to fit into the drill's chuck.

International Textbook Company International Correspondence School Reference Library: Rock Boring, Rock Drilling, Explosives and Blasting, Coal-Cutting Machinery, Timbering, Timber Trees, Trackwork International Textbook Company, Scranton, PA, 1907 A35 p37.

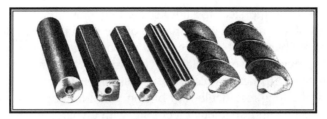

The cross-sections depict five of the principle types of drill-steels used for hardrock and coal mining. From left to right: round hollow steel used in large hammer drills, square hollow steel used in heavy hammer drills and stopers, hexagonal steel used in sinker drills and stopers, cruciform steel used in stopers prior to the late 1920s, and the right two steels are coal augers. Most of the steels feature holes to accommodate a water jet.

Denver Rock Drill Manufacturing Company Drill Steel Denver Rock Drill Manufacturing Company, Denver, CO ca. 1924 [Trade Catalog] p2.

Regardless of the specific type of stock from which a drill-steel had been made, the blacksmith had to confront the problem of sharpening the star bit, and keep the water hole open in the steels used by hammer drills. As with hand steels, the blacksmith had to place the machine steels in the forge fire to heat them to the proper temperature. He simply laid short steels on the forge, but he had to use either a special stand or a long hook suspended from the building's roof rafters to support drill-steels in excess of three feet long. When the blacksmith extracted a steel from the forge with the intent of dressing the bit, he used a tool known as a *swage* or *dressing dolly*, to resurface the star's cutting edges. If the drill-steel arrived in the shop with a chipped or cracked bit, the blacksmith upset the damaged portion by using a chisel to cut it off, then hammered out a new end with enough flare to facilitate creation of a star bit. The blacksmith also ensured that he had centered the star, that the blades were uniform in width, and that the butt of the steel was smooth and symmetrical. After he had dressed the bit, the blacksmith filed imperfections out of the blades, followed by tempering and hardening. All through this process the helper assisted the blacksmith when handling long steels.[24]

Some companies running medium-sized mines supplied their blacksmiths with an appliance known as a *backing block* to ease the difficulties of sharpening unwieldy machine drill-steels. Ordinarily, the blacksmithing team had to act in close concert when sharpening machine steels. The helper leaned the red-hot drill-steel against the anvil located adjacent to the forge and braced it with both hands while the blacksmith dressed the bit with a dolly. However the propensity of the steel to slide, sway, and move under the black-smith's blows, and the giving nature of the shop's earthen floor presented problems that often resulted in poor sharpening. A backing block provided a sound platform for drill-steels, permitting blacksmith teams to better dress bits in less time. Backing blocks consisted of a long rectangular bar of iron, often four-by-four inches in cross-section and up to eight feet long, with divots spaced every half foot. The iron bar was firmly anchored in the ground and it extended outward from the anvil block. To use it, the blacksmith's helper placed the butt of a red-hot drill-steel in one of the block's divots and leaned the steel's neck against the anvil to permit the blacksmith to dress the bit. The backing block provided sound resistance to the blacksmith's heavy blows while holding the drill-steel in place. Each divot in the block accommodated a different length of drill-steel, from two foot starter-steels to ten-foot finishing steels. These ingenious appliances began appearing during the 1890s in the shops of large mines where rock drills were used. Mining companies with sufficient capital purchased factory-made cast-iron models, while penny-pinching outfits engaged their shop workers to forge their own from scrap iron such as salvaged railroad rail.[25]

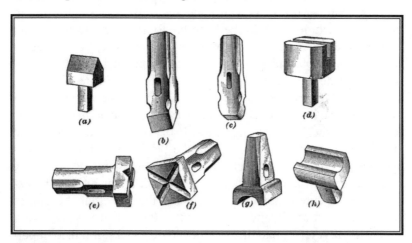

The line drawings depict a set of hand tools that blacksmiths who worked for modestly financed mining outfits used to manually sharpen machine drill-steels. Swages a, d, and h fit into a socket in the anvil, and swages b, c, e, f, and g were affixed onto hammer handles. The blacksmith used each hammer-head against the corresponding anvil swage. Specifically the sooty shop denizen used swages g and h to dress the drill-steel butt, and swages e and f to sharpen the star bit.

International Textbook Company International Correspondence School Reference Library: Rock Boring, Rock Drilling, Explosives and Blasting, Coal-Cutting Machinery, Timbering, Timber Trees, Trackwork International Textbook Company, Scranton, PA, 1907 A35 p36.

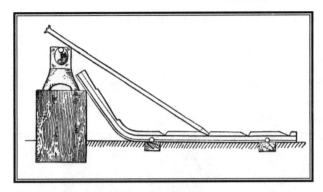

Blacksmiths used backing blocks to brace red-hot machine drill-steels for sharpening. As the profile illustrates, the blacksmith placed the steel's butt, which is the diagonal rod, into a receptacle in the backing block and leaned its neck against the horn of the anvil at left. When his assistant held the steel steady, the blacksmith used a swage to form and sharpen the bit. While mine supply houses offered factory-made backing blocks, many blacksmiths made their own backing blocks out of railroad rail, as illustrated in the profile.

Engineering & Mining Journal Details of Practical Mining McGraw-Hill Book Company, Inc., New York, NY, 1916 p14.

In the first decade of the twentieth century the largest of the western mines, where dozens and even hundreds of miners dulled carloads of drill-steels per shift, attempted to mechanize the sharpening process in hopes of drastically increasing the efficiency of harried shop crews. The well-financed mining companies purchased, seemingly on an experimental basis, compressed air powered *drill-steel sharpening machines,* which had just been released onto the market by manufacturers such as T.H. Proske in Denver and the Compressed Air Machinery Company in San Francisco.[26] The early drill-steel sharpeners, similar in appearance to large horizontal lathes, consisted of a cradle approximately eight feet long and a tall sharpening mechanism that stood on several legs bolted to a substantial foundation. A blacksmith operated the sharpener by clamping a red-hot drill-steel into a small sliding carriage on the cradle, he pushed the steel under the sharpening mechanism, and locked it in place. The shop worker threw a lever that activated a modified piston drill fixed onto the machine's end, which hammered the red-hot end of the dulled steel with a special swage. Most of the early drill-steel sharpeners also featured a second piston drill mounted overhead, which used a special chisel bit to upset the dull steel, should it have any significant defects.[27]

Manufacturers advertised that their sharpeners could streamline the sharpening process and reduce costs. Drill-steel sharpeners were operable by one man; they had the capacity to replace the traditional crew of blacksmith and helper, and with a change of dies they could have been used to sharpen hand-steels and pick tines. Even though the drill-steel sharpeners cost hundreds of dollars, they proved economical and grew in popularity.[28]

Machinery makers began offering practical drill-steel sharpeners around 1900. The shop machinist clamped a red-hot drill-steel in the sliding dolly on the track at right and the two compressed air powered pistons at left worked swages that cut off the old bit, if damaged, and pounded a new bit into form. The intense vibration of the machine required that it be bolted onto a stout concrete foundation. The large and complex machines greatly increased the number of drill-steels one shop worker could handle during a shift, but their high cost relegated them to the shops of heavily capitalized mining outfits.

Mining & Scientific Press April 11, 1903.

Leading rock drill makers, including the Sullivan Machinery Company of New Hampshire, the Ingersoll-Rand Drill Company, and the Denver Rock Drill Company introduced competing units during the early 1910s that had abandoned the lathe-like sliding track and large piston drill swages.[29] The new drill-steel sharpeners instead featured a heavy compressed air-powered clamp capable of holding drill-steels of any length, and they had small, light hammer drills to work the swages. In addition, manufacturers supplied interchangeable dies that permitted shop workers to sharpen any of the varieties of drill-steel types used in the West at that time. The net result of the changes in the form and function of drill-steel sharpeners was a reduction in the amount of floor space they occupied, from at least ten-by-two feet in area to between five-by-two feet and two-and-one-half by two-and-one-half feet. The labor-saving machines primarily made themselves of value to mining outfits because they reduced the time required to sharpen dull drill-steels to less than one minute, with the potential to retouch up to one-thousand-fifteen dull drill-steels in a nine hour shift. It stands as a curious fact that many of these machines had been designed in Denver. Sullivan purchased the Imperial sharpener from T.H. Proske, Ingersoll-Rand used a design manufactured by George Leyner, and the Denver Rock Drill Company produced the third machine.[30]

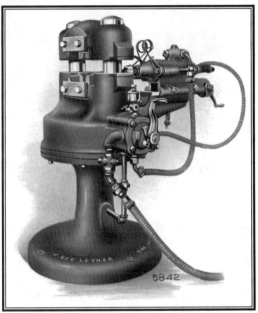

Shortly after 1910 the leading rockdrill makers had revised the drill-steel sharpener, reducing its size and its cost. As a result, a greater number of mining companies employed drill-steel sharpeners to expedite materials handling. The shop worker clamped a drill-steel under the raised head and operated the lever at center to engage the machine.

Ingersoll-Rand Co. Leyner Drill Sharpener "I-R" Model: Instructions for Installing and Operating Ingersoll-Rand Co., New York, NY, 1913 p2.

The reduction of size and price of the new drill-steel sharpeners, and their ease of use made them attractive to a broad spectrum of medium-sized and large mines. Both moderate and well-financed mining companies with an expectation of longevity installed the improved drill-steel sharpeners with increased frequency throughout the 1910s. Most small mining companies, on the other hand, did not purchase drill-steel sharpeners because such outfits lacked available capital, their miners were unlikely to generate enough dull steels to justify the expense, and they did not possess adequate air compressors. Instead, they relied on traditional forge sharpening methods.[31]

The early lathe-like drill-steel sharpeners, used in small numbers during the first decade of the twentieth century, were large, heavy machines subject to extreme vibration. As a result the manufacturers and mining engineers recommended that the apparatuses be firmly

bolted either to heavy concrete or timber foundations. Units manufactured by the Compressed Air Machinery Company and T.H.
Proske were bolted to concrete pylons. The main portions of both
machines were fastened to blocks measuring approximately four-by-
two feet in area and two feet high, and the far end of the sliding
tracks resting on blocks one-by-two feet in area and two feet high.
Other varieties of the lathe-like drill-steel sharpeners such as the
Word model and T.H. Proske's Little Giant were bolted to heavy parallel timbers at least eight feet long and two feet apart, which were in
turn bolted to the shop building's subframe.[32]

A shop worker is operating a Leyner drill-steel sharpener to form the star bits
on hammer drill-steels. The remainder of appliances depicted in the vignette
are typical of well-equipped shops, and they include an oil forge at left, a tri-
pod to support long drill-steels at center, and a large anvil at right. Note that
four bolts anchor the drill-steel sharpener onto a wood foundation.

Ingersoll-Rand Co. Leyner Drill Sharpener iI-Rî Model: Instructions for Installing and Operating
Ingersoll-Rand Co., New York, NY, 1913 p6.

The upright sharpeners, for which no alternative existed by the
late 1910s, became increasingly popular as more mining companies
were able to purchase new units, and used models became available.
Generally, mine machinists and mining engineers found that the
heavily vibrating drill-steel sharpeners destroyed unpadded concrete
foundations over time, and they recommended that upright models
be bolted to timber foundations embedded in the earthen floors of
shops, or to wood footings over concrete shop floors.

Well-financed, large mining companies operating after 1890
appointed their shops with a few additional expensive but efficient

appliances. Professionally-trained mining engineers and shop super-intendents installed one of several specialized furnaces exclusively for heating drill-steels, to supplement the old-fashioned coal forge. A few engineers experimented with coke-fired furnaces that automatically fed the fire with fuel, but oil forges proved to be by far the most popular. Oil forges began appearing shortly after 1905, and they grew in popularity during the 1910s as fuel oil became more common. The devices were about the same shape and size as traditional free-standing forges, and they had fuel lines and hoods with slits designed to admit between ten and twenty drill-steels. Engineers felt the alternative furnaces provided clean heat and permitted a rapid turnover of materials, a necessity at a busy mine.[33]

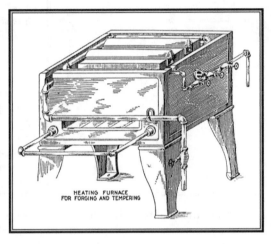

Oil forges did not experience popularity until the 1910s and even then only among well-financed mining outfits because the appliances were expensive, unconventional, and they required a constant supply of fuel oil. Yet for a large mine shop handling numerous drill-steels they proved to be economical.
Engineering & Mining Journal
Oct. 6, 1917.

HEATING FURNACE
FOR FORGING AND TEMPERING

Mining engineers and shop superintendents at large operations installed power hammers to permit a single blacksmith to do some types of fabrication work that usually required a team of two. Shop superintendents overseeing the highly profitable mines of the West installed factory-made steam or compressed air-powered models, which consisted of a heavy plate iron table fixed to a cast-iron pedestal, and a piston hammer that pounded with tremendous force. These hammers were expensive to purchase and transport. They occupied the same area as a drill-steel sharpener and weighed several tons. Many engineers, especially seasoned, self-made individuals like W.S. Stratton and Frank Crampton, were unwilling to spend the capital required to install expensive factory-made hammers, yet they recognized the usefulness of such an appliance. They employed an ingenious alternative that consisted of a worn piston drill fixed onto a stout vertical timber. The old drill stood over a plate iron table fastened onto the top of a timber post several feet high, and when a shop worker threw the air valve open, the drill's piston chuck rapidly tapped the iron table. Usually they clamped a special hammerhead fitting into the chuck to facilitate blacksmith work, and in rare cases the shop superintendent had the drill suspended from a special track hanging from the building's rafters for mobility.[34]

Heavily financed mining companies equipped their shops with costly power appliances that broadened the scope of their work to include machining and light foundry capabilities. The large and smoky shops were appointed with power-driven pipe and rod threaders, metal band saws, drill-steel straighteners, large lathes, hole punchers, spacious drill-steel sorting tables, tool cabinets, plumbing, and turntables at the intersections of mine rail lines.

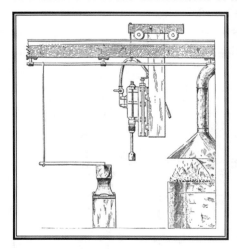

Clever mining engineers adapted old and worn piston drills to serve as vernacular power hammers in mine shops rather than purchase costly factory-made models. Such appliances could only have been used at mines equipped with an air compressor. Contrary to the illustration, in most cases engineers affixed the hammer onto a bracket bolted onto a stout stationary timber within the shop.
Engineering & Mining Journal Details of Practical Mining McGraw-Hill Book Company, Inc., New York, NY, 1916 p18.

Today the forges in the mine shops of the West have grown cold, the machinery is now silent and still, and the ring of blacksmith's hammer can no longer be heard. Despite the removal of tools, appliances, and even entire structures, the visitor to today's historic mine sites often can reconstruct the makeup of the shop based on the material remains. Visitors may find this exercise useful because the shop speaks of a mining operation's capital, productivity, and duration of activity. All of the sizes and duties of the shops discussed above left distinct types of evidence in the forms of artifacts, structural materials, and foundations.

The large, well-equipped shops left the most obvious remnants, and the visitor to a mine site will have to work a little harder to distinguish the remains of lesser shops at small mines. The most apparent shop appliance may be the remains of a vernacular dry-laid rock, wood box, or hewn log forge and an anvil block. Intact forges are almost always square or slightly rectangular and filled with well-sorted gravel. Even though the forge may have collapsed over time, the mound of gravel impregnated with bits of clinker and coal should be apparent. Anvil blocks were typically either timber or hewn log posts eighteen inches high and well-embedded in the ground. The tops of the blocks typically feature four heavy nails, or two crossed iron straps which held the anvil in place.

The plan view depicts the layout of a shop typical of well-financed, modestly profitable mining operations. This shop, at the Delmonico Mine in Cripple Creek, was enclosed in a simple frame building and it included an iron tank forge, a drill-steel sharpener bolted onto a concrete foundation, an anvil, a work bench, and hand tools.

Author.

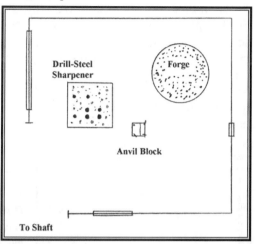

The shop at the Golden Curry Mine in Elkhorn, Montana represents an example of the type of shop erected by well-capitalized mining outfits. The building has been divided into a blacksmith room and a repair room where machine work and carpentry were carried out. The repair room was floored with planking while the blacksmith room featured an earthen floor. Note the two anvil blocks, the backing block, and the power hammer mount.

Author.

A deposit of forge clinker and anthracite coal almost always accompanies the remains of a blacksmith shop. A scattering of clinker not only denotes the shop's existence, but it may reflect the approximate footprint of the building. Blacksmiths usually picked clinker out of the forge as part of their morning ritual, and much of it wound up on the shop's earthen floor. After months to years of compaction by foot traffic, the clinker deposit assumed the footprint of the shop building. The visitor to a mine site must also be aware that blacksmiths made an effort to dispose of clinker, forge-cut iron scraps, bits of industrial refuse, and the upset ends of hand and machine drill-steels outside the shop, resulting in ashy shop dumps. The distinguishing characteristics of a shop dump consist of an amorphous shape, vertical depth of the deposit, the inclusion of heavy industrial artifacts, and relatively light density of building materials such as nails and window glass.

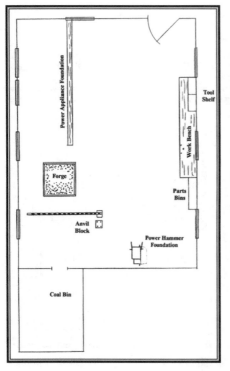

The American Eagle Mine in Cripple Creek was a well-financed operation with a first-rate production-class surface plant. Around 1900 the mine's engineers erected a spacious and mechanized shop adjacent to the shaft, which is depicted by the plan view. The shop included a large wood box forge, an oil forge, a backing block and anvil, a workbench, a tool cabinet, a drill-steel sharpener, and mechanized metal and woodworking appliances. The workbench features bolt-holes for heavy hand tools.

Author.

The visitor to a historic mine may also determine the location, size, and approximate date of a blacksmith shop in some cases by the structural remnants and artifacts left. In most cases mine workers erected frame buildings on dry-laid rock foundation footers that supported the walls. Well-capitalized mining companies may have provided sufficient funding to permit construction of concrete or rock masonry wall footers, either of which may still be visible. After a mining company went broke, laborers dismantled the shop building, leaving a concentration of nails, broken window glass, structural hardware, stovepipes, and in a few cases, electrical insulators where the building stood.

Many of the power shop appliances discussed above left unique evidence. The visitor may encounter sets of timber posts for vernacular power hammers and drill presses. The remains may also feature other power appliance foundations for drill-steel sharpeners, threading machines, and lathes. Artifacts such as canvas drive-belt remnants, belt stitches, and overhead belt shafting reflect the use of power appliances, and air hoses reflect the use of power hammers.[35]

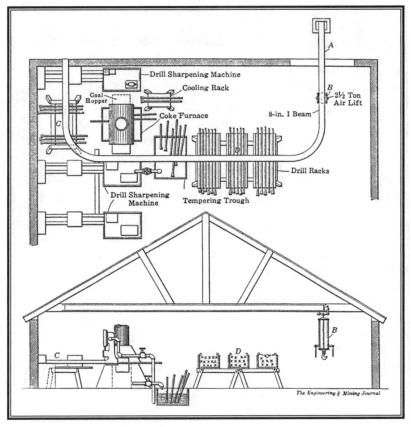

The plan view and accompanying elevation drawings document the spacious and well-equipped shop at Michigan's Champion Mine around 1910. While the shop features a few appliances for general machine work, the facility's primary duty was sharpening drill-steels, and other metal and woodwork were carried out in specialized shops. The building appears to have been approximately 25 by 50 feet in area.

Engineering & Mining Journal May 4, 1912 p881.

Mine shops were gritty, sooty, smoky, dark, and noisy. They reeked of coal smoke and gear oil, and they had the potential to be hellacious in the summer. However, no mine could have functioned without a shop, especially when under the supervision of an experienced mine blacksmith. The blacksmith in the photo is operating a Leyner drill-steel sharpener in a well-appointed shop in Butte, Montana around 1910. Drill-steels are being heated in the large iron box forge to the left, and a hammer drill lies partially disassembled on the work bench at right.

Courtesy of the World Museum of Mining, Butte, MT.

The photo depicts the interior of the well-equipped metal shop at the Gagnon Mine, one of Butte's wealthy operations, and it is representative of the shops erected by well-capitalized and profitable mining companies elsewhere in the West. The roof beams support several overhead drive shafts for belting that powers the lathe at left, the drill-press at right, and another appliance in the right background. To stop a machine from operating, the shop worker pulled a lever that derailed the drive belt off its pulley. A small upright steam engine in the back corner of the shop, at photo center, is turning the overhead drive shaft via additional belting. The shop worker, probably reflooring the cage behind him, posed for this photo around 1910.

Courtesy of the World Museum of Mining, Butte, MT.

$\times$

Production-Class Ventilation

Building a bigger and better shop was not the only worry a mining engineer had to contend with as the operation grew from a small prospect into a paying mine. As miners drove the underground workings deep, ventilation became a genuine issue requiring definite resolution, or the mining crews would not, and at worst could not, go to work. A wagon road graded to the mine afforded the engineer the latitude to effectively address the problem.

The simple wind socks and hand-turned mechanical blowers that sufficed for shallow prospect operations were not effective for the workings of larger mines. The engineer applied several superior methods for providing the miners with fresh air. Ventilation that relied on natural atmospheric currents to replenish breathable air were known as *passive systems,* and those that utilized machines were termed *mechanical systems.* Many mining engineers believed that doing nothing to ventilate mine workings was tantamount to achieving passive ventilation, because they felt that the foul gases naturally dissipated out of the adit portal. In reality, passive ventilation required a genuine incast air current balanced by an outcast current laden with bad air. Multiple mine openings proved to be the most effective means of achieving a flushing current, and temperature and pressure differentials acted as the driving force that moved the air.[36]

In most busy mining districts where operations were spaced closely together, such as at Cripple Creek, mining companies ordinarily at odds with one another cooperated in terms of ventilation and linked their workings together to attain multiple openings. But isolated operations had no neighbors to link their workings with, and instead some mining engineers put miners to work driving air shafts upward from deep within the underground workings. Such shafts improved the air flow through the mine, and served as secondary entrances for the efficient movement of miners and materials. When an adit had been linked to a shaft, the temperature differential between the mine environment and the surface conditions became the greatest mover of air. During the warm months of summer the relatively heavy, cool mine air flowed out of the adit, drawing fresh air down the shaft, and during the frigid winter, geothermally heated air rose up the shaft and drew fresh air in through the adit.[37]

Mines featuring three or more openings to ground-surface presented considerable problems in terms of getting fresh air to miners because air currents short-circuited work areas by following the shortest path through the mine. To address this problem, miners installed air-control doors at strategic intersections within the mine, and at the adit portal, and opening and closing specific doors had the effect of routing the air current through the desired portion of the mine.

Western mining companies preferred passive ventilation for replenishing mine workings with fresh air. The universal preference for natural ventilation is aptly stated in an 1890s mining engineers' textbook. "Whereever it is practical to get along without the use of blowing or suction machinery, the metal miner will inevitably do so."[38] While natural ventilation incurred the high costs of driving necessary workings, over twenty dollars per two to four feet of passage, this type of system required little if any maintenance.[39]

The sollar was a wooden structure built inside of a mine tunnel to exacerbate the natural movement of air currents. Mine carpenters erected cap-and-post timber sets that they had lined with well-fitted lagging, and they built an additional false ceiling to separate the traffic way from an overhead air duct. Fresh air was supposed to flow into the mine through the traffic way and the foul air out the overhead duct. As the reader can infer from the illustration, a sollar consumed immense quantities of lumber, and for this reason, and due to limited effectiveness, sollars saw little use in the West.

International Textbook Company A Textbook on Metal Mining: Preliminary Operations at Metal Mines, Metal Mining, Surface Arrangements at Metal Mines, Ore Dressing and Milling International Textbook Company, Scranton, PA, 1899 A41 p138.

Mining companies that relied on a single opening, and most western mines did just that, were not able to rely on natural ventilation because the underground workings were perpetually isolated and confined. Engineers working under such conditions employed ventilation systems based on either *semi-mechanical* or fully mechanical principles. Semi-mechanical systems relied on two types of wooden structures within a tunnel to enhance the movement of natural air currents. The first consisted of a horizontal partition made of boards known as a *sollar* that divided the tunnel into a traffic compartment and overhead air duct. The divider consisted of a false ceiling of boards nailed along the tunnel's support timbering. Warm, foul air flowed out the overhead passage, drawing cool fresh air in through the traffic-way below. While the sollar did not rely on costly machinery to circulate air, it did require immense quantities of lumber and expert timbermen to make it airtight. Due to high costs and limited effectiveness, sollars saw little use.[40]

*Miners intuitively under-
stood that drilling and
blasting several mine
entrances afforded the best
ventilation. However, three
or more openings allowed
the natural air currents to
circumvent work areas. In
response miners erected air
control doors at strategic
points within the mine to
route the air as they saw
fit. Today's visitor may still
encounter these doors both
underground and at the
tunnel portal.*

Author.

Mining engineers interested in effecting the movement of air while avoiding the costs associated with a sollar attempted to use a *brattice*. Similar to the sollar, the brattice consisted of a vertical wall made of boards or tightly-pitched canvas, dividing a tunnel into a traffic-way and an adjacent air compartment. Theoretically, fresh air flowed in through the traffic-way and foul air exited the adjacent passage. While the costs of installing a brattice were not quite as much as a sollar, it too was rarely used in the West where materials and labor costs were relatively high and the mines short-lived.[41]

A mining engineer in Mexico employed a simple and inexpensive approach for exacerbating the movement of natural air currents in a mile-long tunnel. The engineer instructed his miners to blast a shallow trench in the floor along one of the tunnel's walls for drainage. The miners then covered the channel with stone slabs and chinked the cracks. The mouth of the channel, deep in the mine, collected foul air which flowed out through the stone duct with the drainage water. The water in the channel cooled the air, making it heavy, and the water's flow exacerbated the air's outcast movement. While this ventilation method was not as expensive as brattices and sollars, it required specific geologic and environmental conditions seldom seen in the West, and as a result it too saw little use.[42]

Some mining engineers claimed that the continuous flow of exhaust air from one or two rock drills run by miners provided sufficient ventilation. While this may have been true for short tunnels and drifts, this method proved inadequate in long tunnels that filled with poisonous gases following the end-of-shift blasts. The exhaust from drills was inadequate in volume, and often tainted with oil vapors.

One of the most popular and genuinely effective approaches for ventilating deep tunnels in the West, when natural ventilation was impractical, lay in employing power-driven fans and blowers. As far back as the 1870s mining companies working through shafts on Nevada's Comstock Lode employed belt-driven mechanical blowers to force fresh air into the Hades-like underground. But at that time only well-financed mining companies were able to devote the substantial quantities of capital necessary to install the mechanical components and ducts. By the 1880s mining machinery makers responded to the demand for blowers among mining companies and began offering factory-made systems.[43]

The illustration depicts a large squirrel cage fan designed to be driven by a pulley located on the other side of the shroud. The vanes forming the fan drew air in through the large round opening in the foreground and forced it out a rectangular port at left.

Engineering & Mining Journal March 26, 1910 p674.

During the 1880s mining engineers had a paucity of ventilating machines to select from, but by the 1890s machinery manufacturers offered three basic varieties of blowers in a multitude of sizes. Engineers had termed the first design, which dates back to the Comstock era, the *centrifugal fan,* and miners knew it as the *squirrel cage fan.* This machine consisted of a ring of vanes fixed to a central axle, much like a steamboat paddle wheel. The fan, turning at a high speed, drew air in through an opening around the axle and blew it through a port extending out of the shroud. Inexpensive fans featured canvas or wood veneer shrouds, while efficient and more costly units featured sheet iron shrouds that tightly enclosed the fan.

Manufacturers produced centrifugal fans in sizes ranging from one foot to over ten feet in diameter. The small units were employed for both mining and a variety of other purposes such as ventilating industrial structures, and the largest units were rarely used in the western mines, being employed principally at large coal mines in the East and Midwest. The exact name engineers gave to a centrifugal fan was a function of the direction the outflow port faced. For example, a fan with a port that pointed upward was a *top discharge* fan, and a unit with a port pointing to the tunnel portal was a *front discharge* fan. The second type of fan engineers commonly employed to ventilate tunnels also acted on centrifugal principles, but it consisted of a narrow ring of long vanes encased within a curved cast-iron housing. The *propeller fan,* the third type of blower, was similar to the modern household fan with the addition of a shroud.[44]

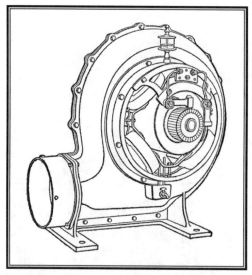

Many mining companies favored the small and curvaceous form of centrifugal blower, like the unit shown. Most of these models were belt-driven, but a few were powered by integral electric motors.

International Textbook Company A Textbook on Metal Mining: Preliminary Operations at Metal Mines, Metal Mining, Surface Arrangements at Metal Mines, Ore Dressing and Milling International Textbook Company, Scranton, PA, 1899 A41 p146.

Mining engineers made use of the most cost-effective power source available in the mining district to drive ventilation fans. Until the 1900s steam engines were ubiquitous, but as the twentieth century unfolded, steam saw heavy competition from electricity in well-developed mining districts, and from gas engines in remote regions. Each of these three motive sources turned the fan via canvas belting.

The way in which mining engineers assembled ventilation systems changed little during the Gilded Age. The blower and drive engine or motor were installed immediately outside of and aligned with the tunnel, and connected to ducting. Most engineers agreed that iron ventilation tubing was superior to canvas tubing for routing the air because it produced less enervating drag on the air current and was cheaper. But unlike canvas, air tended to leak out of the nested joints between the lengths of tubing, which ultimately put a heavy drain on the fan and engine. Medium-sized and small centrifugal fans usually stood upright and were bolted to four anchor bolts, 1/4 to 1/2 inches in diameter, arranged in a rectangular pattern with the motor mount situated nearby. Mine workers usually constructed the foundations either of timber or concrete. Some models of squirrel cage centrifugal fans greater than four feet in diameter were designed to be mounted on their sides and bolted to heavy, square concrete foundations.[45]

Engineers could not agree on whether the fan should blow fresh air into the mine or draw foul air out. Academic engineers understood that the answer depended on the current of whatever natural ventilation the mine might already possess. Forcing air into a mine that featured one opening meant that the noxious gases would flow out of the tunnel or shaft, which was undesirable if it accommodated continuous traffic. If the mine had some natural circulation, then forcing fresh air into the dead-end workings was preferable because the bad gases would have been carried off, probably up a shaft or old stope. Theory aside, most mining engineers used machines to blow air into the mine, which proved to be more energy efficient than trying to draw it out. In a third application, mining engineers had the mine crew place an electric fan deep in the mine and have it push the foul air out of through a line of ducting.[46]

In some western mining districts the local geology was as much of a culprit in rendering mine atmospheres unbreathable as was respiration and the use of explosives for blasting. Cripple Creek serves as one of the more notorious examples. There miners employed a variety of ventilation systems in hopes of flushing foul gases from deep workings. As in most western mining districts, the mining companies of Cripple Creek attempted to link their workings to establish natural ventilation currents, often spending extravagant sums of money and receiving mediocre results. Mining engineer S.A. Worcester spent much time working in the district, and he made several observations regarding these ill-fated attempts:

> "At the Blue Bird Mine a large sum of money has been spent in making a connection 1100 feet long to the Portland Mine. This obtained a strong draft, which, however, weakens or stops entirely in unfavorable weather, so that gas still interferes with working at times. At the Moon-Anchor Mine, an expensive 300-ft. connection was made from the bottom of the 1000-ft. shaft to the adjoining Conundrum Mine, expecting to effect a natural draft, and ventilate both mines. This connection is said to have cost $15,000, and when it was completed no draft whatever resulted. The heavy gas in

> the Conundrum, having about 10% carbon dioxide, seemed
> to prevent circulation as effectively as water."[47]

It is a marvel that miners were able to drive such workings in the heavy, unbreathable air in the first place. To mitigate the intrusion of natural oxygen-displacing gases into Cripple Creek's mines, mining companies attempted to install the best mechanical ventilation systems the Gilded Age had to offer. Worcester further notes:

> "At the Ophelia Tunnel, a blower of 15,000 cu.ft.
> capacity, placed 4000 ft. from the portal of the 7000-ft. tun-
> nel, keeps the outer workings fairly free from gas in good
> weather, but shows no perceptible effect on the remaining
> 3000-ft. of tunnel which is inaccessible, except in most
> favorable weather. At times of low barometer the entire tun-
> nel fills with gas and all work is suspended."[48]

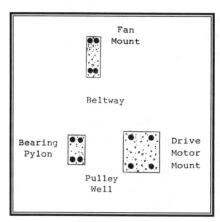

The plan view depicts the foundation arrangement typical of centrifugal fans and their drive motors. The fan body stood on a narrow concrete pad at top, the motor mount lay a slight distance away, and next to it was a pylon that anchored the outboard end of the motor axle. A belt passed from the fan to the motor, where it wrapped around a broad pulley that rotated between the motor mount and the bearing pylon.

Author.

The plan view depicts a second popular arrangement of fan and motor foundations.

Author.

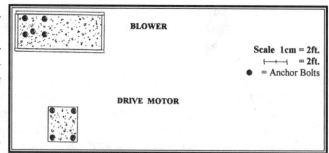

With plenty of good ore in sight but an underground environment too hostile to permit work, it is no wonder that some mining companies were willing to expend considerable capital on ventilation. Worcester identified the main problem in Cripple Creek as the seepage of natural gases from fissures in the country rock and into mine atmospheres. His solution, which proved to be quite effective, was to render the mine workings airtight and pressurize them with powerful, high-speed centrifugal fans of the cast-iron shroud variety, in some cases augmented with air piped from a compressor. The increase in pressure used by Worcester forced the natural gases back into the crevices from whence they came. To enable the mine to be pressurized, Worcester had miners build airtight doors over tunnel portals and shaft collars, and install airtight wood and concrete stoppings in abandoned workings. Further, once Worcester had pressurized the mine, he found that open-ended pipes acted as siphons after blasting, sucking explosives gases out of fouled workings. Such ventilation solutions, which were experiments attempted by professionally-trained engineers, were beyond the limited skills of many self-made engineers and beyond the financial constraints of most mining companies.[49]

Air Compressors

Mining engineers gave heartfelt celebration for the completion of a wagon road to the mine. They fully appreciated the fact that a road afforded them a variety of mechanical luxuries that ordinarily would have been too bulky, heavy, and awkward to carry to remote locations. Academically-trained mining engineers often took a different stance than self-made engineers in terms of installing additional mechanical components to increase ore production and the driving of underground workings. Self-made mining engineers such as Frank Crampton and W.S. Stratton tended to eschew heavy mechanization in favor of continuing the tradition of hand labor, while formally-trained engineers understood and even promoted the seemingly abstract benefits that heavy machinery brought to a mine. One mechanical system capable of stepping up production was the use of rock drills for boring blast holes underground. As early as the 1870s professionally-trained mining engineers convinced mining companies to expend considerable capital on the equipment necessary to run rock drills, while self-made engineers working as late as the 1910s hired miners to drill by hand.[50]

The employment of rock drills greatly expedited the blasting process. Blasting was of supreme importance to mining because it was the prime mover of rock underground. The types of drills manufactured during the 1880s and 1890s permitted miners to bore enough holes to advance a tunnel or shaft approximately three to five feet per shift in extremely hard rock, and improvements in drilling technology made during the 1890s and 1900s allowed miners to make even greater progress. Hand drilling by comparison was slow and laborious, and with it miners typically advanced tunnels and shafts only one to two feet per shift in hard rock. Many formally-educated mining engineers came to the conclusion that the costs of installing and running a compressed air system offset the relatively high labor costs and profitless operating time associated with driving underground workings by hand.[51]

The difference of opinion between formally-trained engineers and self-taught individuals regarding rock drills reaches back in time to the 1870s when the cumbersome and chattering machines were being developed. Commercial rock drills hit the mining industry in the early 1870s, and they were tried on an experimental basis at huge projects such as the Sutro Tunnel driven under Nevada's Comstock Lode, and Colorado's Burleigh Tunnel which penetrated deep mine workings near Georgetown. The western mining industry did not immediately embrace the new drilling technology because it was expensive and unproven, but operators of large mines kept a close eye on the success or failure of the few machines then in use.[52]

At this time, the mining industry was reluctant to expend the large sums of capital required to operate mechanical drills. The early piston drills were mechanically unreliable. Drilling crews constantly broke drill-steels, the chucks on the drills jammed, internal working parts failed, and compressed air plumbing and hoses blew out. As we can surmise, early rock drills made frequent trips to the mine shops, where blacksmiths and machinists had to puzzle out the problems. These issues, in conjunction with the cumbersome sizes and weights of the machines, earned rock drills a bad reputation among miners. Some of those miners in turn became self-taught engineers who influenced their professional progeny. As a result, many self-made engineers who began their careers working underground came to value the skills of the hardworking hand-drilling miner over the abstract idea of increasing production with rock drills, as well as some other forms of mechanization.

The 1880s saw a widespread rise in the numbers of mining companies that introduced drills into their operations. Many reasons contributed to the increased use. First, mining in the West underwent a broad scale geographic expansion, giving rise to an increase in the total number of large mining companies with sufficient financing. Second, the prices of drills and compressors fell while their efficiency and reliability improved. Third, engineers became experienced at installing compressed air systems during this time, and miners became more proficient at using rock drills, improving the economic worth of the machines. During the 1890s mechanical drilling technology became even more accessible to mining companies and at a moderate cost. By the 1900s rock drills had attained a commonplace status at many operations in well-developed districts, and by the 1910s the machines became standard among similar mines and deep prospect operations located in remote areas.

Despite the technological advances of compressors and rock drills, the basic arrangement and form of the overall air system changed little from the 1870s until the 1930s. All air systems consisted of a compressor situated at the top end, a source of power to drive the compressor, an air-receiving tank, plumbing extending underground, and rock drills at the bottom end.

The compressor was the heart of an air system, and because it was the most expensive component, it commanded much attention from the mining engineer. Academic mining engineers viewed air systems in terms of an economic equation that took into account numerous variables. They attempted to base their choice of a compressor on how much air they anticipated a shift of miners to consume during peak drilling times, on the working pressure required to run the drills, how much energy the machine consumed, the quality of the road to the mine, and especially how much available capital they had.

Engineers defined peak air consumption as the air required to simultaneously run all of the mine's rock drills, the air lost through leaks in the plumbing, and numerical constants of efficiency lost at altitude.[53] Engineers working at particularly large mines also had to figure in the use of air hoists and drill-steel sharpeners. Mining and mechanical engineers used *cubic feet per minute* (cfm) as the measurement of air consumption. Not only did the number of drills constitute a variable the mining engineer had to factor into his equation, but the specific types and sizes of drills proved important as well. Mechanical engineers defined the size of a drill not by its weight or its physical size, but by the diameter of the piston that drove the drill-steel into the rock. Light drills, defined as having a piston two-and-one-quarter inches in diameter, consumed approximately seventy-five cfm, while heavy drills, with three inch pistons, consumed one-hundred-twenty-five cfm. For the sake of discussion we can average the two figures and assume that piston drills generally consumed 100 cfm. Therefore, an air compressor producing 300 cubic feet of air per minute could power three rock drills. During the 1880s and 1890s the variety of piston drills was limited, making calculations of air consumption relatively straightforward for the engineer.[54]

All rock drills, from Simon Ingersoll's original contraption to the twentieth century's most efficient machines, operated under heavy air pressure, which, when combined with the cubic feet of air flow-ing through the plumbing, acted as the driving force behind the drills' working parts. The compressed air flowed through the rock drill in an attempt to escape from the pressurized air lines, doing work in the form of moving the parts on its way out of the machine. The types of piston drills used by the mining industry from the 1880s into the 1900s typically operated under sixty-five to eighty *pounds per square inch* (psi), which was the unit mechanical and mining engineers used to measure air pressure.[55]

George Leyner greatly complicated mining engineers' calculations when he introduced his hammer drill in the 1900s, because his large machine, and the spin-off stoper and sinker drills discussed earlier in the chapter, required a slightly higher operating pressure, usually 90 to 110 psi, while consuming differing volumes of air. Leyner's heavy hammer machines consumed at least 100 cfm, while the much smaller and lighter stopers and sinker drills manufactured during the 1900s and 1910s consumed between forty-five cfm and ninety cfm.[56]

While air compressors manufactured between the 1880s and 1920s came in a variety of shapes and sizes, they all operated according to a single basic premise. Compressors of this era consisted of at least one relatively large cylinder, much like a steam engine, which pushed air through valves into plumbing connected to an air receiving tank. The volume of air a compressor was able to deliver (cfm) depended on the cylinder's diameter and stroke, as well as how fast the machine operated. The pressure capacity (psi) depended in part on the above qualities as well as how stout the machine was, its driving force, and on the check valves in the air line. Generally high pressure, high volume machines were large, strong, durable, complex, and as a result, expensive.

The mechanical workings of the air compressors manufactured prior to around 1890 were relatively simple. The two most popular compressor types manufactured during this time were *steam-driven straight-line* and *steam-driven duplex models,* and both styles constituted designs that served the mining industry well for over fifty years. The straight-line compressor, named after its physical configuration, was the least expensive, oldest, and simplest machine. Straight-line compressors were structurally based on the old-fashioned horizontal steam engine, the workhorse of the Industrial Revolution. A mechanical engineer in the eastern states created the straight-line compressor in the 1860s, and machinery manufacturers such as the Clayton Air Compressor Works began making them by the early 1870s. The earliest compressors were no more than a compression cylinder grafted onto the end of a factory-made steam engine, and the compression piston had been coupled directly to the steam piston via a solid shaft. The mechanical action was simple, and these early compressors were inefficient and unreliable. Despite this, they were of the utmost importance and served as a springboard for mechanical engineers to develop better models.[57]

By the early 1880s when large mining companies throughout the West were beginning to experiment with rock drills, field-worthy straight-line compressors had taken form. These machines featured a large compression cylinder at one end, a heavy cast-iron flywheel at the opposite end, and a steam cylinder in the middle, all bolted to a cast-iron bedplate.

During the 1870s and early 1880s mechanical engineers ironed out many of the inefficiencies of these early apparatuses. First, engineers modified the compression cylinder to make it double-acting, much like an old-fashioned butter churn. In this design, which became standard, the compression piston was at work in both directions of travel, being pushed one way by the steam piston and dragged back by the spinning flywheel. In so doing the compression piston devoted one-hundred percent of its motion to compressing air.[58]

The other fundamental achievement attained by engineers during the 1880s concerned cooling. By nature, air compression generated great heat, which engineers found not only fatigued the machine but also greatly reduced efficiency. As a result early compressor makers added a water-misting jet that squirted a spray into the compression cylinder, cooling the air and the machine's working parts. While the water spray solved the cooling issue, it caused corrosion, washed lubricants off internal working parts, and humidified the compressed air, all of which significantly shortened the life of a costly system. By the mid-1880s American mining machinery makers replaced the spray with a cooling jacket consisting of one large void or multiple ports for the circulation of cold water around the outside of the compression cylinder, leaving the internal working parts dry and well-oiled. Mining companies that installed compressors had to include a water system for cooling, which often was no more than a tank connected to the compressor through input and output lines consisting of one to one-half inch piping.[59]

During the early 1880s mechanical engineers made several other significant improvements to the workings of compressors. Engineers found that coupling the compression piston to the steam piston with a solid rod, so that both acted in perfect synchronous tandem, proved highly inefficient. The steam piston was at its maximum pushing power when it was just beginning its stroke, while the compression piston, also beginning its stroke, offered the least resistance. When the steam piston had expended its energy and reached the end of its stroke, the compression piston offered the greatest resistance because the air in the cylinder had reached maximum compaction. Mechanical engineers recognized this wasteful imbalance and designed a breed of straight-line compressors with an intermediary crankshaft, so that when the compression piston had reached the end of its stroke and offered the most resistance, the steam piston was beginning its movement and was strongest.[60]

When rock drilling technology was in its infancy, a few well-financed mining companies in the West attempted to employ a significant number of the machines underground at once. The peak air consumption at these mines often exceeded the capacity of a single large straight-line compressor, which provided up to 1,400 cfm, and as a result mining machinery makers introduced the duplex compressor.[61] The early versions of these large machines were no more than two straight-line compressors arranged side-by-side, sharing a common flywheel and crankshaft. Mining engineers chose one duplex compressor over several straight-line models because duplex units were more energy efficient, they cost less, and they occupied less floor space.[62] However, the relatively high costs of purchasing, installing, and operating duplex compressors made the machine economically unpalatable to most small and medium-sized mines. Rather, machinery manufacturers produced duplex compressors for large, well-funded mining companies that sought long-term savings while maintaining a heavy production schedule.

The illustration depicts a moderate-sized straight-line compressor manufactured around the mid-1880s by the Rand Drill Company. The steam drive cylinder is at left and the compression cylinder is at right. Many western mining companies intent on running up to 6 drills favored these types of compressors because of the modest purchase and installation costs, and they remained serviceable for decades. As the machine operated, the flywheels provided momentum and smoothed out the action.

Rand Drill Company Illustrated Catalog of the Rand Drill Company, New York, U.S.A. Rand Drill Company, New York, NY, 1886 p22.

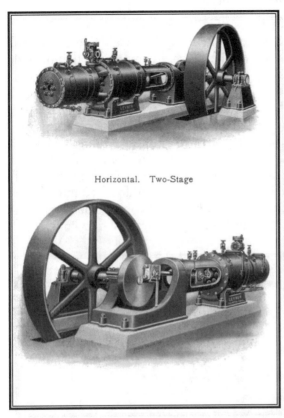

Horizontal. Two-Stage

Some western mining companies installed large straight-line machines that featured an outboard flywheel bearing, as shown by the illustration. These types of compressors, popular between the late 1880s and around 1910, required special foundations that are distinct amid the ruins of mines today. The large flywheel also lent itself well to being driven from another engine via a belt.

J. George Leyner Engineering Works Catalog No. 8 Carson-Harper, Denver, CO, 1906 [trade catalog] p16.

During the late 1880s and early 1890s academic mining engineers fine-tuned compressed air technology used for hardrock mining. During this time they applied cost-benefit analyses and science to effect further improvements that lasted for decades. The most significant advance was the design of compressors that generated greater air pressure than had previously been used by the mining industry. Mining engineers found that they wanted more pressure because it made their drills run faster, enabling miners to bore through more rock than before. Engineers had also found that the pressurization of the maze-like networks of plumbing in large mines placed a heavy burden on compressors. In response to these factors, mining machinery makers began offering straight-line and duplex compressors capable of achieving what the industry termed *multistage compression*.

Mechanical engineers found that attempting to squeeze more pressure out of the conventional 1880s straight-line and duplex compressors required an uneconomical amount of power and energy. Instead, they realized that they could achieve the pressure demanded by mining engineers if they divided the compression between high and low pressure cylinders in several *stages*, instead of in a single cylinder. They designed the low-pressure cylinder to be relatively large, and it forced semi-compressed air into the small high-pressure cylinder, which then highly compressed the air and released it into a receiving tank. Compounding the air compression between several cylinders generated heat which threatened the efficiency that engineers hoped to achieve. Like the simpler single-stage compressors, engineers designed effective cooling apparatuses for the added cylinders, and they also found that chilling the compressed air between stages significantly improved efficiency. The air rarely passed directly from one compression cylinder to the next. Rather, the air exited the machine altogether and passed through an *intercooler*, which essentially was a heat exchanger cooled by circulating water, and then it entered the high-compression cylinder.[63]

During the 1880s and 1890s only well-financed mining companies wishing to power numerous rockdrills spent the lavish sums necessary to install massive duplex compressors such as the model illustrated. Each compression cylinder (at right) and each steam cylinder (at left), and the flywheel bearings were bolted onto individual masonry pylons underneath the plank flooring. Note the potted plants on the windowsill at right.

Rand Drill Company Illustrated Catalog of the Rand Drill Company, New York, U.S.A. Rand Drill Company, New York, NY, 1886 p20.

Mining machinery makers released variations of multistage straight-line compressors with two and even three compression cylinders coupled onto the steam drive piston, and they produced duplex compressors with several multistage cylinder arrangements. The most common multistage duplex compressor was the *cross-compound* arrangement, in which one side of the machine featured the low-pressure cylinder, and the air passed from it, through an inter-cooler, to the high-pressure cylinder on the other side. For mines with heavy air needs mining machinery manufacturers offered a duplex machine with high and low pressure cylinders on both sides, which produced twice the volume of high-pressure air.

After installing straight-line and duplex compound machines at large western mines in the 1890s, technically-trained mining engineers

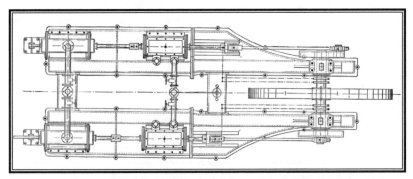

Compare the plan view of the duplex compressor in the diagram with the duplex compressor foundations cataloged several pages below.

Ingersoll Rock Drill Company Catalog No.7: Rock Drills, Air Compressors and Air Receivers
Ingersoll Rock Drill Company, New York, NY, [1887] p34.

The lithograph depicts a small "pony" straight-line compressor manufactured by the Rand Drill Company during the 1880s and 1890s. The steam drive cylinder is at right and the compression cylinder is at left. Small western mining operations located in remote areas appreciated these types of compressors because they were designed to be disassembled into small and easily transported pieces, they were relatively inexpensive, and they could be powered with no more than a portable locomotive or upright steam boiler.

Rand Drill Company Illustrated Catalog of the Rand Drill Company, New York, U.S.A. Rand Drill
Company, New York, NY, 1886 p25.

found that multistage compression was by far the most economical for high pressure, and that standard compression was most economical for conventional pressures. They determined through empirical studies and calculation that single-stage compression was most economical for air pressures up to around 90 psi. Two-stage compression was best for between 80 and 500 psi, and three-stage compression was most economical for 500 to 1,000 psi. Four-stage compression existed, but western mining companies almost never used the process.[64]

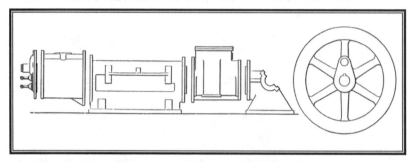

The diagram illustrates how a "pony" compressor disassembles in preparation for being hauled into the backcountry.

Rand Drill Company Illustrated Catalog of the Rand Drill Company, New York, U.S.A. Rand Drill Company, New York, NY, 1886 p24.

In sum, prior to the 1890s medium-sized and large western mining companies employed straight-line compressors, and a few large companies installed duplex models. By the 1890s a few companies with extensive air networks and heavy air demands installed expensive two-stage straight-line units, while a handful of large outfits with substantial capital purchased two-stage duplex compressors. In general, due to high costs and difficulty of transportation, until the 1890s mines in remote districts rarely installed compressors at all.[65]

Once the mining engineer ordered a compressor, the manufacturer shipped it in pieces via rail to the commercial center nearest the mine. There, the mining engineer inspected it, and upon approval, ordered laborers to load it onto heavy freight wagons. Compressors produced during the 1880s and 1890s were conducive to being shipped by wagon because they disassembled into a few large but manageable components. The cylinders unbolted from the bedplate, and the flywheel often came from the manufacturer in half sections. The accompanying steam boilers, while cumbersome, were easier to move around because they came from the maker on wood or steel skids.

Despite mechanical disassembly, the individual components of most straight-line and duplex compressors remained too large to be hauled over the roughest roads and pack trails to isolated mines. Thus, during the mid-1880s the Rand Drill Company advertised a small straight-line compressor designed to be disassembled into pieces light enough to be hauled over relatively treacherous terrain and reassembled with no special hoisting derricks. Other compressor makers followed suit.[66]

Mining engineers realized that multi-stage compression was best for fulfilling the high-volume and high-pressure air needs of advanced and expansive mines. These machines were large, the costs of purchasing and installing them were exorbitant, and they were not easily shipped to remote operations. The illustration depicts a two-stage duplex compressor, which can be identified by the asymmetry of the compression cylinders C and D and the mandatory intercooler E located between the cylinders. The steam cylinders A and B are equipped with Corliss valves, which was the one of most efficient steam engines that the Industrial Revolution had to offer.

International Textbook Company A Textbook on Metal Mining: Steam and Steam-Boilers, Steam Engines, Air and Air Compression, Hydromechanics and Pumping, Mine Haulage, Hoisting and Hoisting Appliances, Percussive and Rotary Boring International Textbook Company, Scranton, PA, 1899 A20 p34.

While the compressor and boiler were en route to the mine, the engineer made the site ready for their arrival. Often the engineer chose to place the new facilities not on the waste rock dump, which could have been soft and subject to settling, but on a cut-and-fill platform adjacent to the adit portal, as compressors required substantial foundations that were most stable when tied into bedrock.[67]

The labor gang excavated pits to bedrock for the compressor foundation and the steam boiler setting. In the event that bedrock lay under thick soil, a rare scenario in the mountainous West, the labor crew excavated deep and broad pits for foundations with large footers, providing them with ample surface area to discourage subsidence.

Erecting a set of foundations that successfully accommodated a compressor and boiler and permitted them to interface with each other, and with additional machines, constituted a major engineering achievement. The first step toward installation was to survey a *master datum line* oriented along the same axis as the tunnel, adit, or the shaft. The engineer and his assistants used the master datum line, usually a cord or chain stretched tight between iron pins, to establish parallel and ninety degree subdatum lines for the precise placement of machinery foundations, for other facilities, and associated buildings. Using rudimentary tools including line levels, protractors and angles, plumb bobs, and measuring tapes, self-taught and formally-trained mining engineers built complex, heavily mechanized industrial plants where there once had been forested mountains or rocky desert hillsides. Engineers working for mining companies with at least modest financing took no chances on making costly mistakes and supplemented the above implements with surveyors' instruments.[68]

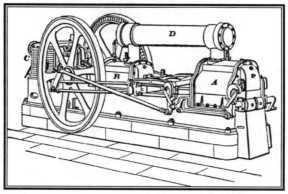

Most mining companies interested in installing costly multi-stage compressors opted for less-extravagant straight-line two-stage units. The machine illustrated consists of a steam drive cylinder C at left, a low-pressure compression cylinder A at right, a high compression cylinder R center, and the mandatory intercooler D.

International Textbook Company A Textbook on Metal Mining: Steam and Steam-Boilers, Steam Engines, Air and Air Compression, Hydromechanics and Pumping, Mine Haulage, Hoisting and Hoisting Appliances, Percussive and Rotary Boring International Textbook Company, Scranton, PA, 1899 A20 p32.

Academic and self-taught engineers each approached foundation construction differently. Academically-trained engineers exercised great care in calculating the stresses to which a foundation would be subjected, and they designed the foundation's size and weight accordingly. Self-taught engineers might have applied a general rule-of-thumb suggesting that the foundation's weight needed to be three times that of the compressor, or they may have responded to their gut feeling as to the best size, and what they had seen at other operations. Ultimately, the foundation reflected the shape and size of the machine.[69]

Once the foundation was designed, a masonry crew began work. Until the 1890s, masons constructed foundations for heavy machinery with one of three methods: a rock masonry core capped with natural concrete, pure rock masonry, or red brick masonry. Red brick generally saw use only in well-developed mining districts where freight costs were low, while local rock was heavily used in remote districts where shipping costs were prohibitive. High quality concrete was difficult to obtain in the West as late as the 1890s, and the lesser grades of concrete were too poor in constituency for lasting industrial work, discouraging the mining industry from using it for machine foundations. By the 1890s improved varieties of concrete had become available throughout the West, and engineers and masons embraced the material because it was easier and faster to work with than brick or stone. However, old traditions died hard among mining engineers, and as a result some operations continued to erect stone and brick masonry foundations for machinery into the 1910s.

The masons began assembling the materials for the foundation, using a wood template sent by the compressor maker. The template showed the exact locations of the anchor bolts, which corresponded to sockets in the compressor's bedplate.[70] Prior to the 1920s mine construction crews used several varieties of anchor bolts in both masonry and concrete foundations to hold fast against extreme tensile and rotative forces. The top ends of anchor bolts were always threaded, while the bottoms featured heads forming either right angles, T's, or square flanges. Some heads had nuts and masonry washers threaded on. Academic mining engineers recognized the last three types of heads as being best for masonry and concrete, but because blacksmiths were able to produce the first type of anchor bolt with little effort and time, many less demanding mining engineers permitted their use.[71] Only the best masons were able to maintain the anchor bolts in exact positions while they built the foundation. Less exacting construction laborers encased the upper portions of the bolts in short sections of pipe to afford them a little flexibility in case they did not match up perfectly with the compressor.

Mining engineers at remote mines, and operations with little capital, sought alternatives to masonry and concrete in hopes of saving money. In many cases small operations bolted their compressors onto timber foundations or drove the anchor bolts directly into bedrock. Engineers recognized such foundations as being informal and temporary, but they acknowledged that some mining operations had no choice due to limited financing. The most popular types of timber foundations consisted of cribbing cells made of timbers pinned together with spikes and bolts, and short cribbing frames bolted onto wood pallets, all of which were buried with waste rock.[72]

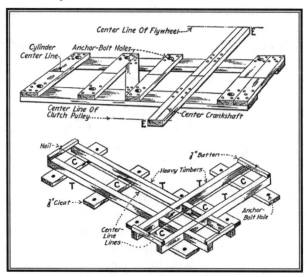

Before masons went to work erecting compressor and hoist foundations using stone, brick, or concrete, the mining engineer presented the crew with a wood template to guide the placement of the machine's anchor bolts. Perfect bolt location was imperative because heavy machines featured sockets in their bedplates with little tolerance for error. The masons placed the template over the beginnings of the foundation, inserted the bolts, and aligned the assembly according to the master datum line. The template on the top might have been for a compressor or gas hoist while the template on the bottom was for a straight-line compressor or a steam engine with an outboard flywheel.
Croft, Terrell Machinery Foundations and Erection McGraw-Hill Book Co., New York, NY, 1923 p247.

Once the construction crew had completed the foundation and the concrete or mortar finished curing, they began moving the compressor for installation. Only large mines in popular mining districts had the luxury of using swing-boom hoisting derricks for lifting the heavy machinery into place. Engineers and laborers who had no access to derricks had to be clever when installing heavy machinery. Construction workers employed several ages-old methods for manipulating the big iron components. One method involved building a dock of timber cribbing between the loaded wagon and the new compressor foundation. The crew rolled the heavy parts from the wagon to the foundation across the platform using either pipes or old boiler tubes as rollers. Workers jacked up the heavy compressor components when they inserted or retrieved the rollers, and when they built up or took down the scaffolding. Once the crew had moved the compressor parts over the foundation, the engineer checked the alignment of the anchor bolts to bolt holes, fine-tuned the machine parts' positions, and instructed the laborers to begin lowering the pieces into place.

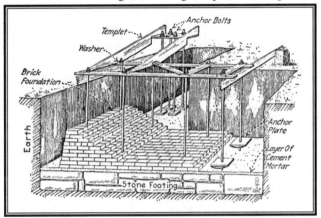

Building a foundation for a heavy hoist or compressor required exacting work. As the diagram depicts, masons laid down a footing of rocks, put the template and anchor bolts in place, and began laying brick, rock, or pouring concrete. In most cases mining companies used rough fieldstones instead of faced stone blocks, as shown in the illustration. The diagram depicts construction of a compressor or engine foundation.

Croft, Terrell Machinery Foundations and Erection McGraw-Hill Book Co., New York, NY, 1923 p311.

If the stone or concrete mason was confident and experienced, he rigidly set the anchor bolts in concrete. If he felt unsure he used pipe casements, and as the compressor components came down, laborers flexed the anchor bolts into the bolt holes. When the compressor parts were poised several inches above the foundation surface, workers laid down a veneer of grout or molten sulfur, filled the encasements around the anchor bolts, and lowered the machine parts onto their resting places. The grout and sulfur acted as a pillow and as cement, filling in minor undulations in the foundation's surface. All was now ready for assembly of the smaller mechanical parts.

Each type of compressor required a foundation with a specific footprint. Straight-line single-stage steam compressors stood on rectangular foundations between approximately nine and twenty feet long and up to three feet wide. Multi-stage straight-line compressors often featured multiple compression cylinders that required special support, and as a result, foundations for these machines tended to be long, up to twenty feet in some cases, consisting of multiple pads studded with anchor bolts. A few types of multi-stage straight-line compressor foundations also featured flywheel wells along their sides or at the end. In rare cases large compressors featured heavy flywheels that required support from an outside offset bearing.

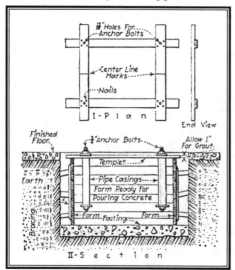

By the 1890s the quality of concrete had attained a level of quality sufficient for use in constructing mine machinery foundations. To build a con-crete foundation masons had to excavate a pit, throw in a dry-laid rock footing, build a wood form, assemble the anchor bolts, casings, and template, and pour the concrete.

Croft, Terrell Machinery Foundations and Erection McGraw-Hill Book Co., New York, NY, 1923 p248.

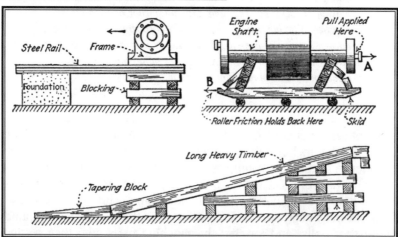

Moving the components of mine machinery onto the foundation and lowering them into place in an era when construction derricks were few required ingenu-ity. The diagrams depict several means of moving heavy machine parts around a mine site. Some miners raised heavy parts up on a timber cribbing scaffold, laid down steel rails and skidded them along as the top two diagrams illus-trate, or they built timber ramps, as the bottom diagram depicts.

Croft, Terrell Machinery Foundations and Erection McGraw-Hill Book Co., New York, NY, 1923 p39.
Croft, Terrell Machinery Foundations and Erection McGraw-Hill Book Co., New York, NY, 1923 p 604.

The foundations for duplex compressors were more complex. Because early small duplex models often consisted of two straight-line units arranged in parallel with a flywheel between, their foundations tended to be a pair of long, narrow pads spaced approximately two to three feet apart. Foundations for large duplex machines installed as late as the 1900s consisted of a symmetrical arrangement of masonry or concrete pylons studded with heavy anchor bolts for each compression cylinder, each steam cylinder, and the flywheel bearings. In addition, duplex compressor foundations featured depressed flywheel wells between the bearing pylons. Small duplex compressor foundations occupied a space approximately eight-by-fourteen feet while large foundations were as large as fifteen-by-thirty feet.[73]

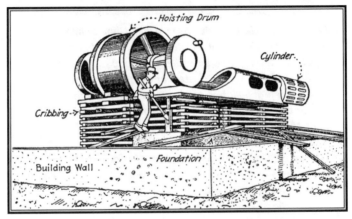

Once the heavy part had been positioned over the new foundation, miners began to lower it by jacking it up and slowly extracting the timber cribbing. Many surface plants in the West were built with the above methods.

Croft, Terrell Machinery Foundations and Erection McGraw-Hill Book Co., New York, NY, 1923 p39.

Mines active between the 1900s and 1920s may feature foundations for compressors, or other machines for that matter, constructed with one of two types of concrete. Prior to the 1910s, masons and construction crews worked almost exclusively with what the building trades termed *natural cement*, which superior *portland cement* almost totally supplanted.[74]

The construction trades, as well as mining engineers, understood that portland cement was far superior to natural cement in longevity, water resistance, and especially strength; yet its high price and scarcity rendered it prohibitive for most uses during the Gilded Age. As a result, natural cement dominated most concrete and masonry work in the United States until the 1910s. Around 1900 the price of portland fell and more cement companies in the United States began manufacturing it. The average price dropped from three dollars per barrel in the 1880s to one dollar-and-forty cents per barrel by 1899, and it continued to decrease after the turn of the century to one dollar and nine cents, where it could at last compete with natural cement.[75]

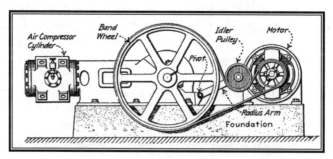

By the 1910s electric belt-driven straight-line compressors began to see widespread popularity among mining companies that had limited funds. Both the motor and compressor were usually anchored to two clusters of anchor bolts set in a common concrete foundation block, as the illustration depicts. The bolt arrangement can potentially mimic the footprint of a two-stage steam compressor. However, foundations for belt-driven straight-line compressors rarely exceeded 12 by 2 feet in area and the bolt groups are often offset. The foundations for two-stage steam compressors were longer and wider, and their bolt patterns are symmetrical.

Croft, Terrell Machinery Foundations and Erection McGraw-Hill Book Co., New York, NY, 1923 p398.

The illustration depicts another popular arrangement for a belt-driven straight-line compressor and motor. In this case the foundation is stepped, and the motor is mounted below the compressor. Mining operations used this type of arrangement from around 1910 into the 1940s, and possibly later.

Mining & Scientific Press Feb. 2, 1913 p13.

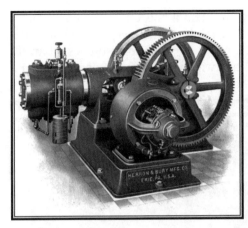

Gear-driven single-stage straight-line compressors were quite rare because the drive arrangement was more costly than belting, and the savings in power did not justify the added expense. Note the L-shaped foundation.

Hardsocg Wonder Drill Company Bulletin A: Hardsocg Wonder Drill, Herron & Bury Compressors Hardsocg Wonder Drill Company, Ottumwa, IA, 1905 [trade catalog] p19.

The transition from natural to portland cement was not instantaneous, and the rate of change varied in different parts of the nation. The East, with its heavy concentration of cement plants, a heavy materials consumption, and a greater competition among cement companies, switched between approximately 1895 and 1910, while the West changed slightly later. Until 1902 almost all of the cement sold on the West Coast was of the natural variety. By 1922, however, the West alone hosted twenty-eight portland cement plants, five of which served Colorado and Utah, and production of natural cement had virtually ceased across the remainder of the nation.[76]

Natural and portland cements encountered at historic mines today exhibit slightly different physical characteristics. Natural cement tends to have a loose hold on aggregate materials and is subject to decay and spauling, especially where it has frequent contact with water. Natural cement may have a white lime precipitate on its surfaces and it also effervesces with hydrochloric acid more vigorously than does portland cement. These characteristics result from 20 percent to 25 percent of the calcium carbonate in natural cement remaining inert because the manufacturers had not chemically balanced the ingredients. Natural cement may also appear off-white to pinkish, or yellow to brownish in color. Portland cement on the other hand, is much stronger, structurally sound, water and weather resistant, and white to gray in color. Its greater strength and lower rate of effervescence with hydrochloric acid results from almost 100 percent of its ingredients being chemically engaged.[77]

Toward the turn of the century, air compressors changed in form and motive source. The titanic steam-driven compressors that once made engineers swell with pride fell out of favor with the mining industry during the 1900s, which began to use smaller and faster models. While the small compressors of turn-of-the-century vintage were not able to generate as much air as their huge cousins, their higher working speeds did grant them a substantial output. Around 1900 many of America's leading mining machinery makers offered a variety of improved steam-driven duplex compressors less than fifteen-by-fifteen feet in area, fitted with a variety of single, double, and triple stage compression cylinders.

Compressor makers had revised the design of duplex units during the 1890s, reducing the size, increasing the working speed, and improving efficiency. The Rand Drill Company Imperial Type 10 represents the general form of these new machines. The model illustrated is a two-stage cross-compound unit, meaning the low-pressure cylinder in photo center partially compressed the air, the air traveled through the intercooler atop the machine and across to the smaller high-pressure cylinder at left. Like some manufacturers, Rand had devised a core bedplate which was capable of accommodating various combinations of single, double, and triple stage cylinders arranged in line or in cross-compound, and powered by steam cylinders, gearing, or belt-drive. Note the plethora of cooling pipes, drain plugs, and oilers on the machine.

Rand Drill Company Imperial Type 10 Air Compressors Rand Drill Co., New York, NY, 1904 p12.

Machinery makers adapted several designs to be run by electric motors and gasoline engines, which were energy sources well-suited for the remote mining districts of the Great Basin and Southwest. Progressive mining engineers working in regions where fuel was costly eagerly experimented with electricity and gasoline, while mining companies in the mountain states where coal and cord wood was more plentiful continued to use steam compressors as late as the 1910s. Gasoline and electric compressors were not accepted overnight. But once they had proven their worth to the mining industry by the 1910s, many companies throughout the West replaced their aging steam equipment with electric machinery.

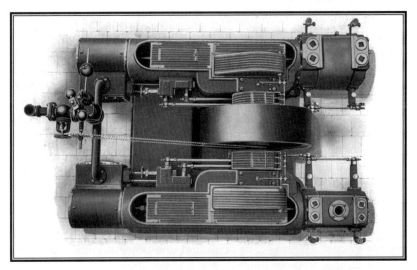

Compare the plan view of Rand's Imperial Type 10 compressor with the U-shaped foundation in the catalogs of compressor foundations several pages below. The unit illustrated is a two-stage cross-compound compressor powered by steam cylinders fitted with Corliss valves.

Rand Drill Company Imperial Type 10 Air Compressors Rand Drill Co., New York, NY, 1904 p2.

Chain drives were quieter and more efficient than belt-drives, but because they were costly few western mining outfits used them to power compressors. The relationship of drive-motor to the duplex compressor illustrated is very similar to the way engineers set up belt-drives. Note the foundation.

Rand Drill Company Imperial Type 10 Air Compressors Rand Drill Co., New York, NY, 1904 p30.

The motor-driven compressors were a take-off of the old belt-driven types in existence since the early 1880s, and the new units were designed to run at the high speeds associated with electric motors. By the late 1890s mining machinery makers offered three basic types of electric compressors, including a straight-line machine that was approximately the same size as traditional steam versions, a small straight-line unit, and a duplex compressor. Mining machinery manufacturers offered their electric compressors with five basic means of coupling to motors; a belt-drive, direct-drive in which the motor was integral with the flywheel, a gear-drive, a chain-drive, and a rope drive. Western mining outfits by far favored the belt-drive because it was a widely understood technology, least expensive, and easiest to install.[78] Gearing and direct-drive compressors were more energy efficient and quieter, but they were unpopular because of their high cost. Gearing also permitted one powerful motor to drive several duplex compressors through a single driveshaft.

Gearing was the most effective means of powering duplex air compressors because it was efficient and quiet, but high costs relegated it to the compressor houses of large and well-capitalized mines under the supervision of progressive engineers. Note how the ends of the motor drive shafts have been supported by bearings fastened onto concrete blocks. The method was commonly employed for drive shafts.

Rand Drill Company Imperial Type 10 Air Compressors Rand Drill Co., New York, NY, 1904 p29.

Motor-driven compressors were ideal for busy mining districts that had electric grids, such as Cripple Creek and Creede, Colorado, Goldfield and Tonopah, Nevada. Further, because motor-driven compressors lacked steam equipment and needed no boilers, they cost less. Of these three basic forms of electric compressors, duplex models, conducive to multistage compression, were most popular among medium-sized and large mining companies. Small mining operations favored the small straight-line units. Due to limited air output compared with a relatively large floor space, the large electric straight-line compressors never saw popularity.[79]

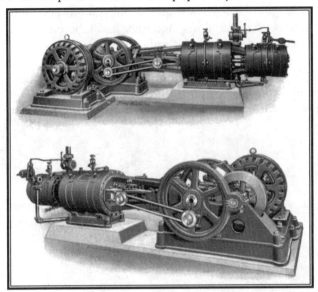

Between the 1890s and around 1920 a few western mining companies installed large belt-driven and geared two-stage straight-line compressors instead of duplex machines. When machinery makers began offering duplex compressors in a multitude of sizes, the meager popularity for straight-line units quickly dropped off.

J. George Leyner Engineering Works Catalog No. 8 Carson-Harper, Denver, CO, 1906 [trade catalog] p18.

Compressor makers also developed economically attractive gasoline units ideal for inaccessible operations. The gasoline compressor, introduced in practical form in the late 1890s, consisted of a straight-line compression cylinder linked to a single cylinder gas engine. Most mining engineers considered gas compressors to be for temporary duty only. Large machines produced up to three-hundred cubic feet of air at ninety pounds per square inch, permitted mining companies to run up to four small rock drills. Prior to the introduction of gasoline compressors, such operations fully expected to employ little or no mechanization and to rely instead on the brawn and skill of old-fashioned hardrock miners.[80]

The noisy machines had needs similar to their steam-driven cousins. Gasoline compressors required cooling, a fuel source, and a substantial foundation, and they came from the factory either pre-assembled, or in large components for transportation into the backcountry. The cooling system often consisted of no more than a continuous-flow water tank, and the fuel system could have been simply a large sheet iron fuel tank connected to the engine with one-quarter to one-half inch metal tubing. Most mining engineers agreed that petroleum compressors required substantial concrete foundations due to severe vibrations, and they had laborers pour a rectangular foundation slightly longer and wider than the machine, and up to three feet high. In a few instances small, poorly funded mining companies bolted the machines to impermanent timber cribbing foundations conforming to temporary-duty characteristics.[81]

Small mining and prospect operations working in woodless and remote areas greatly appreciated petroleum-powered compressors. The little machines, which grew in popularity during the 1910s, were capable of powering up to four heavy drills, and they were relatively easy to pack into the backcountry. The compressor in the photo was probably located at a mine in either California or Mexico in 1899.

Engineering & Mining Journal March 11, 1899 p295.

The brilliant application of electricity and petroleum engines as a motive source around 1900 made compressors affordable for a broader spectrum of mining operations. Mining companies that ordinarily might not have spent the capital installing steam compressors and their trappings were blessed by the introduction of electric power. By contrast, early petroleum engines were unreliable, cantankerous, and incapable of generating much horsepower. As a result petroleum compressors did not see even mild popularity until around 1910.

The types of compressors developed around the turn of the century possessed footprints as distinct as the early steam-driven models discussed above. The high-speed duplex models, whether steam or motor-driven, usually stood on substantial U-shaped concrete foundations featuring totally flat surfaces. Unlike their massive steam-driven counterparts, the small and faster duplex machines consisted of components both bolted to and cast as part of a large common bed plate that had to be anchored to the concrete foundation. The hollow portion of the U in the foundation accommodated the flywheel.

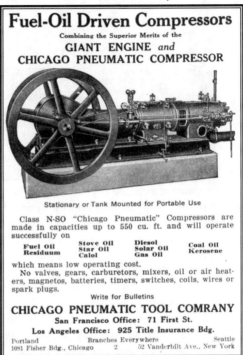
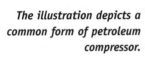
The illustration depicts a common form of petroleum compressor.

Mining & Scientific Press Aug. 5, 1916.

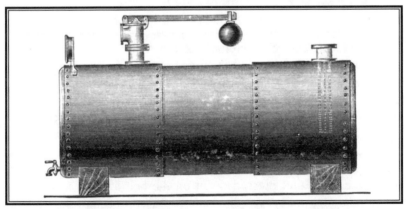

Every properly engineered compressed air system featured a receiving tank plumbed to the compressor. The receiver moderated the flow of compressed air and dampened pulsing caused by the compressor. The size of tank was relative to the output of the compressor and the consumption of air for drilling in the mine. The illustration depicts a horizontal receiving tank.

Rand Drill Company Illustrated Catalog of the Rand Drill Company, New York, U.S.A. Rand Drill Company, New York, NY, 1886 p31.

The lithograph depicts the interior of a spacious, masonry compressor house similar to the costly facilities built by productive and well-funded western mines. The attendant, oil cadger in hand, provides scale for typical straight-line compressors used during the 1880s and 1890s. The lithograph also captures the surface components of compressed air systems, including the incoming steam line, at left, the outgoing air line at right, the receiving tank complete with pressure gauge and blow off valve, and the cooling line descending into the floor from the compression cylinder at right. Compressor attendants had to be careful not to allow loose clothing to snag on the rapidly rotating flywheels. Today all that typically remains of these systems are building and machine foundations, and industrial artifacts.

Ingersoll Rock Drill Company Catalog No.7: Rock Drills, Air Compressors and Air Receivers
Ingersoll Rock Drill Company, New York, NY, [1887] p30.

A distinguishing characteristic of the motor-driven straight-line and duplex compressors manufactured between the 1890s and 1910s, as opposed to their steam-powered cousins, was an adjacent rectangular motor mount featuring four small anchor bolts. Because the drive belt passed from the motor pinion to the compressor's flywheel, engineers had workers construct the motor mount between six and eighteen feet away from the compressor, and offset to accommodate the belt. The use of a drive-belt required a tension pulley, which was an adjustable roller that pressed down on the belt to keep it tight. On small units, the pulley was often bolted onto the compressor frame, while it was anchored to small timber foundations at floor level in association with large compressors. Geared compressors featured mounts for the drive shaft bearings at the open end of the "U".

Petroleum engine compressors required foundations unusually stout in proportion to the size and weight of the machine. Because petroleum compressors vibrated violently, mining engineers and engine makers recommended pouring a heavy rectangular concrete foundation, which may be distinct to visitors at mine sites today. Generally the foundations were seven to ten feet long, eighteen inches to two feet wide, and two to three feet high.[82]

Catalog of the variety of straight-line steam-driven compressor foundations today's visitor is likely to encounter at historic mines that operated between the 1880s and 1900s. While the caption for each foundation specifies a construction material and the type of compressor, in actuality masons may have interchanged stone with brick or concrete.

Author.

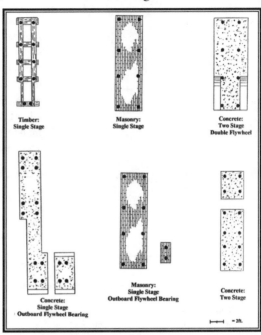

Three samples of straight-line belt-driven compressor foundations today's visitor is likely to encounter at historic mine sites dating between the 1900s and 1920s. By the time electric motors had become popular for powering compressors, construction crews ubiquitously used concrete.

Author

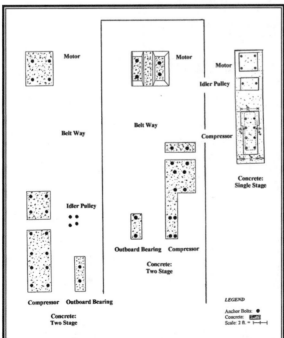

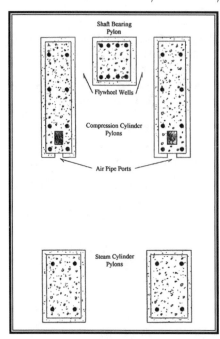

The plan view depicts the foundation for a large duplex steam-driven compressor. The foundation, approximately 25 feet long, was constructed at the Sun & Moon Mine in Colorado's Idaho Mining District. Compare the foundation with the compressor illustrated on page 88.

Author.

Catalog of the variety of duplex steam-driven compressor foundations today's visitor is likely to encounter at historic mine sites. While the caption for each foundation specifies a construction material and the type of compressor, in actuality masons may have used stone, brick, or concrete.

Author.

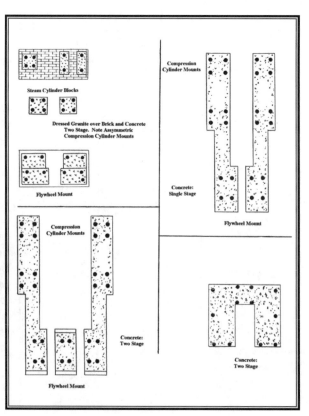

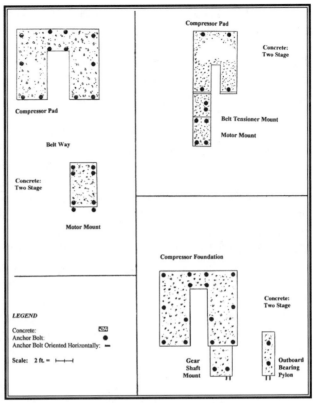

Catalog of general examples of the form of multi-stage duplex belt-driven compressor foundations today's visitor is likely to encounter at historic mine sites active between the 1890s and 1940s. The foundation at bottom right supported a gear-driven compressor, which the reader should compare with the unit illustrated several pages above. *Author.*

While the compressor may be characterized as being the heart of an air system, it was useless without the other components. Mining engineers understood that the output line of a compressor had to be plumbed to an *air receiving tank,* which served to regulate and cool the flow of air, and dampen the pulses created by the compressor's pounding pistons. Mining engineers recommended that the receiving tank be placed no more than fifty plumbing-feet from the compressor, preferably at the adit portal or shaft collar, and that if the lines into the mine were long, that a second unit be located at the other end.[83]

Mining engineers employed either a horizontal receiving tank or a vertical model. Both had to be built to boiler specifications to withstand the tremendous air pressure, and as a result they were made of boilerplate iron assembled with hot-riveted joints. To this end, some engineers working for low-budget operations, adapted discarded boilers to serve as air receiving tanks, because they were capable of withstanding the pressure while costing little.[84] Receiving tanks also featured ovoid pressure hatches that permitted some unfortunate mine worker to clean the black carbon deposits from inside the vessel. The mechanical motion and heat created by air compressors converted lubricating oil into carbon and gas, which proved explosive in the right quantities. The Vindicator Mine in Colorado's Cripple Creek Mining District featured a large receiving tank that accepted air from three gear-driven duplex compressors. Like many large mining companies, the Vindicator was often lax on maintenance, and it neglected to regularly clean the receiving tank. Output of carbon and gases from the three compressors proved too much for the system, and as a result the tank exploded with a terrific crash and shot up into the shaft house rafters once in 1918, and again in 1920.[85]

During the 1880s academically-trained mining engineers found that the type of piping used for air lines proved to have a significant impact on the efficiency of the air system. Professionally-trained engineers determined that forcing large volumes of air through small pipes created enervating friction, which impeded compressor efficiency. Academic engineers found that using pipes with four to six inch inside-diameters to assemble trunk lines greatly decreased energy loss, and they had pipemen install distribution lines ranging from one to two inches in diameter into workings where miners were drilling. Self-made mining engineers, on the other hand, installed air plumbing in response to a combination of the cost of the piping and how they had seen other operations install air systems.[86]

Moving the Materials of Mining: Transportation Systems

Miners and engineers alike had first-hand experience of literally turning mountains inside out in pursuit of riches. Great tonnages of waste rock and ore flowed out of the mine, and substantial quantities of timber, tools, equipment, and men flowed in. Such mass exchanges of heavy materials, a hallmark of productive mines, required a transportation system better and more lasting than the light-duty conveyances typically used by prospect operations.

Steel ore cars and flatcars on miniature gauge rail lines proved to be the best solution for the transportation issue, and this system served western mining for over one hundred years. The specific type of rail system installed by an operation reflected the experience and judgment of the engineer, the financial status of the company, the extent of the underground workings, and whether the mine produced much if any ore. Like most mine plant components, rail systems proved to be full of complications and nuances that drew different responses from self-made and from professionally-trained mining engineers.

Nearly all deep prospect operations and paying mines used ore cars on steel rails spiked 18 inches on-center to move the materials of mining. Miners pushed the cars by hand at small mines while large operations used animal power or mechanical locomotives. The illustration depicts two ore cars parked under the loading chutes associated with an overhead stope. Loading stations were an important part of a mine because they facilitated the transfer of ore into a vehicle, they provided a place for equipment storage, and they served as a work area for miners to prepare explosive charges. The photo was taken in the underground museum at Creede, Colorado.

Author.

The basic rail system used in nearly all western mines was fairly simple and straightforward. This plant component typically consisted of a main rail line that extended from the areas of work underground out to the waste rock dump. Miners in the underground drilled, blasted, and shoveled the resultant shot rock into an ore car. A miner then pushed the loaded car out of the mine onto the edge of the waste rock dump where he discharged the car's contents. As the drilling and blasting crew advanced the workings, they laid rails in the new space to facilitate bringing the car close for loading. Large mining companies generated enough rock and ore car traffic underground to warrant hiring mine laborers known as *muckers* specifically to load empty ore cars, and *trammers* who pushed full cars out of the mine and sent empty ones back in. Productive mines and deep prospect operations usually had rail spurs extending off the main line underground to other headings in feeder drifts and crosscuts where more drilling and blasting teams were at work. Spurs also branched off into stopes and ore bin stations where muckers loaded empty ore cars with payrock. Substantial mines with extensive surface plants also featured spurs off the main line on ground-surface that extended to different parts of the waste rock dump, to a timber storage area, and to the mine shop. Many large mines built special stake-side, flatbed, and latrine cars for the coordinated movement of specific materials and wastes.

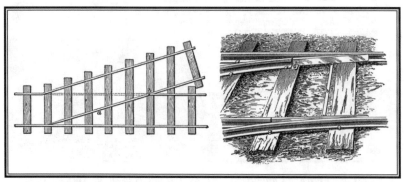

Small mines with rail lines that featured spurs and forks required the installation of switches. The limited volume of traffic and light-duty nature of the ore cars permitted the switches to be simple and inexpensive. Further, today's visitor may encounter intact switches above and below ground, as well as their pieces and parts. At left is an example of the most common switch used in the West. Rail a pivots on a hinge pin b that rests in a socket in the cross-tie. A miner merely had to swing the rail according to the fork on which he wished his car to go. Legends claim that seasoned mine mules kicked the switches into place with their hooves. The diagram at right illustrates another common light-duty switch.

International Textbook Company International Library of Technology: Hoisting, Haulage, Mine Drainage International Textbook Company, Scranton, PA, 1906 A48 p44, p36.

Western mining companies often had different outlooks on just how to improve a mine's transportation system as the operation expanded. Most people in the mining industry understood that wooden cars and strap rails were temporary, and most mining engineers and superintendents hoped to see their replacement at some point in a mine's life. However, the cost of hauling tons of rails, spikes, and ties across the West proved so economically distasteful to mining companies as late as the 1880s that even highly productive operations, especially in the Great Basin, used perishable wood until the mines ceased life. In general, however, by the 1880s most mining operations spent the capital necessary to install steel rails, and by the 1890s most of these operations were also using steel ore cars.[87]

The general rule of thumb followed by both self-made and professionally-educated engineers working at small operations was to spend little capital improving the rail system, provided it already consisted of steel components. Typically, at small and medium-sized operations miners installed lines consisting of rail no heavier than sixteen pounds-per-yard spiked at eighteen-inch gauge to ties spaced every three feet. Further, the ties usually consisted of hewn logs or salvaged lumber ranging from two-by-four to six-by-six inch stock. In efforts to save money, miners often reused the ties several times. This type of line existed throughout the West because it accommodated hand-pushed ore cars and it was relatively inexpensive to build. Many self-made engineers saw no reason for installing heavier rails and possibly a broader gauge as the mine underwent significant expansion. Academically-trained engineers working at large mines, however, had miners replace light-duty rails with heavier, lasting rails as the iron succumbed to corrosion, or they had miners replace only the trunk portions of the main rail line with heavy rails, leaving alone the ends and spurs.[88]

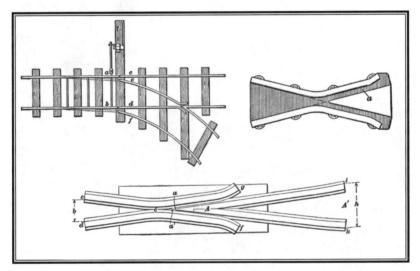

Productive mines featured rail lines that required heavy-duty switches capable of locking in position. The illustration at left shows a sample switch activated by a lever that locked in place. Complex rail networks featured multiple lines that crossed over each other. Rail frogs, at right and bottom, facilitated the cross over, and they were stable and lasting enough to withstand locomotives.

International Textbook Company International Library of Technology: Hoisting, Haulage, Mine Drainage International Textbook Company, Scranton, PA, 1906 A48 p39, p37, p36.

Many professionally-educated mining engineers understood that hiring miners to hand-tram single ore cars was the most cost-effective means of transportation at small and medium-sized operations. But at large mines, where high volumes of materials had to be efficiently handled over great distances, they strongly recommended the use of ore trains pulled by a motive source greater than one or two struggling miners. Prior to the 1890s, when progressive engineers began experimenting with compressed air and electric locomotives, mining companies seeking to move large tonnages of ore and waste rock turned to the use of draft animals. In extremely rare cases a few highly productive mines spent enormous sums of money developing large haulage tunnels to accommodate small, low-clearance steam engines to pull ore trains. Mining companies found that this transportation system presented problems as great as the advantages it provided. Most mining engineers condemned steam locomotives for their exhaust gases, the heavy costs involved in enlarging the underground workings to adequate sizes, and in laying heavy-duty railroad lines.

Few mining companies in the West were expansive enough or generated ore in sufficient tonnages to justify the high costs associated with installing the system necessary to run an electric locomotive. The few mines that warranted such a drastic improvement in their transportation systems realized a decrease in the cost per ton of rock hauled out. Operation of an electric locomotive required specialized labor positions including a locomotive driver and an electrician.

International Textbook Company International Library of Technology: Hoisting, Haulage, Mine Drainage International Textbook Company, Scranton, PA, 1906 A56 p21.

As hardrock mining matured throughout the nineteenth century, miners learned that mules were the best animals suited for work underground because they were reliable, strong, of even temperament, and intelligent. Mining engineers, seeking to put science and calculation behind the use of mules to improve efficiency defined sixteen pound rail as being best in conjunction with small ore trains because it resisted wear. Underground workings capable of accommodating a mule had to feature either an area spacious enough to turn the animal around, or they had to have a loop-circuit.[89]

The electric locomotive, termed an *electric mule* in some mines, arrived in the West during the 1890s. Mining engineers working for coal mines in the East and in the Appalachians introduced the first electric locomotives in 1887 or 1888 to move the immense volumes of the fossil fuel produced by coal companies. The early machines consisted of a trolley car motor custom-mounted onto a steel chassis, and it took its power from overhead *trolley lines* strung along the mine's ceiling.[90]

Under the advice of engineers, some western mining companies opted for compressed air locomotives, which operated according to the same mechanics that drove steam engines. Mining engineers felt that small compressed air locomotives, such as the unit illustrated, were more cost-effective than their electric cousins because they were able to negotiate the 18 inch gauge rail lines that wended their ways through the tortuous drifts endemic to western metal mines.

International Textbook Company International Library of Technology: Hoisting, Haulage, Mine Drainage International Textbook Company, Scranton, PA, 1906 A55 p35.

The spread of the electric mules to hardrock mining in the West proved slow. Locomotives required special mechanical and electrical engineering, which was largely unavailable during the 1890s and 1900s. In addition, electric mules were too big for the tortuous drifts typical of most metal mines, and they required considerable capital to purchase and operate. During the first decade of the twentieth century the electrical system necessary to power a locomotive included a steam engine, a generator, electrical circuitry, plumbing for the engine, and an enclosed building. The system alone cost around $3,100, and a small locomotive cost an additional $1,500.

Further, an electric locomotive cost approximately $7.50 per day to operate. A mule, on the other hand, cost only $150 to $300 to purchase and house, and between sixty-cents and $1.25 to feed and care for per day.[91]

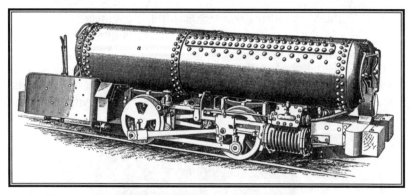

Expansive mines with lengthy tunnels employed large compressed air locomotives that were efficient for pulling long trains through the main haulage way.
International Textbook Company International Library of Technology: Hoisting, Haulage, Mine Drainage International Textbook Company, Scranton, PA, 1906 A55 p33.

Upgrades to the rail line necessary to accommodate a heavy locomotive included additional costs. Mules were able to draw between three and five ore cars that weighed approximately 2,500 pounds each, and for this, sixteen pound rails spiked to eighteen-inch gauge proved adequate. But electric locomotives and their associated ore trains usually weighed dozens of tons, and as a result they required broad tracks consisting of heavier rail. Mining engineers recommended that at least twenty-pound rail spiked twenty-four inches apart on ties spaced every two feet be laid for small to medium-sized locomotives. Heavy locomotives required rail up to forty pounds per yard spiked to thirty-six inch gauge. The reason for the heavy rails and closely spaced ties was that the heavy machines pressed down on the rail line and perpetually worked uphill against the downward-flexed rails. This wasted much of the locomotive's power and energy, and engineers sought to minimize the sag with stiff rails on a sound foundation of closely spaced ties. In addition, light rails presented a greater chance of derailment, which proved to be a logistical and economic disaster in the confines of a haulage tunnel.[92]

Some academic mining engineers criticized the fact that electric locomotives were tied to the fixed route defined by the trolley wires. To remedy this problem, electric machinery makers introduced the storage battery locomotive around 1900, which had free rein of the mine's rail lines. Despite its independence, very few western hardrock mining companies employed battery-powered locomotives during the Gilded Age because they were costly, they required a recharging facility, and they too were physically inhibited by the mine's tight passageways. While they worked for coal mining, they were not appropriate for the piecemeal work endemic to western hardrock mining.

A few prominent academic mining engineers, including Herbert C. Hoover, espoused the compressed air locomotive, which saw limited use in western hardrock mines beginning in the 1890s. This interesting contraption consisted of a compressed air tank fastened to a miniature steam locomotive chassis. Eastern mining engineers disliked the locomotives, criticizing their need for an expensive three to four stage compressor, valves and fittings for charging the tank with air, and a limited range of travel. Western mining engineers rebuffed the complaints, claiming that the machines were well suited for western mines. The locomotives were able to negotiate tight passageways, they had plenty of motive power, and they spread fresh air wherever they went. Some of the machines were able to operate on the ubiquitous eighteen-inch rail gauge, and they did not require complex electrical circuitry. However, compressed air locomotives were expensive, costing as much as their electric cousins, and they required a costly compressor capable of delivering air at pressures of 700 to 1000 pounds-per-square-inch. However, most metal mines large enough to warrant a locomotive employed a high-pressure compressor to run rock drills and other air-driven machinery anyway. Such arguments were a moot point for most western mines, which continued to rely primarily on the efforts of struggling trammers, and occasionally on mules for moving the heavy materials of mining.[93]

Today most historic mine sites in the West feature only vestiges of the rail systems that miners used to move the mountains of ore and waste rock. Yet, if visitors to such sites take a close look, they may encounter enough evidence to reconstruct a mine's transportation system. Often labor gangs disassembled rail lines following a mine's abandonment, but they inevitably left small items behind. Miniature rail spikes, track bolts, rail connector plates, and ties featuring spike holes all serve as evidence that a mine site had a rail line. Track bolts and connector plates are distinct types of artifacts, the bolts featuring an ovoid shank immediately below a dome-like head, and the connector plate having four corresponding ovoid holes for the bolts. This design permitted miners to insert bolts into the connector holes binding two rails together and to thread nuts on.

The visitor to a historic mine site may be able to estimate gauge and duty of a rail line by measuring the space between sets of spike holes in abandoned wood ties, and by examining the sizes of the ties, connector plates, and remaining spikes. Ties made of six-by-six inch timber stock, spikes longer than four inches, heavy connector plates and bolts, and a rail gauge twenty-four inches or wider may reflect the use of a locomotive. If a mine site features evidence of such a rail line in combination with the foundation for a particularly large duplex compressor, the mining company may have used a compressed air locomotive. Mine sites featuring rail lines comprised of sixteen to twenty pound rails spiked at twenty-inch gauge may have employed a mule to pull ore cars, while lines consisting of a smaller gauge and lighter rail probably relied on the labor of men to move ore cars.

Surface plants typically included several dead-end rail lines for dumping waste rock. The end of the line usually stood on several trestle piers, also known as bents, that afforded plenty of vertical space for the accumulation of waste rock under-neath. Over time, waste rock engulfed the pier, leaving only the tips of two log or timber posts showing. At left is a typical trestle associated with rail lines on waste rock dumps.

International Textbook Company A Textbook on Metal Mining: Preliminary Operations at Metal Mines, Metal Mining, Surface Arrangements at Metal Mines, Ore Dressing and Milling International Textbook Company, Scranton, PA, 1899 A42 p17.

Whether the cars were pushed by miners or pulled by locomotives, once they had been shuttled onto the surface of the waste rock dump, a surface laborer uncoupled them for dumping. He pushed each car to the end of the rail line, which stood on a trestle extending outward from the shoulder of the waste rock dump. The worker released the ore car's catch-lever and emptied the unwanted contents. Load after load, the buildup of waste rock filled the area around the trestle, creating a lobe, or tongue of rock that projected out from the main portion of the dump. The mining company, understanding that trestles' lives were limited, often directed laborers to build the structures out of expendable salvaged lumber or hewn logs to save money. Trestles for waste rock usually consisted of two or more *piers* also known as *bents*, and each pier was made with two or more posts known as *pilings*, tied together with *diagonal braces* and a *cap timber*. Because miners often spaced the piers, at most, three feet apart, they simply spiked the rails directly to the cap timbers and laid planking for a walkway. Productive mines also featured well-built trestles that supported a spur rail line extending to an ore bin or ore sorting house, where trammers pushed cars loaded with payrock.

Ore Storage

The presence of an ore bin is a hallmark of a productive mine. Sloped floor ore bins, such as this two-cell structure in Colorado's Red Mountain District, were best for ore storage but they were most expensive to build, and they saw construction in the West beginning in the 1870s.

Author.

While capitalists, mining engineers, and miners often held differing opinions as to how to set up and run a mine, all were in agreement that the primary goal was the production of ore, and lots of it. Most western hardrock mines, of course, were failures, producing little or no payrock. A few operations, however, proved to be profitable, and a tiny fraction made millionaires of their owners. Those mines with any measurable output of payrock usually included an ore storage facility as part of their surface plants, and operations that produced either miniscule amounts of payrock or were *dry holes* almost never featured ore storage facilities. Like most of a mine's facilities, ore bins and ore sorting houses reflected the financial state of the mining company, the mine's volume of production, and the type of ore the miners drilled and blasted.

Ore bins were functionally different from *ore sorting houses,* and the mining engineer based his choice on which structure the company erected on the type of ore being mined. Free gold, tungsten, and copper usually occurred in veins and masses that were fairly consistent in quality and rock type, and they warranted storage in an ore bin. The quality and consistency of silver and telluride gold, on the other hand, varied widely in any given mine, and they required sorting, separation from waste rock, and rudimentary concentration in an ore sorting house. Both types of structures required a means of inputting ore from the mine, and a means of extracting it for shipment to a mill for finer concentration.

Mining engineers recognized three basic types of ore bins: the *flat-bottom bin,* the *sloped-floor bin,* and a structure which was a hybrid of the two known as a *compromise bin.* Flat bottom bins, which generally consisted of a flat floor, high walls made of heavy planks, and a louvered gateway in one wall, had a greater storage capacity per square-foot than the other two types of structures. However, laborers had to stand on the pile of shifting payrock and work in choking dust to shovel it out into a waiting wagon or railroad car. Sloped floor bins, on the other hand, were expensive to build as they required proper engineering, but they were conducive to automatically unloading the ore, which naturally flowed out of the structure through chutes. Compromise bins combined the above two designs, half of the floor being sloped and half being flat, to create a bin which automatically unloaded when full, and required shoveling when almost empty.[94]

Mining companies with substantial capital backing and heavy ore production often erected large, sloped-floor ore bins. These structures were lasting, strong, and had a look of permanency, solidity, and they inspired confidence. Such bins were uncommon in the West prior to the 1880s due to a general lack of capital, but they became more popular over the ensuing decades. Well-built sloped-floor bins, which cost more than twice to build than flat-bottomed bins, typically consisted of a heavy post and girt frame made with eight-by-eight inch timbers sided on the interior with two-by-six to two-by-twelve inch planking. The structures generally stood on foundations of posts tied to heavy timber footers placed on terraces of waste rock. The biggest and best ore bins stood on concrete or stone masonry footers. To ensure the structure's durability in the onslaught of the continuous flow of sharp rock coming from the mine, construction laborers often armored bin floors with salvaged plate iron. Mines of a small order used sloped-floor bins consisting of a single-cell, for example twenty-by-twenty feet in area. Large productive mines erected long structures that included numerous bins to hold different grades of ore or batches of payrock produced by multiple companies of lessees working within the same mine.

Mining companies with limited financing and minor ore production erected flimsy, flat bottom bins because such structures were inexpensive to build. Rarely did these ore storage structures attain the size and proportions of their large sloped-floor cousins because the walls were not able to withstand the immense lateral pressures exerted by the ore. Flat-bottomed bins had to contend with pressures on all four walls, while sloped-floor bins directed the pressure against the front wall and the diagonal floor.

By nature of their function, ore bins and ore sorting houses had to be linked to the mine via a rail line for the input of fresh ore, and they had to provide for the removal of stored ore. Trammers and miners filled ore bins by pushing loaded ore cars across a small trestle, and over the bin. To facilitate a level rail connection, the rim of the ore bin had to be at the same elevation as the tunnel portal or shaft collar. As a result, mining engineers usually located sloped-floor bins, which tended to be rather tall, out on the flank of the mine's waste rock dump. Many flat-bottom bins, and some small, poorly built, flimsy sloped-floor bins, were located at the toe of the waste rock dump on stable ground. Trammers loaded ore into these structures by dumping the rock into a long chute. Prior to the 1900s, some mining companies extracting very limited quantities of ore, countersunk small flat-bottomed bins into the waste rock dump near the adit portal. Such bins, often no more than twenty-by-twenty feet in area, were accessed by a mine rail spur curving off the main line, and the trammer merely pushed a loaded ore car to the bin's edge and disgorged the car's contents.[95]

Ore sorting houses were generally more complex and required greater capital and engineering to erect than ore bins. The primary functions of ore sorting houses were both the concentration and the storage of ore. The earliest ore sorting at western mines did not occur in structures at all. Rather, in silver districts such as the Comstock and Tybo in Nevada, miners simply dumped loaded ore cars out onto the waste rock dump where surface workers handpicked the ore by daylight. Such crude methods proved to be very slow, sloppy, limited in production, and subject to the weather, so mining engineers moved ore sorting indoors for greater efficiency and year-round operation.[96]

In keeping with gravity-flow engineering typical of mining, engineers usually designed sorting houses with multiple levels for input, processing, and storage. These structures usually featured a row of receiving bins located at the top level, a sorting floor under the receiving bins, and a row of holding bins underneath the sorting floor. Receiving bins always had sloped floors, and in most cases the holding bins below did too. In the cold and windy mountain states, a cupola sheltered the top level, and the sorting floor was fully enclosed and heated with a stove. The structure usually stood on a foundation of heavy timber pilings and hewn log cribbing walls.[97]

Like the processes associated with ore milling, mining engineers utilized gravity to draw rock through ore sorting houses. The general path the ore followed began when the drilling and blasting team, the mucker, or shift boss, all working underground, characterized the nature of the ore they were extracting. They communicated their assessment of the ore's quality to the trammer via a labeled stake, a message on a discarded dynamite box panel, or a tag. The trammer subsequently hauled the loaded car out of the mine and pushed it into the sorting house. He emptied the car into one of several bins. *High-grade ore* went into a small and special ore bin at one end of the structure, *run-of-mine ore*, which was not particularly rich but required no sorting, went into another bin at the opposite end of the structure. *Mixed ore* that was attached to or combined with considerable waste rock went into one of several receiving bins located in the center of the long ore sorting house. When released from the car, the mixed ore slid across a heavy grate, known as a *grizzly*, at the bin's bottom. The principle behind the grizzly was that the rich portions of telluride and silver ores fractured into *fines*, and the large cobbles that remained intact through blasting, mucking, and unloading contained waste rock that needed to be *cobbed*, or knocked off, by surface laborers. The valuable fines dropped through the grizzly directly into holding bins at the bottom of the structure, while the waste rock-laden cobbles rolled off the grizzlies and into chutes that fed onto stout tables on the sorting floor. There, the laborers worked by daylight admitted through windows, and by kerosene or electric lighting to separate the ore from waste.

On the sorting floor, laborers, wearing slouch hats and gloves, hovered over a row of around two to six iron-clad sorting tables. A few workers would loosen the stoppings on the chutes holding mixed ore until their tables were full of cobbles, then they sorted through the rock, occasionally using a hammer to knock off waste. The workers would drop the recovered ore through unguarded openings in the floor where it fell into the holding bins below. They tossed waste rock onto the floor or swept it into short-handled shovels that they emptied into ore cars parked on a rail line inside the structure.

Full ore bins were a measure of a mine's success. But to make the ore pay, it had to be transferred into wagons or railroad cars for shipment to a mill. Most mines, even in well-developed districts, were not productive enough to warrant direct rail access, and they had to be served instead by teamsters. To collect a load of payrock from a mine, a teamster maneuvered a stout wagon on the road directly along side the ore bins. When a wagon had been positioned, a mine laborer opened the gates on the chutes which allowed the ore to pour forth. The types of chutes included louvered boards, iron gates raised by gearing, or a pivoting gate that opened when a laborer pulled down on a long lever. The louvered plank stopping proved to be most popular in the West because it cost least and was easiest to install.

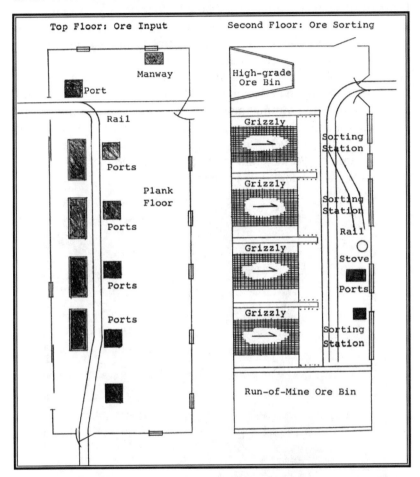

The plan views depict the top two floors of the sorting house at Cripple Creek's Strong Mine, built around 1900. The ore sorting process began when a miner pushed an ore car along the track leading into the ore input floor illustrated at left. If the rock contained high-grade ore then he dumped it into the port at upper left, where it accumulated in the bin labeled "High-grade" shown on the second floor. If the ore was average in quality the miner dumped it into the port at bottom left, where it dropped into the bin labeled "Run-of-Mine". If the ore was mixed with waste he deposited it into one of the center bins. The rock fell onto a grizzly, the telluride-bearing fines dropped through and the waste-rock laden cobbles accumulated at one of the sorting stations. Miners separated the waste and pitched it into an ore car parked on the interior rail line. Mining companies applied this same process for silver ore.

Author.

The layout and nature of access roads for large and productive mines differed from the roads at small mines. Most medium-sized and large mines were served by broad roads forming a circuit, or they featured spacious flat areas that granted a teamster plenty of room to pull his rig underneath the ore chutes and turn around once the wagon had been filled. Such traffic control facilitated the efficient movement of entire wagon trains. Inefficient dead-end roads, on the other hand, often served small mines. Where possible a wise teamster turned his wagon and team around and backed up into the loading area, and when turning room did not exist he had to unhitch his team while mine laborers manually turned the wagon around and loaded it with ore. Such roads were not intended to maximize the flow of materials. Rather, mining companies graded them in hopes of minimizing the expenditure of capital.

Interior of the sorting floor in the sorting house of the Gold Dollar Mine in Cripple Creek around 1910. Miners dumped ore from cars into one of several bins behind the high wall at left. The telluride-bearing fines dropped through a heavy screen while waste rock-laden cobbles rolled onto the sorting tables where the workers are standing. The laborers cobbed off waste and dropped the recovered ore through ports in the floor at their feet. To improve vision, engineers designed ore sorting houses with skylights, as shown, supplemented by electric lights.

Mining & Scientific Press August 23, 1913 p297.

Large and productive mines always hoped for rail service, because trains hauled much more ore for less money than wagons. However, except for operations in wealthy mining districts with developed rail networks, such as Cripple Creek and Butte, only mines rich enough to attract capitalists on the scale of George Graham Rice and mining engineers the caliber of Herbert C. Hoover were directly served by railroad lines. Even without direct rail service, the mere presence of a railroad in a mining district benefited all operations because the costs of shipping ore and the prices of machinery and other goods dropped significantly.

A train of two ore wagons at the World Museum of Mining in Butte, Montana.

Author.

With a few exceptions, ore bins and ore sorting houses serve as indicators denoting a mining operation as having been productive. However, the presence of an ore bin cannot answer the question of whether a mining company sank more money into the ground than it took out. In other cases a few mining companies, such as George D. Roberts' State Line Mine, erected ore bins to fraudulently give the public the impression that the mine was producing ore.

Overall, intact ore storage structures are a rarity in the West today. Instead, the visitor to a historic mine site is often faced with remains. Large ore bins and ore sorting houses were usually complex buildings made of heavy timbers and lumber fastened with large nails, bolts, mortise and tennon joints, and iron tie rods. As a result, even after the structures had been disassembled or demolished, they usually left distinct traces. The most common evidence left by an ore bin or sorting house following its removal consists of foundation remnants, including log or timber pilings, log cribbing walls, dry-laid rock walls, and eroded terraces of waste rock which supported the head or toe of the ore structure.

Flat-bottom ore bins may be somewhat more distinct that the clusters of timber pilings vaguely denoting the location of a sloped-floor bin. The remains of a flat-bottomed bin may appear as an open-topped wood box embedded in the edge of a waste rock dump. Often the remains of flooring and plank walls are visible, but in instances where the building materials have been removed, the bin location may appear merely as a rectangular depression with a flat floor.

Because ore bins and ore sorting houses were materials intensive, they usually left a distinct artifact assemblage. Invariably laborers dismantling an ore bin left behind relatively large quantities of intact and fragmented hewn logs, heavy timbers, and other types of dimension lumber in the approximate place where the structure stood. The laborers also left hardware such as lag bolts, heavy nails, large-diameter construction washers, iron brackets, and iron tie rods. The former locations of ore sorting houses, where mine workers spent a day's shift, may also feature food items, stove parts, and small industrial items.[98]

Aerial Tramways

In the West, prospectors discovered a good many productive mines in impossible terrain. Some of the locations were so inaccessible that pack trains proved to be the only viable means of transporting in the materials of mining and hauling out ore. Eleven burros or donkeys were required to pack one ton of ore, and for many mining operations this was not cost-effective. In some cases mining engineers spent lavish sums of capital to build circuitous wagon roads in hopes of mitigating transportation problems. However, the steep and winding wagon roads proved to be only somewhat better than pack trails, economically squelching what could have otherwise been a highly profitable operation.[99]

Western mining companies began experiencing these transportation-related problems as early as the 1860s when prospectors found tantalizing deposits of gold and silver in the rugged Great Basin and Rocky Mountains. At that time mining engineers dreamt of fanciful solutions to move great tonnages of ore to points of rail shipment, or directly to local concentration mills. One such engineer and mining machinery maker in San Francisco, Andrew S. Hallidie, was the first to turn fantasy into reality. Hallidie, son of steel wire maker Andrew Smith, was born in London in 1836, and there he learned bits and pieces of his father's trade. Hallidie also picked up machining skills from a brother. With the potential to capitalize handsomely on the California Gold Rush, Andrew Smith and his son traveled to the southern Mother Lode in 1852 to inspect a mining proposition in which they were offered stock. Smith and his son had a falling out and young Andrew elected to remain in California, where he changed his name to Hallidie. As a metalworker he gained a favorable reputation when he demonstrated his engineering and construction talents by building a wire suspension bridge and a flume for a placer mine. In 1856 he began making wire rope for placer mines in San Francisco.[100]

Combining his superior wire rope, his engineering skills, creativity, and knowledge of European mining technology, Hallidie hit upon an invention that solved the transportation problems presented by the West's high mountains and impassable winter snows. In the late 1860s he developed and patented the first practical aerial tramway. Hallidie's apparatus, modified by the ski industry during the twentieth century into today's chairlift, conveyed ore buckets suspended from an endless loop of his wire rope over the most hostile terrain. Hallidie's system consisted of a series of strong wooden tram towers featuring cross-members tipped with idler wheels that supported the lengthy loop. The towers were built to heights dependent on the relief of the terrain in efforts to keep the pitch of the tramway consistent. The loop of rope passed around large sheave wheels at top and bottom stations, and Hallidie ingeniously designed the system to move under gravity. The loaded buckets gently descended downslope, pulling the light empties back up to the mine.[101]

Hallidie's design changed little from the 1870s until the 1910s. Empty buckets entered the upper tram terminal located at the mine, workers loaded them with payrock, and once full they whisked around the sheave wheel and continued down to the bottom terminal. When the bucket entered the bottom terminal a steel guide rail upset it, dumping the contents into a receiving bin as it passed around the bottom wheel. A few feet farther on a group of laborers may have been busy loading empty buckets with dynamite, drill-steels, food, and forge coal to supply the miners at work high above.

At the cost of $5,000 per mile during the 1870s, only the wealthiest mining companies were able to afford aerial tramways. Still, rich mines under the direction of progressive management and competent engineers in Frieberg, Nevada, Ophir and Silver Plume, Colorado, and in Utah dared to install the capital-intensive systems during the 1870s and 1880s. These companies, in turn set a precedent for a wide-spread employment of aerial tramways in the Great Basin and mountain states in subsequent decades.[102]

By the 1880s, enough mining companies had installed Hallidie aerial tramways to enable academic engineers to evaluate their economic worth and performance. The mechanical wonders remained unrivaled for moving large volumes of ore across untraversable terrain, but they possessed several undeniable limitations. The tramways had distance and elevation limitations of two miles and 2,500 feet, respectively. Longer circuits required very expensive transfer stations. Because the buckets were fixed to the wire rope, they had to be filled and unloaded while in motion, limiting their load and giving them the greater potential to wreak havoc with the system. When the grade traversed by Hallidie's tramway was less than fourteen feet rise per one-hundred feet traveled, an expensive steam engine had to power the rope. Last, because the system relied on one rope to both carry and move the buckets, the weight capacity of each bucket had to be curtailed to minimize strain and the cataclysm of breakage.[103]

A line of pyramid tramway towers stand in echelon along the fall line of a steep hillslope at the Mary Murphy Mine in Colorado's rugged Collegiate Mountains. Both sides of the towers feature two cables, and each tower had two cross-members, indicating that a Bleichert Double Rope Tram served the mine.

Author.

With these problems in mind, Theodore Otto and Adolph Bleichert, two German engineers, developed an alternative system first employed in Europe in 1874. The *Bleichert Double Rope* tramway utilized a *track rope* spanning from tram tower to tram tower, and a separate *traction rope* that tugged the ore buckets around the circuit. The track rope was fixed in place and the buckets coasted over it on special hangers featuring guide wheels. The traction rope was attached to the ore bucket's hanger via a mechanical clamp known as a *grip*. Like *Hallidie Single Rope* systems, *Bleichert Double Rope* tramways relied on top and bottom terminal stations where the buckets were filled and emptied, and they too usually ran by gravity.[104]

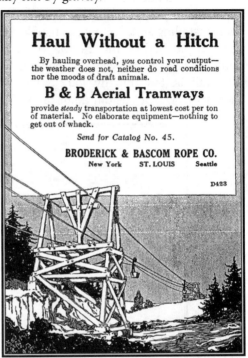

The advertisement depicts a series of through towers built for a Bleichert Double Rope Tram somewhere in the West. Through towers were the most expensive type of tower, but they were also the most structurally sound.
Mining & Scientific Press Jan. 19, 1913 p33.

Bleichert's system offered advantages that endeared the systems to academically-trained technology-loving mining engineers. First, even though Bleichert systems were up to 50 percent more expensive to erect than Hallidie tramways, they proved to be better for heavy production because they were able to handle greater payloads which resulted in a higher income for the mining company. In addition, the grip fastening the buckets to the traction rope was releasable, permitting workers to manually push the buckets around the interior of the terminal on hanging rails and fill them at leisure without spillage. The double-rope system also permitted the entire tramway circuit to be extended up to four miles in length and almost to any height.[105]

Mining companies began experimenting with Bleichert Double Rope systems in the 1880s, ten years after Hallidie began manufacturing his marvelous aerial tramways. Due to superior performance, the popularity of Bleichert systems eclipsed the less expensive Hallidie tramways by the 1890s, when the use of tramways in the West surged. Still, many western mining companies with limited production and moderate amounts of capital continued to install Hallidie systems after the turn of the century.[106]

Purchasing and installing aerial tramways was beyond the rough-and-ready skills of self-made mining engineers. The systems were complex and very expensive, they required economic and engineering calculations, and a state of mind bordering on the experimental and progressive. Even professionally-trained mining engineers installing aerial tramways usually required direction from engineers dispatched by the tramway maker. While mining companies purchased basic tramway components from manufactures such as A.S. Hallidie & Company, Park & Lacy, and Bleichert & Company, each setup was a custom affair tailored to a specific mine's needs.

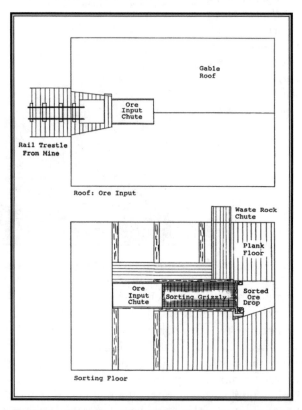

The plan view depicts the top two levels of the Bleichert tramway terminal at the upper haulage tunnel of Colorado's Pennsylvania Mine. A trammer pushed a loaded car across a high trestle and disgorged the contents into an ore chute that descended into the terminal. The ore slid down the chute and rolled across a wire mesh grizzly located on the top floor. The silver-bearing fines dropped through the mesh and accumulated in the holding bin on the second floor, while the cobbles stopped at the ore sorting station. Laborers recovered ore from the waste rock and dropped the value rock through a port in the floor where it too accumulated on the second floor.

Author.

Mining engineers and the builders of tramways assembled systems from fairly standardized technology, but rarely were two setups alike.

Tramways offered mining companies the advantage of producing ore in economies of scale, but perhaps greatest of all, in the eyes of engineers, was the statement such a system made about the engineer himself, and about the mining company that backed him. The constant aerial parade of loaded ore buckets inspired management and investors alike, it spoke of prosperity and wealth, and was a mechan-

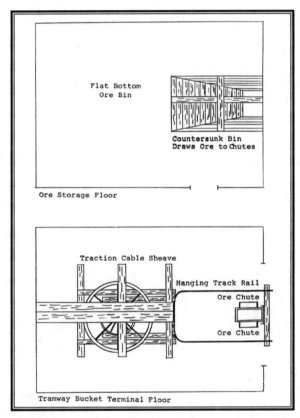

The plan view depicts the bottom two levels of the tramway terminal at Colorado's Pennsylvania Mine. The second floor consisted of a large holding bin for ore, and the first floor featured the equipment associated with the tramway. A tram bucket entered the first floor through a doorway in the building's upper right corner where a worker uncoupled it from the traction rope, which passed around the large sheave. He stopped the bucket underneath an ore chute, filled it, and rolled the bucket around to the other side of the track where a second worker reconnected it to the traction rope. Note that the sheave is encased in a heavy timber framework. On Hallidie systems the sheave was exposed on three sides to permit the buckets, which were fixed to the rope, to swing around.

Author.

ical fascination. The action inside the loading and unloading terminals was no less inspiring. A busy crew of mine workers uncoupled every bucket as it arrived, they pushed the empties over the hanging rail to the ore chutes where another worker filled them, and workers recoupled the buckets onto the ever-moving rope for the ride down.

Engineers used four basic types of tramway towers for both Bleichert and Hallidie systems. These included the *pyramid tower,* the *braced hill tower,* the *through tower,* and the *composite tower.* The pyramid tower consisted of upright legs that joined at the structure's crest. The through tower resembled an A-shaped headframe consisting of a wide rectangular structure stabilized by fore and back braces, and the tram buckets passed through the framing. Composite towers usually had a truncated pyramid base topped with a smaller frame supporting a cross-member. The braced-hill tower was similar to the through tower, except it had exaggerated diagonal braces tying it into the hillside.

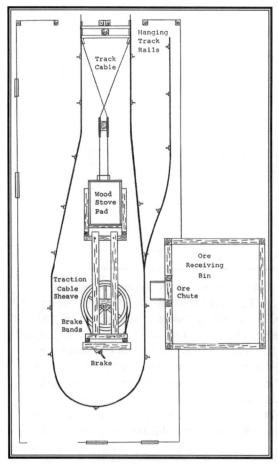

The plan view illustrates the upper terminal for the Bleichert tramway that served Colorado's Mary Murphy Mine. Tram buckets entered the structure through the doorway in the upper left corner where a worker uncoupled them from the traction rope. He rolled the buckets along the hanging rail to the other side of the terminal where a second worker filled them at the ore chute, then reconnected the buckets to the traction rope for the trip down. Note the right rail spur for keeping extra buckets. Engaging the handbrake encircling the sheave had the capacity to stop the tram in the event disaster struck.

Author.

Interior view of the tram terminal at the Mary Murphy Mine. The tram buckets entered the doorway in the left background. Compare with the plan view.

Author.

Towers for both Bleichert and Hallidie systems required stout cross-members to support the wire ropes at a distance that permitted the buckets to swing in the wind and not strike the towers. Hallidie systems, with their single wire rope and fixed buckets, needed only one cross-member that featured several idler wheels or rollers. Bleichert systems, on the other hand, required a stout cross-member at the tower top to support the stationary track cable, and a second cross-member three to seven feet below to accommodate the moving traction rope. The second cross-member almost always featured either idler wheels or a broad steel roller.

Engineers found great challenge in attempting to design tram towers. They had to minimize the quantity of construction materials, yet create a tall structure that resisted a complex interplay of forces. Building a sound structure that met the last criterion was perhaps the most rigorous engineering goal. Tram towers had to withstand the downward pressure exerted by the weight of the cable and ore buckets, the horizontal forces created by starting and stopping the system, and lateral forces created by windshear on the towers, cables, and buckets. This last force was not to be underestimated in the rugged and mountainous West, and it had caused a number of major malfunctions at mines.

The choice of tower form and spacing was a function of topography, local weather, and the pitch of the line. Pyramid and composite towers could have been built higher than the other types, and they were the least costly. Hillside towers were best for very steep terrain, and they, as well as through towers, gave greatest stability during severe weather. Engineers recommended using steel beams for construction, but this was far too expensive, and most mining companies in the West built with timber ranging from six-by-six to ten-by-ten inch stock, fastened with bolts. Where the bucket line traversed forested hillsides, laborers had to cut a path through the trees.

Tramway terminals presented engineers with no fewer design problems than did the towers. Terminals had to be physically arranged to permit the input and storage of tons of ore from the mine, they had to facilitate transfer of the payrock into or out of the tram buckets, they had to resist the tremendous forces put on the sheave wheel by the traction rope, and in the case of Bleichert systems, they also had to anchor the track cables. Mining engineers designing small-capacity tramways attempted to solve all of the above problems literally under one roof, while the terminals for large-capacity tramways were enclosed in more complex buildings.

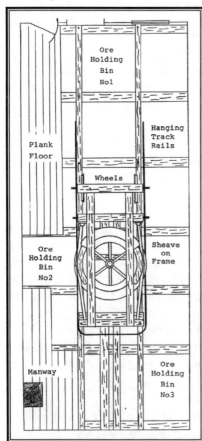

The plan view illustrates the bottom terminal for the Bleichert tramway that served Colorado's Mary Murphy Mine. Tram buckets entered the structure through the doorway in the upper left corner where a worker uncoupled them from the traction rope. He rolled the buckets along the hanging rail and stopped them over one of the three ore bins, depending on the grade of ore in the bucket. He released a catch lever, emptied the bucket, rolled the vehicle to the other side of the terminal where a second worker reconnected it for the return trip to the upper terminal. The timber frame encasing the sheave featured railroad car wheels so the carriage could have been moved to adjust the tension in the traction rope.

Author.

The view depicts the sheave carriage inside the Mary Murphy's bottom tram terminal. Most Hallidie and Bleichert tramways featured similar carriages to adjust the ropes. Compare with the diagram above.

Author.

Regardless of the type of tramway a mining company had installed, special accommodation had to be made for the sheave wheels in both terminals. They had to resist the significant horizontal forces of keeping the traction rope taut. The sheave in the top terminal was usually fixed onto a heavy timber framework anchored to bedrock and partially buried with waste rock ballast. Typical sheave wheels, six feet in diameter for small systems and twelve feet for large systems, featured a deep, toothed groove for the rope, and they were fixed onto a heavy steel axle set in cast-iron bearings bolted to the timbers. The teeth in the groove gripped the rope in the event that a terminal worker had to throw the brake and stop the system. Brake levers were very long to provide great leverage, and they were located either on a catwalk or adjacent to the wheel at ground level, both of which afforded the straining worker a view of the system he was attempting to stop. The lever controlled heavy wooden shoes that pressed with much force against a special flange fastened to the sheave wheel. These brakes may seem dubious, but they were reputed to easily bring to a halt entire lines of full buckets.

At the bottom terminal the sheave had be moveable to take up slack in the rope line. In many cases the wheel was fastened onto a heavy timber frame pulled backward by adjustable anchor cables or threaded steel rods. The wheel carriage also featured hardware that automatically upset the ore buckets, and they emptied their precious contents into an ore bin underneath the terminal. The terminals for Bleichert systems included anchors for the track cable, room for workers to uncouple the buckets from the traction rope, and hanging rails for directing the buckets. Nearly all of the Bleichert tramway terminals built in the mountainous West had to be enclosed to provide shelter for the crew, however meager, in the face of frigid winds during winters. Hallidie systems, which automatically loaded and emptied the buckets, did not always require this added expense.[107]

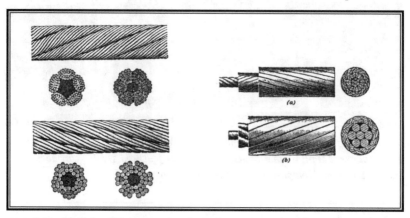

At left is a sample of the type of cable typically used with mine hoists, and at right is a sample of the cable used in association with tramways. Note that the tram cable is smooth to facilitate easy and consistent movement over rollers and around sheaves.

International Textbook Company International Library of Technology: Hoisting, Haulage, Mine Drainage International Textbook Company, Scranton, PA, 1906 A51 p5.

Mining engineers with foresight and access to capital not only took delight in the fact that tramways were powered by gravity and required no costly drive mechanisms, but they went an extra step and attempted to make the continuously moving mechanical wonders do work for the mine. Besides using the empty buckets to deliver supplies to lofty mine workings, clever engineers added gearing to the rotating sheave wheels that turned electrical generators. The problem with harnessing tramway power, claimed mining engineers, was that interruptions of tramway service stopped the generator or any other machinery so connected, and as a result, they could not use tram power to run crucial plant components.[108]

Today, intact tramways are almost nonexistent in the West. Yet, by examining the remains of towers and tram terminals, today's visitors may be able to reconstruct the type of system that served the mine. In many cases the sheave was centrally located in the terminal. If the sheave has been removed by salvage operations, its origi-

nal position may be demarcated by heavy cantilevered timber framing braced with diagonal beams designed to counter the rope's horizontal pull. Bleichert tram terminals often surrounded the sheave
with the timber framework, but Hallidie systems had to have clearance on both sides and the rear of the sheave to permit the buckets
to swing around. Also, because Hallidie buckets were fixed to the
rope, the ore chutes had to be very close to the bucket pathway. On
the other hand, workers uncoupled and rolled Bleichert buckets
across hanging rails, which had been routed under ore chutes. As a
result, the ore chutes at Bleichert tram terminals were distant and
often behind the sheave wheel at locations impossible for Hallidie
buckets. The hanging rails were usually bolted to ceiling beams, and
while the rails may have been removed, the visitor inspecting a tram
station may be able to locate the pattern of telltale bolt-holes in the
undersides of some of the remaining beams.

*The traction ropes for Bleichert tramways had to be well-anchored to withstand
the tremendous forces of wind and weight. The photo illustrates the connection
used at the Mary Murphy Mine between the smooth track cable at right and a
rough conventional cable at left. The tape measure in the photo has been
extended three feet for scale.*

Author.

As grand a solution as Hallidie and Bleichert tramways were for
facilitating the procession of ore from a mine, they were too big and
expensive for many small operations that possessed only modest
amounts of capital. Yet, rugged terrain and locations high on the sides
of mountains presented no less a problem for these limited but profitable operations. The high relief and steep slopes also provided an

answer to their dilemma of access. Rather than install the large and efficient but costly tramways relished by academic engineers the likes of T.A. Rickard and Herbert C. Hoover, the smaller companies strung up *single* and *double-rope reversible* aerial trams.

In addition to hauling ore, tramways served as an important mode of travel for men and materials from a mine's base camp up to lofty and remote workings. Unregulated for safety, a ride in a tram bucket could have been hair raising, but it saved miners having to ascend several thousand feet on foot over the course of hours.

Courtesy of the Colorado Historical Society, Denver, CO.

Well-engineered single-rope trams typically consisted of simple components. A fixed line extended from an ore bin located high up at the mine down to another ore bin below. A hoist at the mine wound and unwound a second cable that pulled a bucket. The cost of installing such a tramway was very low, but many engineers scorned them because these conveyances were slow and inefficient, relying on one vehicle moving back and forth between the bins.[109]

The primary materials a mining company needed to build a single-rope tramway were abundant and inexpensive. A mining outfit had to procure two lengths of cable, lumber, a hoist, and a vessel hung from a pulley. In many cases mining companies, especially those with limited resources, purchased used steam or gas hoists, and impoverished outfits used prospectors' hand-cranked windlasses or crab winches. The lines that mining operations strung up may have been retired hoist cables, and the bucket possibly fashioned from an ore car body, but proper ore buckets were preferred. The mining outfit often anchored the hoist high up at the mine to a sound timber foundation that they tied into bedrock, and they often anchored the ends of the track cable to bolts. Double-rope tramways featured two buckets counter-weighting each other. Ramshackle though they might be, miners working high up on mountainsides would have agreed that the single-rope aerial tramways saved them immense effort at bringing down their precious payrock and sending up drill-steels, dynamite, and the occasional passenger.[110]

CHAPTER 4

GEAR OIL & STEAM POWER: SURFACE PLANTS FOR SHAFTS IN THE GILDED AGE

The Joe Dandy Mine in Cripple Creek represents the typical surface plant arrangement for shaft mines. The headframe stands over the shaft, and the hoist and hoist house, in the foreground, are directly aligned with the headframe and have been located on solid ground a short distance away. The steam boiler stood adjacent and right of the hoist house, the shop, visible immediately above the hoist house, has been situated near the shaft, and the ore sorting house, left of the headframe, has been built also near the shaft. The waste rock dump lies beyond the complex.

Author.

Historic mine shafts in the West today inspire a sense of mystery, wonder, and danger among the general public. The casual observer may view the curious remains of these hardrock operations as no more than a deep hole in the ground surrounded by piles of barren and scorched waste rock. Unknown to many people, however, shafts required support from substantial and complex surface plants, which had to meet the same five fundamental needs of underground operations as adit mines. The plant had to provide an entry underground, ventilation, equipment maintenance and fabrication, transportation in and around the mine, and waste rock storage. As part of the transportation infrastructure, surface plants for shafts also included a *hoisting system* to move materials up and down the vertical passage. The hoisting apparatuses that served shafts during the nineteenth century usually consisted of an engine, a cable, a power source, a headframe, a vehicle, and a means of communication between the hoist operator and the miners below. The complexities of erecting and operating these facilities, as well as driving vertical underground workings, required a much greater degree of mechanization, engineering, and capital than adit mines. In this chapter we will explore and analyze the surface plant components typically associated with shafts.

Shafts underwent a process of discovery, promotion, and development nearly identical in sequence to the events that led to the opening of adits. W.S. Stratton's discovery and development of the fabulous Independence Mine in Cripple Creek is fairly representative of the process. After receiving favorable assay reports from samples Stratton had broken off a granitic dike near what became Victor, Colorado, he sank a shallow prospect shaft to ascertain the mineral body's worth below ground. Under Stratton's experienced and guiding hand the Independence claim underwent a series of improvements, contingent on the injections of borrowed capital. In keeping with traditional mine development, the old prospector first installed a temporary plant, which he hoped would be followed by a production-class plant, provided sufficient ore lay underground. Ore sufficient to make Stratton a millionaire did lie underground, and he designed a surface plant to facilitate light production. Top-notch professionally-trained engineers upgraded Stratton's facilities into a massive permanent plant following his sale of the mine to British interests. Most shafts were not nearly as rich as the Independence, and many more were not even marginally profitable. As a result, they were abandoned in early stages of development, ranging from shallow operations to what mining engineers knew as *deep prospects.*

Pursuing a mineral lead under foot required that the party of prospectors erect a temporary surface plant to support work underground. Popularly known among mining men as a *sinking plant,* temporary plants for prospecting ranged in quality from simple and labor-intensive facilities to substantial and mechanized operations. All varieties, however, shared elements of impermanence, simplicity, portability, a low cost, and unanimous definition among professionally-trained mining engineers as not being suited for ore production.

As with adit mines, miners organized the surface plant components according to a few general patterns. Most basic prospect plants consisted of an entry underground, specifically a shaft collar, a work shop where a blacksmith maintained and fabricated tools and hardware, a ventilation system, a means of waste rock disposal, transportation through the plant, and a hoisting system.

The first arrangement pattern took form when prospectors began to drill and blast into bedrock in pursuit of ore. As they sank their exploratory shaft, they set up a manual *windlass* directly over the shaft to raise waste-rock-laden buckets, and they dumped the vessels downslope from the shaft collar. Over time the waste rock formed a semicircle around the shaft, or a raised pad encompassing the shaft when the terrain was relatively flat. The party of prospectors may have used picks and shovels to create a cut-and-fill earthen platform adjacent to the shaft where they could set up a temporary field forge for sharpening drill-steels, and possibly a small wall tent as living quarters and for materials storage.

Once the prospecting party had demonstrated existence of ore and aroused interest from a mining company, the company engineer, or superintendent acting as such, expanded the surface plant to facilitate deep prospecting. For the second arrangement pattern, surface laborers graded relatively large cut-and-fill earthen platforms on firm ground adjacent to and upslope from the shaft where they placed the hoist, other machinery, and the shop. They installed the ventilation system adjacent to the shaft, and erected a headframe. The mine's transportation system, usually a rail line, extended away from the shaft collar and terminated on the waste rock dump's edge, and two spurs may have veered off to the shop area and to a timber stockpile. Miners continued the practice of dumping waste rock away and downslope from the shaft, while grading it flat in attempts to maintain a working surface.

Clustering the components of the hoisting system and the shop near the shaft collar reduced costs and eased the logistics of surface operations. Arranging the components close made sheltering and attending the machinery easier, and it saved miners time and labor when they handled tools and hardware. In harsh mountain climates shelter was often necessary to run the mine, and mining companies enclosed the core plant components in large frame shaft houses. When the engineer began building the surface plant, he oriented the machinery and buildings according to sub-datum lines taken off the rectangular shaft sides, and he almost without failure directly aligned the hoist with the shaft on the master datum line.[1]

Where the miners had proven the existence of spectacular ore reserves and the engineer decided to upgrade the surface plant with production-class components, he added additional facilities of the types installed at profitable adit mines, and they were arranged according to the above described pattern. The engineer had construction gangs erect an ore bin on the shoulder of the waste rock dump, grade a wagon road that accessed the plant core, and another road that passed along the downslope edge of the ore bin. The plant upgrade often included a better ventilation system, enlarged shop facilities, and a compressor, much of which may have been enclosed in the shaft house. The miners and surface laborers extended the original mine rail line outward from the shaft by adding several spurs that curved toward different parts of the waste rock dump, to the shop, and to the ore bin. In the event the engineer upgraded the hoisting system, he usually placed the new hoist farther from, but still in alignment with, the shaft and abandoned the original hoist. Like adit mines, the surface plant also often included an outhouse, a water tank located upslope, an office, and possibly bunkhouses and a cookhouse situated away from the mine. These patterns of arrangement changed little from the 1870s until the 1940s.

Surface Plants for Prospect Shafts

M any shaft mines in the West began as small and humble prospect operations. When a party of prospectors thought they had a promising lead, they had to decide whether to drive an exploratory adit or sink a prospect shaft to explore the geology at depth. Many chose to labor at the latter, mostly because shafts were conducive to tracking the ore body downward, as well permitting prospectors to drive periodic lateral drifts and crosscuts for horizontal investigation.

Sinking a small, confined prospect shaft was no easy task. In his seminal western work *Roughing It,* Mark Twain provides us with a realistic if not pessimistic account of the hardship some prospectors realized in undertaking such work:

> *"We decided to sink a shaft. So, for a week climbed the mountain, laden with picks, drills, gads, crowbars, shovels, cans of blasting powder and coils of fuse, and strove with might and main. At first the rock was broken and loose and we dug it up with picks and threw it out with shovels, and the hole progressed very well. But the rock became more compact, presently, and gads and crowbars came into play. But shortly nothing could make an impression but blasting powder. That was the weariest work! One of us held the iron drill in its place and another would strike with an eight-pound sledge - it was like driving nails on a large scale. In the course of an hour or two the drill would reach a depth of two or three feet, making a hole a couple of inches in diameter. We would put in a charge of powder, insert half a yard of fuse, pour in sand and gravel and ram it down, then light the fuse and run. When the explosion came and the rocks and smoke shot into the air, we would go back and find about a bushel of that hard, rebellious quartz jolted out. Nothing more."* [2]

Vivid portrayals of this kind illustrate why parties of prospectors easily gave up their search for wealth when they tasted the backbreaking labor required to open a mine. Yet the lust for striking bonanza ore was strong in many prospectors, and as is borne out by the remains in mining districts today, wealth-seekers sank substantial but profitless holes in the ground.

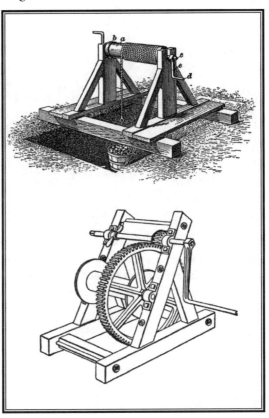

At top is a windlass, which was a simple, inexpensive, and easily transported manual hoist ubiquitous among prospectors throughout the West. Nearly all shafts were opened with this device. The crab winch below offered a pulling power and depth capacity greater than the several hundred pounds and 100 foot limitations of the windlass. However, because they required a foundation, prospectors rarely used crab winches. The model shown has been equipped with a brake.
International Textbook Company International Library of Technology: Hoisting, Haulage, Mine Drainage International Textbook Company, Scranton, PA, 1906 A50 p2.

Sinking a shaft required a sequence of events that were fairly uniform across the West. Laborers first excavated a broad pit through soil down to bedrock, and they began drilling blast holes arranged in a special pattern designed to shatter the rock in a rectangular form. Drilling by hand was difficult work, but drillers had an easier time swinging their hammers downward when sinking shafts than they did while driving horizontal holes in adits. Shoveling the shattered rock off the pit bottom after the miners had shot the round, however, was awkward and frustrating. The confined space made maneuvering a shovel difficult, and cleaning the ragged floor of fractured gravel proved impossible. As the miners sank the shaft, they quickly passed the point where they could heave the rock out with a shovel. To continue work, the miners had to resort to a hoist and bucket. Most small operations employed a *hand windlass,* which was an ages-old manual-powered winch consisting of a spool made from a lathed log fitted with crank handles. The handles also acted as an axle permitting the spool to rest in slotted boards bolted onto a heavy wood frame placed over the shaft.

Prospectors have set up a windlass over the beginnings of the Combination Shaft in Goldfield, Nevada in 1906. The frame that the windlass stands on and the raised wooden collar are typical of prospect shafts.

Mining & Scientific Press Oct. 6, 1906 p413.

When a miner at the shaft bottom finished filling an ore bucket with shot-rock, he signaled his partner at the top who began cranking the windlass and slowly raised the bucket. Once the bucket had reached the shaft collar, the topman undertook the delicate task of wrestling the load of rock, one-hundred pounds or more, onto the shaft rim without letting even a small rock fall, which would have bounced down the shaft walls and probably injured or killed his partner. The topman stood on the wood planks providing firm footing around the shaft rim. He wedged a board under the windlass handle to stay it, and then grabbed the bucket by its bail and tugged it upward. Once the vessel was on firm ground, the topman unhooked it from the rope and dragged it to the edge of the waste rock dump where he spilled its contents. The topman and his partner repeated this process until they had cleared the shaft of all shot rock, then they proceeded to drill and blast another round.

Prospectors sinking inclined shafts had the option of using what mining engineers termed a *geared windlass* or *crab winch,* which offered a greater pulling power and depth capacity. Geared wind-lasses cost much less than other types of mechanical hoists, and they were small and light enough to be packed into the backcountry. The winch was not easily used at vertical shafts because the rope spool and hand-crank fitted onto a frame which had to be anchored onto a well-built timber structure.

Where the shaft penetrated horizons of soil and decaying bedrock, the walls usually required support. Prospectors working in the West favored two basic types of timbering. The first method, known as *closed cribbing*, consisted of hewn logs or three-by-six dimension timbers deeply notched at the corners so they nested together without gaps, while *open cribbing* featured shallow notches permitting gaps between the timbers. Closed cribbing prevented infiltration of loose material, but it required more logs or timber, greater work, and incurred a higher expense than open cribbing. As miners sank the shaft deeper they added timber sets upward, as well as downward, to retain the ever-growing waste rock dump that began encompassing the shaft collar.[3]

The party of prospectors set up a small field forge near the shaft after they had determined that they were going to put some time into the claim. They placed the forge, be it a freestanding pan or gravel-filled rock enclosure, either on the waste rock dump or on a cut-and-fill earthen platform adjacent to the shaft. Except for a wind sock or a hand-turned blower for ventilation, little else comprised the surface plants typical of shallow shafts.

Hoisting Vehicles and Shaft Form

Experienced prospectors and mining engineers recognized that crude prospect shafts were inadequate for anything other than a cursory examination of the geology underground. In instances where a prospecting outfit strongly suspected or had confirmed the existence of ore, they sank a better, more formal shaft that was conducive to deep exploration and even, the outfit hoped, ore production. Between the 1880s and 1920s mining engineers were critical distinguishing between temporary shafts and production-class shafts. Engineers placed great importance on the shaft because it served as the conduit between the underground workings and the ground surface, accommodating the exchange of materials required for development and ore production.

The preference among prospectors and engineers for a rectangular shape remained unchanged throughout the Gilded Age, and it was standard for several reasons. First, such shafts were cheaper to sink and easier to timber than circular ones. Second, western miners inherited the rectangular form from traditional Cornish mining practices.[4]

During the 1860s and 1870s mining engineers began to understand that the clearance within a mine shaft directly influenced a mine's level of production. Small shafts limited the quantity of ore that could be hauled out per vehicle trip, and large shafts facilitated production in economies of scale. Mining in Michigan, California, and on the Comstock during the 1860s had set a precedent in which companies working in vertical shafts aspired to install steam-powered hoisting systems that used a *cage* as the hoisting vehicle. Mining engineers defined a shaft large enough to accommodate a cage to be production-class.[5]

A mining industry institution for over one-hundred years, the cage consisted of a steel frame fitted with flooring for crews of miners coming and going on and off shift, and rails to accommodate an ore car. Nearly all cages used in the West featured a stout cable attachment at top, a bonnet to fend off falling debris, and steel guides which ran on special fine-grained four-by-four inch hardwood rails. After a number of gruesome accidents in which hoist cables parted, such as when a cage loaded with nineteen miners dropped to their deaths into the sump of Montana's Anaconda shaft, mining machinery makers began installing special safety-dogs on cages designed to stop an undesired descent. Usually the dogs consisted of toothed cams that were controlled by springs kept taut by the weight of the suspended cage. If the cable broke, the springs retracted, closing the cams onto the wood rails.[6]

Cages proved to be highly economical because mining companies did not have to spend time transferring ore and waste rock between various vehicles. A miner or trammer merely had to push on an ore car filled at some distant point in the mine, and another worker retrieved it at the surface. However, cages presented mining companies with several drawbacks. One of the biggest problems lay in drilling and blasting a shaft large enough to accommodate the vehicle and support timbering. In addition, mining companies had to spend considerable capital installing fine-grained hardwood guide rails, which had to be bolted the length of the shaft in two perfect lines.

The composition of production-class shafts evolved during the 1850s and 1860s. Between the 1830s and 1860s American miners imitated the Cornish and usually arranged their shafts to consist of a single large compartment.[7] However, during the 1860s and 1870s mining companies began dividing production-class shafts into a combination *hoisting compartment* and *manway*, also known as a *utility compartment*. By the late 1860s mining engineers in the West had recognized the value of *balanced hoisting*. This system relied on the use of two shaft vehicles counterweighing each other, so that as one vehicle rose the other descended. The use of one hoisting vehicle to raise ore had become known as *unbalanced hoisting*, and while this system was very inefficient in terms of production capacity and energy consumption, was the less costly to install. Balanced hoisting required a shaft featuring two hoisting compartments and it necessitated a double-drum hoist, which constituted a considerable expense. But the hoist only had to do the work of lifting the ore, and as a result this system was energy efficient and provided long-term savings. Wealthy mines anticipating production over an extended period of time spared the expense to install a balanced system.[8]

Western mines used three other types of hoisting vehicles during the Gilded Age, in addition to the cage. The first was the old-fashioned *ore bucket*, the second was the *ore bucket and crosshead*, and the last was the *skip*. The ore bucket that had endeared itself to the western mining industry became known as a *sinking bucket* because its shape and features were well suited for the primitive conditions typical of mines under development. Sinking buckets had convex sides that prevented the rim from catching on obstructions such as timbers, and permitted the vessel to glance off the shaft walls while being raised. Manufacturers forged a loop into the bail to hold the hoist cable on center, and the bottom of the bucket featured a ring so the vessel could be upended once it reached the surface. As late as the 1930s some mining companies continued the ages-old practice of using ore buckets instead of cages, and some employed a straight-sided variant known as the *Joplin Bucket*, named for its prevalence in the lead and zinc mines near Joplin, Missouri. To expedite the production of rock while continuing to use ore buckets as their principal hoist vehicles, mining companies discovered that the vessels could easily be mobilized underground when unhooked from the hoist cable and placed on flatcars. They became, in essence, substitutions for ore cars. While ore buckets did not have the same compartment restrictions as skips and cages, most outfits in the West that used them followed the conventions of shaft sizes and configurations recommended by mining engineers.[9]

Companies engaged in deep shaft sinking took great risks when they attempted to use free-swinging ore buckets. The vehicles had a propensity to act like pendulums, striking the shaft walls and posing the very real threat of dumping their loads of rock and tools on the miners below. William Langdon and Wilbur Parffit experienced this problem firsthand while they were mucking shot rock out of the bottom of the Prince Albert Shaft in Cripple Creek in 1917. They had finished filling an ore bucket and signaled the hoistman to raise it out of the shaft. It began swinging on its ascent and caught on some timbering high up, letting two fifteen-pound rocks fall. The projectiles ricocheted down the shaft and one struck Langdon in the head, completely crushing his skull.[10] To prevent such accidents, some mining companies used a hybrid hoist vehicle that consisted of an ore bucket suspended from a frame that ran on the same type of guide rails used by cages. The frame, known as a *crosshead,* held the ore bucket steady and provided miners with a dubious platform to stand on during ascents and descents in the shaft. Many small, poorly financed, and marginally productive mining companies in remote locations favored this type of hoisting vehicle.

Cornish mining engineers in Michigan copper mines developed the *skip* during the 1840s and 1850s for haulage in inclined shafts. The typical skip consisted of a large iron box on wheels that ran on a mine rail line. Skips had little deadweight, they held much ore or waste rock, and because they ran on rails they could be raised quickly. They were also easy to fill and empty. The hoistman on the surface lowered the skip to a shaft station where a miner dumped rock into the vehicle. The hoistman then put on steam and raised the skip into the headframe where it was automatically upset and disgorged its contents. Skips were similar in size to cages and they ran in shafts of approximately the same area.

During the 1890s mining engineers began to recognize the skip as superior to the cage for ore production in vertical shafts. Skips were lighter than cages because they did not have the combined dead weight of the vehicle and an ore car, and the reduced weight resulted in energy savings. They could be filled and emptied quickly, resulting in a rapid turnover of rock. For these reasons, large western mining companies began replacing cages with skips for use in vertical shafts shortly after the turn of the century. The changeover proceeded slowly through the 1900s, accelerated rapidly during the 1910s, and by the 1930s most western mines used skips. Of course, mining companies were able to switch skips and cages at the shaft collar when necessary by unhooking the hoist cable and pulling the vehicle off of the guide rails.[11]

Mining engineers recognized that cages and skips were the vehicles of choice for productive mines, and they relegated ore buckets, used throughout the West for all classes of operations into the 1870s, to a status of shaft sinking and minor ore production. Cages and skips permitted faster hoisting speeds than with free-swinging ore buckets, from 300 to 400 feet per minute up to 3,000 feet per minute in deep shafts, and with them, mining companies hauled great tonnages of rock from underground.[12]

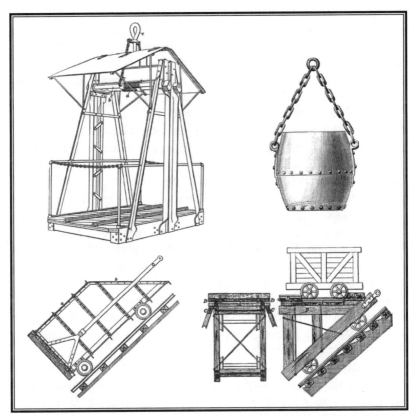

The illustration depicts the principle hoisting vehicles used in western mines. At top left is a cage, ubiquitously used by production operations in vertical shafts for raising full ore cars and sending down miners and supplies. The cage ran on 4x4 timber guides bolted the length of the shaft. Top right is a sinking bucket used by operations with limited funding for prospecting and minor production. Most buckets featured an iron bail with a forged loop instead of the chain bail. In the early years of hardrock mining engineers devised a wheeled platform for raising loaded ore cars out of inclined shafts, which they had termed a cage. It quickly fell out of favor when the skip, bottom left, came into being. The skip contained more rock for less vehicle-weight, holding the contents more securely, and proving inexpensive.

International Textbook Company International Library of Technology: Hoisting, Haulage, Mine Drainage International Textbook Company, Scranton, PA, 1906 A53 p9.

Ingersoll Rock Drill Company Catalog No.7: Rock Drills, Air Compressors and Air Receivers Ingersoll Rock Drill Company, New York, NY, [1887] p64.

International Textbook Company A Textbook on Metal Mining: Steam and Steam-Boilers, Steam Engines, Air and Air Compression, Hydromechanics and Pumping, Mine Haulage, Hoisting and Hoisting Appliances, Percussive and Rotary Boring International Textbook Company, Scranton, PA, 1899 A23 p79, 87.

By the 1880s engineers declared typical prospectors' shafts, usually four-by-six feet in-the-clear or smaller, to be substandard and for shallow exploration only. The premise was that such a shaft was too small to accommodate the preferred production-class hoist vehicles and a utility compartment. During the 1700s and early 1800s, miners in Cornwall extracted copper and tin from shafts that were an almost inconceivably small two-by-three feet to three-by-four feet in area, and the large shafts they used were still a relatively small six-by-eight feet. American mining engineers working in the nineteenth century West considered these sizes to be unacceptable. During the 1870s mining engineers began defining production-class shafts as needing hoisting compartments that were at a minimum four-by-four feet in-the-clear to accommodate cages. By the late nineteenth century the definition expanded as a result of the introduction of larger cages. Mining engineers felt that a four-by-five foot hoisting compartment was better suited for ore production, and five-by-seven feet was best, because it permitted the movement of larger ore cars and machines, which resulted in the extraction of greater tonnages of payrock.[13]

Between the 1870s and 1890s a few particularly large western mining companies spent enormous sums of capital sinking four compartment shafts, two of which were for hoisting, and the other two for utility lines and the rods of the mighty Cornish pump. The pump compartment was often the same size as the hoisting compartments, and it housed the reciprocating wood beamwork that extended from the engine at the shaft collar down into the shaft sump where lay the pump mechanism.

The Cornish pump, brought by Cornish miners at first to Michigan, saw instant success in the West during the 1860s in the well-capitalized and deep mines of California's Mother Lode and on Nevada's Comstock. Mining engineers subsequently adapted the technology to other water-saturated districts such as Eureka, Nevada and Central City, Colorado. The pumps first used in the Comstock were virtually indistinguishable from contemporary models plunging away across the Atlantic, but the design used in other western districts at a later date had been modified to suit the available machinery.

The Locan Shaft in Nevada's Eureka Mining District serves as a fine example of the Cornish pump technology of the 1870s and 1880s. Instead of the huge upright beam engine that traditionally powered Cornish pumps, the Eureka Consolidated engineers installed a conventional horizontal steam engine, similar to those used by industries throughout North America. The engine drove a large bull-gear that transferred motion to the pump's walking beam, and the gear also slowed the movement down. The walking beam pivoted up and down on a heavy axle. One end connected to the rods extending down the shaft and the other end featured counterweights to balance the rods. The foundation featured deep brick-lined wheel wells for the engine's flywheel, gearing, and pump bob counterweight. The shaft house enclosed the assemblage of machinery and carpenters surrounded it with plank flooring. Like other well-engineered operations, the Locan Shaft had two compartments for balanced hoisting, a separate pump-way for the rods, and a utility compartment.[14]

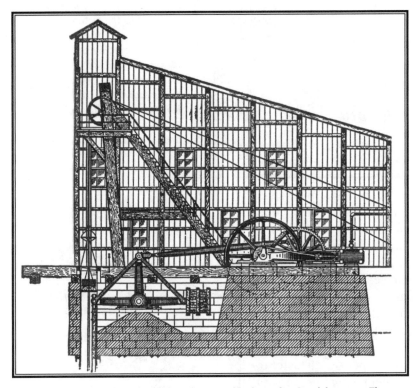

The profile reveals the principal surface mechanism of a Cornish pump. The powerful horizontal steam engine at right turned a large bull gear visible behind the engine's spoked flywheel. The bull gear featured an eccentric crank that imparted a push-pull motion to the heavy horizontal rod extending to the left of the gear. The rod was in turn connected to the pump bob, which is the triangular frame mounted in the pit in the massive stonework. The pump bob acted in a fashion similar to a seesaw, raising and lowering the solid timber rod that extended into the bowels of the mine. The rod hangs off the left side of the bob while a counterweight to balance the rod hangs off the bob's right side. The headframe can be seen towering over the shaft. Cornish pumps were a custom affair, two rarely being identical.

International Textbook Company A Textbook on Metal Mining: Preliminary Operations at Metal Mines, Metal Mining, Surface Arrangements at Metal Mines, Ore Dressing and Milling International Textbook Company, Scranton, PA, 1899 A42 p3.

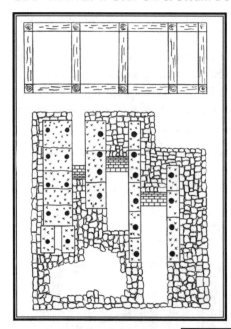

The plan view depicts the Cornish pump foundation at the Locan Shaft in Nevada's Eureka district. The foundation consists of an expansive and high rock footing that features brick-lined wells for the steam engine's flywheel (left), the bull gear (center), and pump bob (right). The engine was anchored to the rectangular stone pad at left, and the other stone pads supported massive bearings for the gearing and pump bob axles. The shaft, located adjacent to the foundation, consists of two hoisting compartments, a pump-way, and a utility compartment lined with 12x12 timbering.

Author.

Whether a mine shaft featured an extra compartment for the rods of a Cornish pump, or whether the shaft consisted of a single hoisting compartment and a manway, it usually had to be timbered. Though western mining companies found the expense distasteful, timbering was necessary to anchor the wood guide rails required to run cages or crossheads, and to support the shaft where it penetrated unstable ground. Timbering a shaft proved to be an engineering challenge that miners did not take lightly, lest structural defects cause a deadly wreck with a cage. The best system proved to be *shaft-set* or *square-set*

The block diagram depicts the shaft form that predominated in the West prior to the late 1870s, when hardrock mining was still in a nascent state. Early mining engineers divided shafts into two to three compartments uniform in size for hoisting. They often lined the shaft walls and built the dividing bulkheads with closed cribbing.

International Textbook Company A Textbook on Metal Mining: Preliminary Operations at Metal Mines, Metal Mining, Surface Arrangements at Metal Mines, Ore Dressing and Milling International Textbook Company, Scranton, PA, 1899 A40 p30.

timbering, which consisted of open cubic frames bracing the shaft walls.

Construction of the timbering began at the shaft collar when miners or timbermen built a timber frame known as the *bearer*, with cross-members dividing the hoisting and utility compartments. Experienced timbermen often assembled the framing with square-notched joints, iron tie rods, and bolts, while mine workers with less skill simply abutted the timbers. The workers sawed vertical posts and cross-timbers to size on the waste rock dump, and sent the prepared material down the shaft. The miners assembled the timbers in the shaft to form an open cube, which they hung from the previous set using square-notching and iron tie rods. In places where the shaft penetrated loose ground, miners lined the outside of the timbering with vertically-nailed planking to stabilize rocks. Last, to ensure that each shaft set was perfectly aligned with the one above, the miners or timbermen hammered wooden

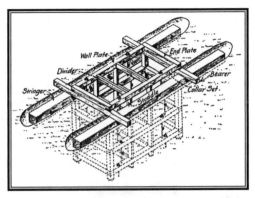

During the 1860s Western mining engineers began to use shaft-sets, also known as square-sets, to support the shaft walls and cage guide rails. Timbermen built a collar around the shaft consisting of heavy timber stringers and bearers, and hung the subsequent sets from the initial structure. They installed dividers between the compartments, which contributed to the skeleton's overall strength. This form of timbering predominated in the West by around 1880. Mining engineers often incorporated the stringers into the foundation for the headframe.

Engineering & Mining Journal Sept. 20, 1913 p546.

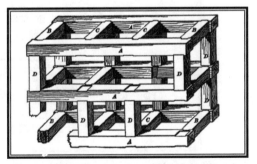

The illustration provides a close view of the shaft-set timbering typically used for production-class shafts in the West. Note how timbermen have cut square-notched joints into the horizontal wall plates to facilitate tight assembly of all the timber members. In the event that the shaft penetrated loose ground, the mine's timbermen lined the outside of the timber sets with board lagging.

International Textbook Company A Textbook on Metal Mining: Preliminary Operations at Metal Mines, Metal Mining, Surface Arrangements at Metal Mines, Ore Dressing and Milling International Textbook Company, Scranton, PA, 1899 A40 p25.

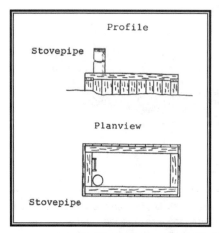

Profile

Stovepipe

Planview

Stovepipe

Prospect shafts were much simpler and smaller than the production-class shafts shown above. The illustrated shaft consists of a single compartment 3 by 7 feet in the clear. It features shaft-set timbering, and it contains a ladder and ventilation duct. The prospect operation, active in Goldfield, Nevada around 1905, relied on a horse whim for hoisting. In many cases the collars of such shafts have given way, leaving a deep conical depression today.

Author.

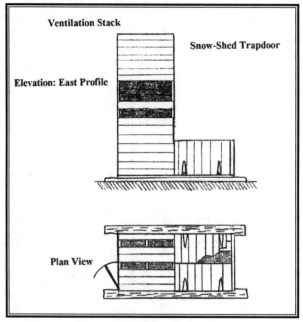

Ventilation Stack

Snow-Shed Trapdoor

Elevation: East Profile

Plan View

The profile and plan view illustrate several features common to the collars of moderately deep prospect shafts. First, the miners built a snowshed trap door over the hoisting compartment to deflect rocks when they emptied the ore bucket at the shaft collar. Second, they built a ventilation stack over the utility compartment, which consists of a plank-lined booth with a door and a port for ventilation tubing. Further, the miners divided the hoisting and utility compartments with a solid wood bulkhead and lined the shaft with well-fitted lagging. They arranged for a mechanical blower to force air into the stack, which created a down-cast current in the utility compartment, forcing foul gases to rise out of the hoisting compartment as an up-cast current. The illustration was taken from a prospect shaft on the fringes of the Cripple Creek district.

Author.

wedges between the timbers and the shaft walls in accord with measurements taken from a plumb bob. This also had the effect of making each set weight bearing and it helped the timbering resist swelling of the rock walls. Where the shaft sets had to merely anchor the cage guide rails, mining engineers felt that the five-by-five inch timbers were sufficient, while large multiple compartment shafts in swelling ground required timbering made of up to sixteen-by-sixteen inch stock backed by three-inch thick lagging. Some mining companies that used free-swinging ore buckets had their miners nail vertical skid boards to the exposed shaft sets to prevent the rim of the bucket from snagging. It is not hard to imagine the immense volume of timbering that mining companies put into their shafts, as well as the heavy expenses they incurred periodically replacing it.[15]

Horse Power to Steam Power:
Hoisting Systems for Prospect Shafts

A well-engineered shaft by itself was not enough to facilitate the exchange of men and materials into and out of deep workings. Mining outfits had to install a hoisting system, which proved to be important because it both governed the quantity of rock that the company extracted from underground during a shift, and the depth to which the company could have worked. Here, we look at the types of systems employed by prospect operations, which differed from those used by profitable mining companies.

Typical hoisting systems installed by western prospect operations consisted of a hoist, a headframe, a power source, and a hoisting vehicle. These components shared fundamental relationships with each other, and they interfaced with the other facilities comprising the prospect outfit's surface plant. For example, the type of hoist an engineer selected influenced the type of headframe, the power source, and the transportation system he subsequently installed. Yet, the greatest factors that overshadowed the types of plant facilities an engineer installed included the financial state of the mining company, the operation's physical accessibility, the operating time period, and the quantity of proven ore.

Self-educated mining engineers often took a different approach toward equipping deep prospect shafts than did their professionally-trained counterparts. For example, cautious self-educated engineers may have installed a succession of increasingly complex hoisting systems as work progressed deeper, while the professionally-trained engineer may have opted to install a large and powerful system at the outset of work.

Whether under the guidance of conservative self-taught engineers or extravagant professionally-trained engineers, nearly all prospect shafts began as small holes served by a windlass. Prospectors working in shallow shafts before the 1920s favored windlasses for their simplicity, great ease of portability, and low pur-

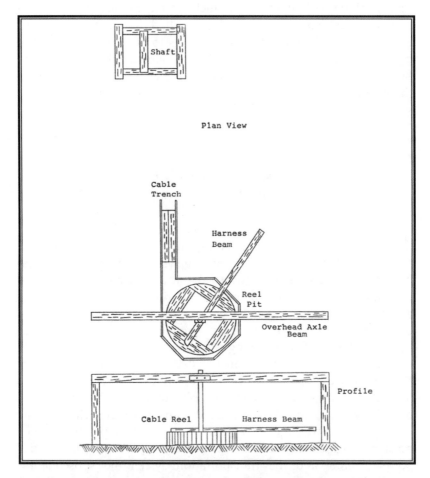

The plan view (top) and profile (bottom) depict the typical malacate horse whim used throughout the West for prospecting. The cable reel, constructed of wood, rotated in a pit lined with plank retaining walls. A trench extends from the pit toward the shaft, and it accommodated the hoisting cable and control linkages. The prospectors laid planks over the trench for the draft animal. The reel's axle has been pinned to a timber footer buried in the ground and to an overhead beam. Prospectors in the West also used an antiquated version in which the reel had been positioned high above the ground surface.

Author.

chase price. Factory-made models fetched twelve to eighteen dollars. Windlasses became ineffective in the hands of all but the most robust prospector when depths exceeded 75 to 100 feet, and when the payload exceeded several hundred pounds. These limitations proved frustrating, because nearly all-promising prospect shafts extended much deeper than one hundred feet, ultimately requiring the installation of a costly mechanical hoist.[16]

The remains of a factory-made horizontal reel whim in Nevada's Como Mining District. In keeping with the typical application of horse whims, a short-term prospect operation installed the whim to serve a shallow shaft that was accessed only by a pack trail. The remains visible consist of a cable reel, an attached harness beam, and a band brake at the reel's base. Note the dry-laid rockwork in the background built to keep material off of the animal path that encircles the reel.

Author.

When a party of prospectors followed a mineral body that gave every indication of extending deeper than the limitation of a windlass, they may have attempted to upgrade their system with a horse whim. Through the 1860s the western mining industry accepted the *horse whim* as being the state-of-the-art hoisting technology for both prospecting and ore production. But by the 1870s practical steam hoists were finally coming of age, and horse whims declined in status. As steam hoists improved and dropped in price through the 1870s and more mining operations were able to install them, professionally-educated and progressive mining engineers deemed that horse whims were well suited for backcountry prospecting, but they were too slow and limited in lifting power for ore production. By around 1880 medium-sized and large western mining operations had fully embraced steam hoists, and engineers' perspective on horse whims became more rigid.[17] The problem with horse whims, according to mining engineers, was that they had a load capacity of around 800 pounds, a depth limitation of 300 feet, and a painfully slow hoisting speed of fifty to eighty feet per minute. However, they were ideally suited for work at remote locations because they were light and could be transported on mule-back, they were easily disassembled, and inexpensive. Horizontal reel whims weighed between 600 and 800 pounds and cost as little as $150, while geared whims weighed twice as much and cost a little more. The draft animal that labored in the harness constituted a renewable energy source, requiring only local feed and water. Because of these factors, western prospect operations employed whims into the 1910s.[18]

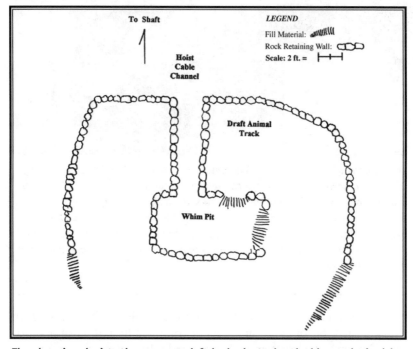

The plan view depicts the remnants left by horizontal reel whims today's visitors are likely to encounter at historic mine sites. The whim's reel was fastened to a foundation in the pit at center, and the hoist cable passed through the trench to the shaft. Prospectors erected rockwork to retain the circular draft animal track surrounding the whim pit. The whim remnant in the illustration was located in Nevada's Austin Mining District.

Author.

Mining companies and prospect outfits employed several varieties of horse whims. The simplest and oldest version consisted of a horizontal wooden drum or reel directly turned by a draft animal. While this model was a modification of the Cornish device, Hispanic miners christened it the *malacate* (mal-a-ca-tay), which stuck in western mining lingo. Early malacates featured a wooden cable drum, a stout iron axle, and bearings fastened onto both an overhead beam and a timber foundation. Prospectors usually positioned the drum so that it rotated in a shallow pit that they lined with either rockwork or wood planking. The cable extended from the drum through a shallow trench toward the shaft. It passed through a pulley bolted to the foot of the headframe, then up and over the sheave at the headframe's top. The draft animal walked around the whim on a prepared track, and the party of prospectors usually laid a plank over the cable trench for the animal to walk across. The controls for the malacate consisted of brake and clutch levers mounted to the shaft collar, and they were connected to the apparatus by wood or iron linkages that passed through the trench. The controls were placed at the shaft collar so the hoistman could communicate with miners in the shaft and keep an eye on the hoist system, as well as manipulate the ore bucket when it arrived on the surface.[19]

The geared horse whim experienced immense popularity among the West's small prospect operations between the 1880s and 1910s. Geared whims were more expensive and heavier than reel whims, and the gearing offered greater mechanical advantage for lifting more weight from deeper shafts.

Author.

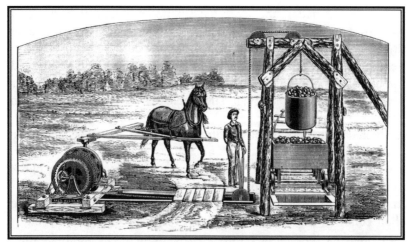

An artist's rendition of a complete horse whim system. While most of the components of horse whim systems have been removed from today's prospect shaft sites, visitors can use this illustration as a template for interpreting the remains they may encounter. In many cases prospectors placed the whim in a shallow pit connected to the shaft by the cable trench shown. The cable extended from the whim to a pulley at ground level, then up and over the sheave atop the small headframe. The trench also accommodated the control linkages, which may still be evident at the shaft collar.

Ingersoll Rock Drill Company Catalog No.7: Rock Drills, Air Compressors and Air Receivers
Ingersoll Rock Drill Company, New York, NY, [1887] p60.

By the late 1870s mining machinery makers began offering factory-made horse whims which were sturdier and performed better than the custom-made units. The Risdon Iron Works in San Francisco began manufacturing the Common Sense Horse Whim. Similar in form to the malacate, Risdon's machine consisted of a spoked iron cable reel mounted on a timber foundation that miners embedded in the ground. To raise a bucket of rock, a mine laborer enticed the draft animal to walk around the track, winding the rope until the vessel had ascended to the top of the headframe, the hoistman threw out the clutch and applied the brake. The brake system consisted of a wood-lined strap that squeezed a flange under the cable reel. The clutch mechanism was ingenious. When the hoistman pulled the clutch lever, a small hinged iron framework popped up from the top of the reel, catching the harness beam. These *horizontal reel horse whims* remained popular among poorly funded prospect operations into the 1890s.[20]

The *geared horse whim* appeared in the West during the 1880s and remained popular among prospect operations, especially in the Rocky Mountain states, into the 1900s. The machine consisted of a cable drum mounted vertically onto a timber frame, and a beveled gear that transferred movement of the draft animal. Sold by the Ingersoll Rock Drill Company, Park and Lacey, and F.M. Davis, geared horse whims were supposedly faster and could lift more weight than horizontal reel models. Geared units featured controls and cable arrangements like the other types of whims, and the drum and gearing were bolted onto a timber foundation buried in waste rock.[21]

In the early 1890s the Cripple Creek Mining District was just beginning to see development, and the lack of capital and poor state of transportation access made the horse whim an ideal hoisting system for the conditions. The photo captures the mighty Buena Vista Mine, which became a heavy producer, before the property was purchased and upgraded with a production-class plant advanced enough to make any mining engineer proud. The arrangement of the components can serve today's historians as an example for interpretation of extant sites.

Courtesy of Colorado Historical Society, Denver, CO.

In keeping with the themes of impermanence and limited budgets, mining operations erected small and simple headframes in conjunction with horse whims. Prospectors favored using either a tripod, tetrapod, or a small four-post derrick that was just wide enough to straddle the shaft. The primary stresses on these headframes were vertical, consisting of the combined weight of the loaded hoisting vehicle and the cable played down the shaft. The lack of other stresses permitted prospect operations to erect structures that were simple and unique to horse whims. The headframes were made of light-duty materials, and did not need backbraces, a firm foundation, or other structural elements necessary for power hoisting. Further, prospectors often used hewn logs up to twenty-five feet long that they cut at little cost, which could have been easily disassembled and moved to other prospect sites.[22]

Prospect operations working in deep shafts began to use steam hoists in large numbers by around 1880. These systems were beyond the means of simple, poorly financed partnerships, nor were they easy to transport deep into the backcountry. Steam hoists and their associated boilers required much capital, and knowledgeable men to install and operate. Deep prospect operations equipped with mechanical hoists usually fell under the auspices of organized mining companies staffed by jacks-of-all-trade miners.

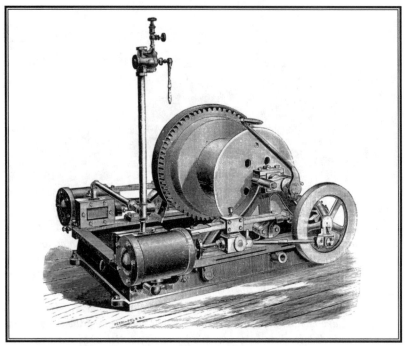

The single drum geared duplex steam hoist, popularly known as the single drum steam hoist, revolutionized shaft mining because it permitted companies to raise greater weight in less time from deeper shafts than before. The installation of such an apparatus and the associated steam boiler required capital and engineering. The front of the illustrated hoist, manufactured during the 1880s, is at right and the hoist's rear where the hoistman stood to operate the controls is at left.

Ingersoll Rock Drill Company Catalog No.7: Rock Drills, Air Compressors and Air Receivers
Ingersoll Rock Drill Company, New York, NY, [1887] p62.

Steam hoisting systems required a relatively substantive infrastructure. Physically, they consisted of a heavy hoist and boiler, cable, pipes, a headframe, and foundations. Because of the numerous heavy components, the steam system had to be planned, engineered, and the claim made ready with a road . The mining company had to provide a reliable source of fuel and soft water for the boiler. Investors and company management often expected deep prospect operations to work year 'round until they found ore or until the money ran out, necessitating the construction of a shaft house to fend off the elements.

During the 1800s steam hoisting technology was in a developmental stage and mining machinery makers devised several variations of the duplex steam hoist. The hoist previously illustrated was the most common form seen in the West, but some mining companies used models with offset cylinders, as illustrated here. These unconventional models disappeared during the 1890s. The illustration depicts the hoist's rear where the hoistman stood to operate the machine.

Engineering & Mining Journal August 20, 1892 p177.

Prospect operations throughout the West active after 1880 typically used *geared single-drum duplex steam hoists*, known simply as single-drum steam hoists. These hoists, "made by the mile and cut off by the yard", became the ubiquitous workhorse for shaft mining. Single-drum steam hoists consisted of a cable drum, two steam cylinders flanking the drum, reduction gears, a clutch, a brake mechanism, and a throttle. The steam pistons turned the drum through the gearing, and hoist makers found that they needed to stagger the movement of the pistons so that when one was just beginning its stroke, the other was half-way through, ensuring that the engine could not stop on top-dead-center, which was a hoistman's nightmare. Top-dead-center meant that the piston stopped at the absolute end or absolute beginning of its stroke, leaving no clearance for the admission of steam and either the exhaust or the steam valve open. The only way to remedy the problem was to physically turn over the engine, which was difficult when geared to a hoist.[23]

Single-drum hoists featured durable and simple controls, and they were easy to use. All hoists had a clutch that uncoupled the drum from the drive shaft, and they also had a reverse link that permitted the engine to run backward. The reverse lever moved a rod that switched the positions of the exhaust and steam valves, closing one if it had been open and opening the other if it had been closed. While the reverse link proved invaluable for slowing the hoist and vehicle during long descents in deep shafts, hoists also featured a mandatory brake. Mining machinery makers equipped their hoists with one of three basic types of brakes: the *block brake,* the *post brake,* and the *strap brake.* The block brake consisted of a single wood shoe that pressed against the bottom of the drum flange, which not only put a heavy strain on the drum bearings but also quickly wore out the shoe. Because of its superior performance, hoist makers favored using the strap brake, which was a steel band with a wood or asbestos pad that wrapped around the drum flange. They also made heavy use of the post brake, which featured two heavy shoes designed to squeeze the drum.[24]

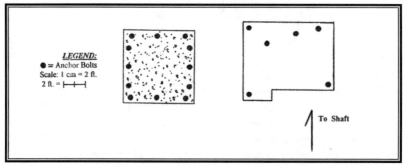

LEGEND:
● = Anchor Bolts
Scale: 1 cm = 2 ft.
2 ft. = ├──┼──┤

To Shaft

The line drawings depict the footprints of foundations for the steam hoists commonly employed by prospect operations and small, marginally profitable mines in the West. The foundation at left anchored the typical duplex geared steam hoist, while the foundation at right supported a single drum hoist with either one or two offset steam cylinders. Note the pattern of anchor bolts for both foundations. In keeping with sinking-class equipment specifications, the foundations are usually less than 6 by 6 feet in area.

Author.

The hoist came from the manufacturer in easy-to-assemble pieces which facilitated hauling it to a mine site. To install the hoist, the mining engineer had a labor crew grade a platform, then build a stout timber or concrete foundation using the same materials and methods we examined for air compressors. Once this had been done, the engineer directed the crew to first install the hoist's bedplate, which they moved into place and lowered over the anchor bolts. They subsequently moved the remaining pieces into place and put the hoist together. Like the plants erected at tunnel mines, the engineer arranged the hoist and boiler according to a master datum line taken from the orientation of the shaft, and he always placed the hoist in alignment with the shaft.

Mining engineers selected the specific model and size of hoist according to the budget the company granted them, and on the speed and depth they anticipated working. Nearly all of the sinking-class hoists that engineers selected for deep prospecting had bed-plates smaller than six-by-six feet in area and were driven either by gearing or by a *friction drive* mechanism. Friction drives consisted of rubber rollers which pressed against the cable drum's flanges, and while they cost less than geared hoists, they were slow and apt to slip under load. Both types of hoists had limited strength, which was often less than forty horsepower, a slow speed of 350 feet per minute, and a payload of only several tons.[25]

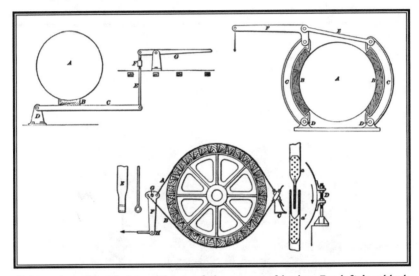

Hoist makers fitted hoists with one of three types of brakes. Top left is a block brake, top right is a post brake, and bottom center is a band brake. Block brakes proved inadequate and were rarely used.

International Textbook Company A Textbook on Metal Mining: Steam and Steam-Boilers, Steam Engines, Air and Air Compression, Hydromechanics and Pumping, Mine Haulage, Hoisting and Hoisting Appliances, Percussive and Rotary Boring International Textbook Company, Scranton, PA, 1899 A23 p50, p51, p52.

Steam boilers were a necessary component of nineteenth century power hoisting systems. While specific designs of boilers evolved and improved, the basic principle and function remained unchanged. Boilers were iron vessels in which intense heat converted large volumes of water into steam under great pressure. Such specialized devices had to be constructed of heavy boilerplate iron riveted to exacting specifications, and they had to arrive ready to withstand neglect and abuse. The problem that boilers presented to mining companies was that they were cumbersome to transport and required engineering to install. The mere thought of attempting to maneuver even a small boiler deep into the backcountry was enough to convince many smaller companies with less resources to continue using traditional horse whims.

During the 1880s the *Pennsylvania boiler,* the *locomotive boiler,* and the *upright boiler,* also known as the *vertical boiler,* quickly gained popularity among the West's prospect operations. These boilers were well-suited to the mining West because they were self-contained and freestanding, ready to fire up, and able to withstand mistreatment. Because these boilers were designed to be portable at the expense of fuel-efficiency, mining engineers declared them fit only for sinking duty.

By the late 1870s, if not earlier, mining operations throughout the West used locomotive boilers to power sinking-class hoists, as well as small compressors. Manufacturers mounted locomotive boilers on skids to facilitate portability. The model illustrated appears to be wood-fired, and the ashes probably dropped through an opening in the bottom of the firebox directly onto the ground. Note the water level sight tube and pressure gauge above the fire door.

Rand Drill Company Illustrated Catalog of the Rand Drill Company, New York, U.S.A. Rand Drill
Company, New York, NY, 1886 p45.

In general, all of the above sinking-class boilers consisted of a *shell* that contained water, *flue tubes* extending through the shell, a *firebox* inside the shell at one end, and a *smoke manifold.* When the *fireman* stoked a fire in the firebox, he adjusted the *dampers* to admit enough oxygen to bring the flames to a steady roar. The *flue gases,* which were superheated, flowed from the fire through the flue tubes, imparting their energy to the surrounding water, and they flowed out the smoke manifold and up the *smokestack*.

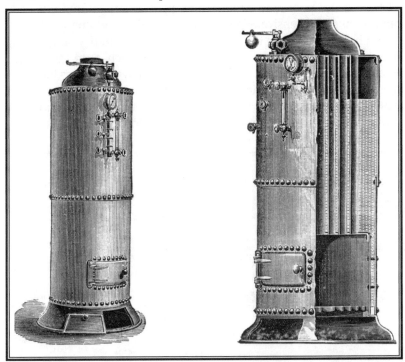

Durable, inexpensive, but highly inefficient, upright boilers had the capacity to power sinking-class steam hoists or other mine machines such as blowers and small compressors. The great ease of portability rendered upright boilers popular among small mining and prospecting operations working in remote areas. However, because upright boilers could not generate a substantial head of steam they saw limited application. Note the water level sight tube and pressure gauge above the fire door.

Rand Drill Company Illustrated Catalog of the Rand Drill Company, New York, U.S.A. Rand Drill Company, New York, NY, 1886 p47.

Great danger lay in neglecting the water level. An explosion was imminent if the flue gases contacted portions of the shell that were not immersed in water on a prolonged basis. Usually the front of the boiler featured a *glass sight tube* much like the level indicator on a coffee urn. When the water level began to get low, the fireman turned the valve on the main that had been connected to the boiler, or he operated a small hand pump if the plumbing had no pressure. Boiler tenders, often also serving as hoistmen, usually kept the boiler three-quarters full of water, the dead space being necessary for the gathering of steam. When the fire grew low the boiler tender opened the *fire door*, the upper of two cast-iron hatches, and threw in fuel. Self-made and professionally-educated mining engineers recognized that cord wood was the most appropriate fuel in remote and undeveloped mining districts, because poor road systems and great distances from railheads made coal too expensive. However, coal was the most energy-efficient fuel, a half ton equaling the heat generated by a cord of wood. Mining operations close to sources of this fossil fuel, such as in the eastern Rockies and New Mexico, preferred it. Expansion of rail lines into mining districts throughout the West lowered the costs of transportation, which made the use of coal economical for increasing numbers of prospect operations through the nineteenth century.

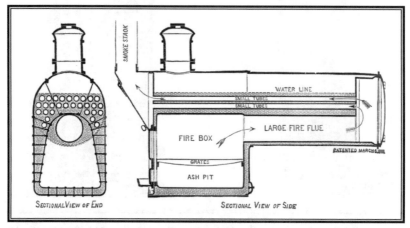

The cut-away view illustrates the components of the Pennsylvania boiler. The fireman put fuel on the fire which combusted on the grates in the firebox. The resultant ashes dropped through and accumulated in the ash pit below. The superheated flue gases follow the path defined by the arrows through the bottom tunnel, then rise and reverse direction and travel through the flue tubes, and escape the boiler out the smokestack. The illustration shows how a jacket of water surrounded the firebox, ash pit, and flue tubes. Pennsylvania boilers were mounted onto timber skids like locomotive boilers, and while they began appearing the West in the 1880s they did not become even mildly popular until the 1890s due to a high initial cost. Mining companies operating in remote areas used Pennsylvania boilers to power small production-class machines instead of conventional return tube units.

Rand Drill Company *Illustrated Catalog of the Rand Drill Company, New York, U.S.A.* Rand Drill Company, New York, NY, 1886 p46.

During the 1880s mining companies came to appreciate the utility and horsepower of the locomotive boiler. The locomotive boiler, so named because railroad engine manufacturers favored it, consisted of a horizontal shell with a firebox built into one end and a smokestack projecting out of the other end. Nearly all of the models used in the West stood on wood skids and were easily portable, but some units required a small masonry pad underneath the firebox, and a masonry pillar supporting the other end. Locomotive boilers were usually ten to sixteen feet long, three feet in diameter, and stood up to six feet high, not including the steam dome on top. These workhorses, the single most popular sinking-class source of steam into the 1910s, typically generated from to thirty to fifty horsepower, which was enough to run a sinking-class hoist. Larger locomotive boilers were available, but were rarely used.[26]

Firing a boiler with coal on a sustained basis generated a dark, scoria-like, light-weight and vitreous residue known as boiler clinker that the boiler tender usually threw out on the waste rock dump. Similar in appearance to forge clinker, boiler clinker dumps include relatively large clasts like those in the photograph. The pocket knife provides scale.
Author.

Upright boilers were the least costly of all boiler types. They tolerated abuse well, and they were the most portable. Because upright boilers could not generate the same horsepower as locomotive or Pennsylvania units, they could not power large hoists, or several machines at once. Upright boilers consisted of a vertical water shell that stood over a firebox and ash pit that had been built as part of a cast-iron base. The flue tubes extended upward through the boiler shell and opened into a smoke chamber enclosed by a hood and smokestack, which appeared much like an inverted funnel. The flue gases' path directly up and out of the firebox made these steam generators highly inefficient, and the rapid escape of gases and quick combustion of fuel caused great fluctuation and inconsistency in the pressure and volume of steam. The short path for the gases and intense fire also put heavy heat stress on the boiler's top end, causing it to wear out quickly and leak, and the firebox and doors also saw considerable erosion. However, upright boilers required little floor space, little maintenance, and were so durable that they almost could be rolled from site to site. Many remote prospect operations suffering from limited capital saw great advantage in vertical boilers, and consequently the apparatuses enjoyed substantial popularity.[27]

The third basic type of sinking-class boiler that western prospect operations used in noteworthy numbers was the Pennsylvania boiler. Like the other portable boilers, the Pennsylvania boiler featured an enclosed firebox that was surrounded by a jacket of water. The flue gases traveled through a broad tunnel in the shell, they rose into a small smoke chamber, then reversed direction and traveled back toward the front of the shell through flue tubes, and finally escaped through the smokestack. The Pennsylvania boiler, which originated in the Keystone State's oil fields, proved to be remarkably efficient and saw use at a number of western mining operations.[28]

Prospect operations engaged in deep subsurface exploration high in the mountains or deep in the desert employed highly versatile, mobile, and modestly priced donkey hoists. As the illustration depicts, the durable machines were self-contained on a common bedplate, and because they were heavy when assembled, they required no anchor foundation. A few impoverished operations used donkey hoists for light ore production. When the hoistman disconnected the clutch, he could have used the drive-belt pulley at right to power other machines such as ventilation blowers while the steam engine idled away.

Ingersoll Rock Drill Company Catalog No.7: Rock Drills, Air Compressors and Air Receivers Ingersoll Rock Drill Company, New York, NY, [1887] p54.

Developed in Scotland for maritime purposes, the Scotch marine boiler was the least popular sinking-class steam generator in the West. Scotch marine boilers consisted of a large-diameter shell enclosing the firebox, and the path for the flue gases was similar to that of the Pennsylvania boiler. While this type of boiler was one of the most efficient portable units, it never saw popularity in the West primarily because convention dictated the use of the other types, and because it was heavy, large, and difficult to haul to remote locations.[29]

A significant number of deep prospect operations in the West found horse whims inadequate, but could not, or would not, come up with the capital necessary to install a conventional steam hoist and boiler. During the late 1870s machinery manufacturers introduced a revolutionary type of hoisting system that met the needs of these small operations. The *steam donkey hoist,* so named for its broad utility, consisted of either a small single cylinder or duplex steam hoist and an upright boiler mounted onto a common wood or steel frame. While donkey hoists were not manufactured exclusively for mining, being heavily used for logging and in freight yards, they endeared themselves to western prospect operations. The durable machines withstood mistreatment, they were relatively inexpensive, they did not require much site preparation, and they could literally drag themselves around the landscape. In addition, donkey hoists did not require a deep understanding of engineering, and nearly anyone on the payroll of a mining company could have operated one.

Mining engineers recognized donkey hoists as meeting only sinking-class specifications. The machines possessed slow hoisting speeds, the boilers offered poor fuel economy, and the hoists had limitations of up to an 8,000 pound payload and a 1,000 foot working depth. Preparing a donkey hoist for use was extremely easy once it had been brought to the prospect shaft. A crew of laborers graded a platform a short distance from the shaft and placed the donkey hoist on it. Usually the sheer weight of the machine was enough to keep it in place during operation, but in many cases prospect operations staked down the rear as a safety precaution. Like all steam hoists, donkey hoists required sources of fuel and water.[30].

Prospect operations seeking riches deep in the backcountry, especially in the arid Great Basin and Southwest where wood and water were scarce, were reluctant to spend the capital required to install steam equipment. The problems they faced were two fold. Not only did these operations have to ship and install a hoisting system, which was arduous, but they also had to continuously feed it fuel and water, which proved costly. In the early 1890s the Witte Iron Works

Company and the Weber Gas & Gasoline Engine Company both began experimenting with a new hoisting technology that alleviated many of the fuel and water issues faced by remote prospect operations. Witte and Weber almost simultaneously introduced the first practical petroleum engine hoists. These innovative machines were smaller than many steam models, they required no boilers, and their concentrated liquid fuel was far easier to transport than wood or coal.

Petroleum hoists, commonly referred to as gasoline hoists, came into being during the early 1890s, but due to their temperamental nature and severely limited performance they did not become popular until the mid-1900s, and even then only among small prospect operations in remote and arid regions. The model illustrated, probably made by Fairbanks-Morse, consists of a large single cylinder petroleum engine at left, a central crankshaft fitted with a pinion gear and flywheels in the center, and the cable drum fitted with a bull gear at right. The hoistman operates a clutch with one hand, a throttle with the other hand, and a brake with his foot. The form of the illustrated hoist, all the components being mounted onto a common frame, was by far most popular in the West.

International Textbook Company International Library of Technology: Hoisting, Haulage, Mine Drainage International Textbook Company, Scranton, PA, 1906 A50 p31.

Despite the potential advantages of petroleum hoists, the western mining companies working in arid and woodless districts did not immediately embrace them. Steam technology, the workhorse of the Industrial Revolution, dominated in the mining industry through the nineteenth century for several reasons. Many mining companies and engineers, such as Frank Crampton and W.S. Stratton, were by nature conservative and stayed the course with steam into the 1910s. During this time gasoline engine technology was relatively new and had not seen widespread application, especially for hoisting. The few operations to employ petroleum hoists during the 1890s found the engines to be temperamental, and their performances were limited. Further, petroleum hoists were slow, possessing speeds of 300 to 400 feet per minute, they could not raise much more than 4,500 pounds, and their working depth was limited to less than 1,000 feet. For these reasons professionally-educated mining engineers felt they were barely adequate for sinking duty, and conversion from steam took approximately fifteen years.[31]

A few mining machinery makers offered gasoline hoists that deviated from convention. These machines consisted of an independent cable drum that had been geared to an adjacent engine. The combined footprints of the cable drum and engine had the potential to mimic an electric hoist, However, in keeping with gas engine technology, construction workers mounted these petroleum hoists onto heavy concrete foundations.

Mining & Scientific Press Jan. 19, 1918 p2.

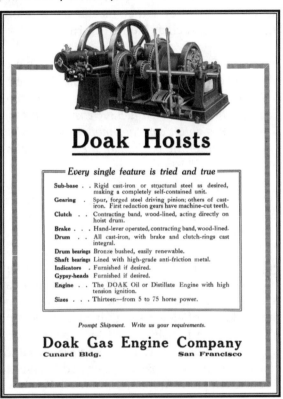

Doak Hoists

Every single feature is tried and true

Sub-base . .	Rigid cast-iron or structural steel as desired, making a completely self-contained unit.
Gearing .	Spur, forged steel driving pinion; others of cast-iron. First reduction gears have machine-cut teeth.
Clutch . .	Contracting band, wood-lined, acting directly on hoist drum.
Brake . . .	Hand-lever operated, contracting band, wood-lined.
Drum . .	All cast-iron, with brake and clutch-rings cast integral.
Drum bearings	Bronze bushed, easily renewable.
Shaft bearings	Lined with high-grade anti-friction metal.
Indicators .	Furnished if desired.
Gypsy-heads	Furnished if desired.
Engine . .	The DOAK Oil or Distillate Engine with high tension ignition.
Sizes . . .	Thirteen—from 5 to 75 horse power.

Prompt Shipment. Write us your requirements.

Doak Gas Engine Company
Cunard Bldg.   San Francisco

The petroleum hoists seen among western prospect operations were similar in form to the old-fashioned and well-loved steam donkey hoists. The engine, a large single cylinder oriented either vertically or horizontally, was fixed to the rear of a heavy cast-iron frame and its piston rod connected to a heavy crankshaft located in the frame's center. Manufacturers located the cable drum, turned by reduction gearing, at front, and the hoistman stood to one side and operated brake and clutch levers, and the throttle. Because the early petroleum engines were incapable of starting and stopping under load, or of being reversed, they had to run continuously, requiring the hoistman to delicately work the clutch when hoisting, and to disengage the drum and lower the ore bucket via the brake. Miners truly placed their lives on the line when riding an ore bucket controlled by a petroleum hoist.

Western prospect operations began showing interest in petroleum hoists during the late 1890s because the small sizes and light weights of the machines made the apparatuses easy to ship. Also, petroleum fuel cost much less to pack to a prospect site than coal or wood, and the purchase price of the hoists was modest. By the 1900s professionally-trained mining engineers granted the hoists recognition for the ability to play an effective role in deep prospecting at remote sites. However, their means of operation bothered some mining engineers, such as Herbert C. Hoover:

> *"Gasoline hoists have a distinct place in prospecting and early-stage mining, especially in desert countries where transport and fuel conditions are onerous, for both the machines and their fuel are easy of transport. As direct gas-engines entail constant motion of the engine at the power demand of the peak load, they are hopeless in mechanical efficiency."* [32]

Despite running at full throttle much of the day, many early petroleum hoists consumed at most ten gallons of gasoline, diesel, or kerosene per ten hour shift, which cost a total of approximately two dollars. By comparison, a cord of wood, most of which a locomotive boiler consumed during a shift, also cost around two dollars where cut, but then it had to be shipped to the prospect site, raising the total cost to as much as ten dollars. Some prospect operations under the guidance of clever self-trained engineers put the constant running of the hoist engine to efficient use by adding a pulley to the flywheel, which then powered air compressors and shop appliances via canvas belting. This partially negated Hoover's criticism of the inefficiencies of petroleum hoists. [33]

The western mining industry did not truly embrace petroleum hoists for deep prospecting, let alone minor ore production, until the late 1900s. By the 1910s gas engine technology had improved and hoistmen understood how to operate the machines without stalling them, and even to work the throttle to maximize efficiency. Petroleum hoists had made such an impact by this time that steam hoists were fast becoming obsolete among remote prospect operations in the Great Basin and Southwest. But in areas where cord wood or coal was plentiful, the transition from steam to petroleum hoists was slower. Competition from electric power, and the availability of inexpensive used steam equipment salvaged from defunct operations, suppressed the conversion to petroleum hoists. [34]

Nearly all mechanical hoisting systems in the West required that the prospect operation erect a headframe over the shaft. The purpose of the headframe was to support and guide the hoist cable into the workings, and to assist in the transfer of rock from, and supplies into, the hoisting vehicle. Professionally-educated mining engineers recognized six basic structural forms of headframes, including the tripod and tetrapod used with horse whims, the two-post gallows, four and six-post derricks, and the A-frame. Of the six varieties, prospect operations used the two-post gallows and small four-post derricks in conjunction with power hoisting, and outfits working in inclined shafts erected small A-frames.[35]

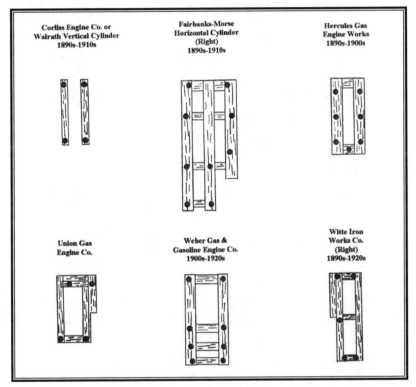

Illustrated is a catalog of the sinking-class timber foundations for petroleum hoists used by small western mining companies between the 1890s and 1920s.

Author.

The *two-post gallows* was one of the most common headframes used throughout the West, and self-made and professionally-educated engineers unanimously agreed that it was best for prospecting. The variety used by small operations usually consisted of two upright posts, a cap timber and another cross-member several feet below, with diagonal braces, all standing at most twenty-five feet high. The cap timber and lower cross-member featured brackets that held the sheave wheel in place. The gallows portion of the structure stood on one end of a timber foundation that crews built equal in length to the head-frame's height. The diagonal backbraces extended from the posts down toward the hoist, where they were tied into the foundation foot-ers. The foundation, made of parallel timbers held together with cross members, rested on the ground, and it straddled the shaft collar.

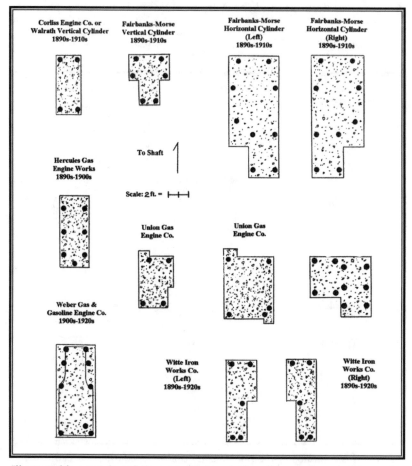

Illustrated is a catalog of the concrete foundations for petroleum hoists used by small Western mining companies between the 1890s and 1920s.

Author.

The four-post derrick erected for prospecting was similar in height, construction, and materials to two-post headframes, and it too stood on a timber foundation. The A-frame was based on the same design as the two-post gallows. The difference was that the A-frame featured fore and aft diagonal braces to buttress the structure in both directions. A-frames were not erected directly over an inclined shaft, rather they were placed between the hoist and shaft so that the angle of the cable extending upward from the hoist equaled that extending down the incline shaft.

Clever mining engineers used the constant running of a petroleum hoist's gas engine to advantage by engaging it to power other mine machines via a drive belt. The plan view depicts the arrangement of foundations for a hoist, compressor, and a drive shaft for a canvas belt. The hoist, which constantly ran, was bolted onto the large concrete block at right, and the engineer fastened a special drive shaft onto its flywheel. Bearings, bolted onto the two concrete pylons, held the drive shaft steady as it rotated at high speeds. A canvas drive belt passed from a pulley fixed to the shaft to a large flywheel on the compressor, which was anchored to the foundation at the bottom.

Author.

Drive Pulley Well

Drive Shaft Bearing Pylons

Hoist Foundation

Compressor Foundation

The photo captures a 50 horsepower Fairbanks-Morse petroleum hoist at the Richmond Mine in Eureka, Nevada. We are looking at the rear right corner of the hoist.

Author.

Designing and building these types of headframes was well within the grasp of both professionally-educated and self-made engineers such as W.S. Stratton and Frank Crampton. The common features shared by the above structures included: a small size, simplicity, minimal use of materials, ease of erection, and portability of materials. For comparison, a two-post gallows frame twenty feet high cost as little as fifty dollars and a slightly larger structure cost $150, while a production-class A-frame cost $650, and a production-class four-post derrick headframe cost up to $900. Sinking-class head-

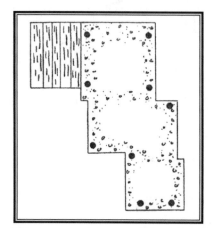

The plan view depicts the foundation associated with large Fairbanks-Morse petroleum hoists.

Author.

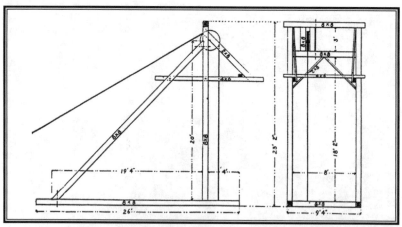

Prospect operations working in both vertical and inclined shafts usually erected a two-post gallows headframe to guide and support the hoisting cable, and to facilitate emptying the ore bucket. Two-post gallows headframes were ubiquitous across the West and were favored because of their relative ease of erection and light use of materials, which translated into low costs. A horizontal beam is visible in the profile near the headframe's top, which held the dumping chain for emptying ore buckets.

Engineering & Mining Journal March 7, 1903 p366.

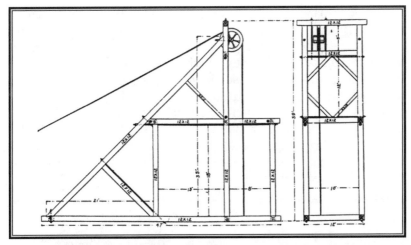

The illustration depicts a large two-post gallows erected for deep shaft sinking. This structure and the headframe illustrated above lack the guide rails necessary to accommodate a cage, indicating that the mining company used an ore bucket.

Engineering & Mining Journal March 7, 1903 p366.

frames had to withstand only a few basic stresses. The most significant consisted of the *live load,* created by the weight of a full hoist vehicle and cable, the *braking load,* which was a surge of force created when the hoistman quickly brought the vehicle to a halt in the shaft, and the *horizontal pull* of the hoist.[36]

To counter these forces, mining engineers had workers typically build their headframes of eight-by-eight timbers, and they installed diagonal backbraces to counter the pull of the hoist. Usually carpenters assembled the primary components with mortise-and-tenon joints, one-inch diameter iron tie rods, and timber bolts. Professionally-trained mining engineers specified that the diagonal backbraces were most effective when they bisected the angle of dead vertical and the angle formed by the hoist cable ascending to the top of the headframe. By tying the backbraces into the foundation between the shaft and hoist, engineers determined that the total horizontal and vertical forces put on the headframe were equally distributed among both the vertical and the diagonal posts. When a mining engineer attempted to find the mathematically perfect location for a hoist after erecting such a headframe, he merely had to measure the distance from the shaft collar to the diagonal brace, double the length, and built the hoist foundation. Most western prospect operations followed this general guideline and arranged their hoisting systems accordingly, but a few poorly educated engineers strayed and gave the diagonal braces either too much or too little of an angle.[37]

While sinking-class headframes tended to be relatively small, assembling and erecting the structure proved to be no easy task. The challenge began when a wagon or a pack train of mules delivered a load of timbers to the mine site, and the construction crew had to sort them by length. When possible, the mining engineer preferred to use full-length timbers for the vertical posts, but hauling twenty-five foot beams through the backcountry to a remote site proved arduous, and in some cases the engineers had to settle for shorter segments. In the event the mining company could not have purchased full-length timbers, the carpenters had to splice several short pieces together using fish joints or scarf joints, bolts, and heavy plate washers.

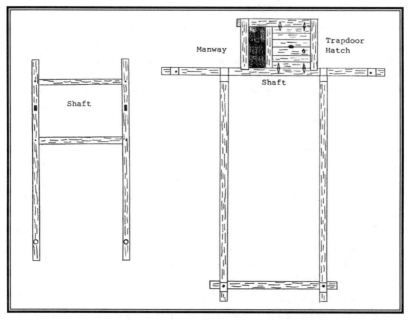

Plan views of timber foundations for two-post gallows headframes that today's visitor may encounter at small historic mines. The foundation at left supported a simple structure, and the foundation at right supported a large headframe with side braces. The backbraces were bolted to the footers at the bottom, and the hoist was located farther down, out of view. The diagrams are to scale.

Author.

After the construction gang sorted the timbers and laid them on wood blocks, some of the workers began cutting them to exact length while others hewed pins and sockets for the mortise-and-tenon joints. Other laborers operated heavy hand-turned carpenters' drills to bore holes for the iron tie rods that the mine blacksmith forged. The first portion of the headframe that the crew assembled after preparing the materials was the foundation frame. The laborers may have used a

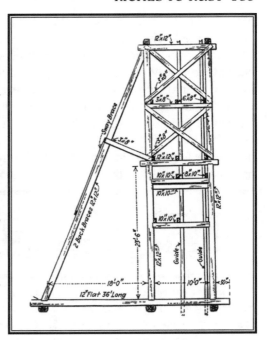

Profile of a four-post derrick headframe. While four-post derricks were stronger than two-post gallows frames, they were not as popular for sinking because they were much more costly to build. Note the simple timber foundation, and the guide rails for either a skip or an ore bucket and crosshead.
Engineering & Mining Journal
Dec. 28, 1912 p1215.

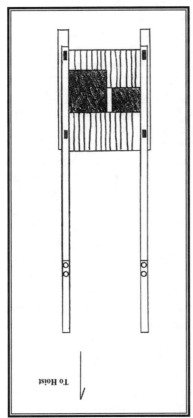

The plan view illustrates a foundation for a sinking-class four-post derrick headframe, as identified by the four mortise-and-tenon sockets at each corner of the shaft, which has been covered with decking.

Author.

block-and-tackle or a draft animal to skid the long timber footers into place on the leveled waste rock pad surrounding the shaft. Teams of laborers struggled with the shorter members and dropped them into place, then slowly pushed the mortise-and-tenon joints together, slid the iron tie rods through the holes, and tightened the nuts. In some cases headframes were assembled with simple square-notched joints instead of mortise-and-tenon joints, which proved to be easier to assemble but not as durable. At some mines the workers buried the headframe foundation in waste rock ballast for stability, while at other operations they placed the frame on the surface of the dump.

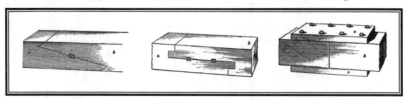

When mining companies were not able to haul full-length timbers into the backcountry for building headframes and associated foundations, carpenters used one of a variety of joints to assemble long beams from short members. The left and center images depict two types of scarf joints, and the right image shows a fish joint, also known as a butt joint.

International Textbook Company A Textbook on Metal Mining: Preliminary Operations at Metal Mines, Metal Mining, Surface Arrangements at Metal Mines, Ore Dressing and Milling International Textbook Company, Scranton, PA, 1899 A42 p44.

The labor gang repeated the steps they had just completed to put together the gallows frame. After it too had been assembled, the crew made ready to erect the structure and fasten it to the foundation. Attempting to raise a twenty-five foot-long timber frame weighing several tons from the horizontal to the vertical presented great engineering and physical challenges, and it was very dangerous. First, the construction crew had to find a capable motive source. Usually, small mining operations resorted to teams of draft animals, a crab winch, or the mine hoist. Several laborers pinned the legs of the gallows frame in place to prevent them from skidding when the top of the structure was raised, and several other workers built a small A-frame between the motive source and the prostrate gallows to give the rope lifting advantage. When the construction boss had given the signal, the workers began either cranking on the winch or hustling the draft animals; the traction rope came under force and the top of the gallows frame began to rise. After it had passed the critical angle, the A-frame supporting the rope fell away and the gallows rose by sheer force of the traction power. If the workers had planned everything correctly, the mortise-and-tenon joints between the gallows frame and footer frame assembled themselves, and several laborers lashed down guy ropes from anchors to the standing structure. The last step in building the headframe involved raising the diagonal backbraces into place, which could have been accomplished using block-and-tackle or with a team. While the structure was in the dangerous state of standing only by the support of guy ropes, workers climbed to the top to guide the backbraces into place and fasten the bolts. Once all of the joints had been checked, the bolts tightened and re-tightened, the headframe, to which many miners entrusted their lives, was ready for service.[38]

In 1906 miners working the Combination Shaft in Goldfield, Nevada were so eager to begin extracting ore that they operated the mine before a proper hoist house could be built to shelter the new Fairbanks-Morse gasoline hoist and a shop. The arrangement of the components for the hoisting system in the photo is representative of the surface plants associated with western prospect shafts, and it serves today's visitors as a model they can refer to. At some point the mining company would have dumped waste rock fill around the shaft to create a level pad, and subsequently erected either a hoist house around the hoist, or a shaft house enclosing all of the plant components that are visible. The pipe extending up from the hoist is for exhaust. Rather than descend the shaft on rickety wooden ladders, miners usually went to work by riding the ore bucket into the depths of the mine, exemplified by the miner in the ore bucket in the photo.
Mining & Scientific Press Oct. 6, 1906 p413.

After the mining engineer had the hoist and boiler installed, the headframe erected, a road graded to the site, and a water pipeline laid, the last step he undertook to ready the plant for deep prospecting was to enclose the hoist and boiler in a building. Small prospect operations working in severe climates warranted being enclosed in a *full shaft house*, but out of financial constraints and a desire for simplicity, min-

Mining was a dirty industrial business where the costs of operating were as important as the necessity of producing ore. The hoist houses of most small mining and prospect operations were rough, noisy, and perpetually in slight disarray. The photo captures the surface crew in the shaft house of the Monongahela Mine above Boulder, Colorado around 1900. The interior is typical of companies with limited capital operating in remote regions. The hoist, center left, appears to be a sinking-class steam model powered by an upright boiler, center. The blacksmith shop is partially visible at the far left, as is one of the headframe's diagonal backbraces. The shaft house floor is earthen, and the structure consists of corrugated steel siding nailed to a 2x4 frame. The mountains of Boulder County get very cold and windy during the winter, necessitating the use of a woodstove, bottom center, in addition to the blacksmith forge and boiler for heat. The men at right are probably engineers or superintendents.

Courtesy of the Carnegie Branch Library for Local History, Boulder Historical Society Collection

ing engineers directed mine workers to build a simple frame *hoist house* for the hoist, boiler, and shop. These structures, built in a style that can be referred to today as *western mining vernacular*, were usually no more than plain plank or board-and-batten sided frame walls with a gabled or shed roof, and no ornamentation. These simple buildings rarely stood on a formal foundation. Rather, the walls stood on dry-laid field stone footers at best, and usually the construction crew merely placed them on the ground. The engineer designed the hoist house large enough to accommodate the hoist and a temporary blacksmith shop in one room, and the boiler in another room. In the more temperate Southwest the boiler may have been sheltered merely by a shed roof, or it may have been totally in the open. By the 1900s deep prospect operations began using small straight-line air compressors, which they usually placed adjacent to the hoist.

The interiors of the hoist houses typically built by prospect operations were at least as rough as the exterior. For best use of space the engineer arranged the building to feature the hoist at one side, the hoistman's plank-decked platform at rear, and the shop on the opposite side. Additional space may have been used for equipment storage, heavy carpentry work, or to accommodate a water tank. Except for the hoistman's platform, hoist house floors were usually earthen, and the building featured several windows and doors, and a port for the hoist cable.[39]

The Barbee Lease, located on the west side of Beacon Hill in Cripple Creek, was a small capital-starved operation that produced ore around 1900. The arrangement of the surface plant components is typical of small marginally productive western mining operations. Everything about the surface plant speaks of a want of capital, such as the tiny hoist house, the use of wheelbarrows instead of ore cars, the small sinking-class headframe, and the flimsy flat-bottomed ore bins. Were it not for the presence of ore bins, it would have been fair to assume that this operation in fact failed to strike ore and died in infancy.

Courtesy of Colorado Historical Society, Denver, CO.

After the surface plant had been completed and the engineer gave the go-ahead for the driving of exploratory workings underground, the mine crew divided into support roles. First were the miners who worked underground drilling, blasting, and filling the ore buckets. They pushed full buckets to the shaft station and exchanged the loaded vessels for empty ones. When the miners had hooked a full bucket onto the hoist cable, one of them pulled the signal bell cord and gave the hoistman clearance to raise the vessel out of the shaft. The signal bell system was a miners' Morse code, and each shaft station, the headframe, and the hoist house featured bells linked to a common cord used for communication.

The hoistman, also known as an engineer, operated the hoist and sent the ore bucket to its specified destination. When he received the bell signal, he threw on the steam, the hoist's pistons wound the drum, and wisps of steam escaped the exhaust pipe projecting from the hoist house. After several minutes of operation, the ore bucket rose with a bang through the trap door covering the shaft. Most shaft collars featured gabled, or angled trap doors to deflect material and keep it from falling onto the miners far below. Some doors opened automatically when the rising bucket struck them, while others featured a pull cord operated by the *topman*, who tended to the surface work. As the bucket rose through the trap doors, the topman watched its progress, and when it had attained a specific height in the headframe, he rung the bell to stop the hoist.[40]

The shaft house of the COD Mine above the town of Cripple Creek, as it appeared around 1895, is a stereotypical example of the sort of structure erected by small modestly-financed western prospect and mining operations, between the 1870s and 1890s. The shaft house encloses a short headframe, a steam hoist and boiler, and the shop. The architectural style can be described as vernacular and the building rests on mother earth. Highly productive mining companies quickly outgrew such confining shaft houses and either built a larger structure, or enclosed all of the surface plant components in individual structures.

Courtesy of Colorado Historical Society, Denver, CO.

The next task was to empty the bucket of its contents and send it back down to be filled again. The topman climbed onto the headframe, or used a long pole to hook the *dumping chain* onto the bottom ring of the bucket. Then, he signaled the hoistman to slowly lower the bucket, and as the vessel descended toward the ground the dumping chain became taut and upended the bucket. The rock slid out, hopefully into a waiting ore car. Mining operations usually anchored the chain onto one of the headframe's cross-members, or onto a special beam that projected outward from the structure. Sinking-class two-post gallows headframes used in the Rocky Mountains often featured a special plank-dumping apron that deflected rock away from the shaft into an ore car when it spilled out of the bucket. When the bucket had emptied itself, the topman signaled the hoistman, and the team reversed the process, sending the bucket back down for another load.[41]

Large prospect operations also kept on staff a fireman to tend the boiler. Otherwise the hoistman had to keep it fueled and watered. The crew also included a blacksmith and the mining engineer, who, judging from the examples set by W.S. Stratton and Frank Crampton, worked elbow-to-elbow with the miners and surface laborers.

Miners found that transferring the rock from the hoist vehicle at inclined shafts to be slightly easier. Under the scrutiny of the topman, the hoist operator slowly pulled the vehicle up the A-frame's timberwork, where special guide rods automatically upset the vessel and poured the rock into a waiting ore car, or into an ore pocket affixed to the headframe. The topman then pushed the freshly-loaded car across the mine rail line to the edge of the waste rock dump where he emptied it.

While the surface plants that incorporated horse whims, donkey hoists, and fixed steam hoists differed in performance and levels of capitalization, they were all transitory and impermanent. As result, when a mining company went bankrupt or gave up the search after encountering nothing but barren rock, they removed the equipment as quickly and easily as when they had brought it to the mine site. As a result, unproductive prospect shafts remaining in the West today rarely feature intact structures and machines for visitors to see. But upon close inspection the visitor may encounter subtle evidence hinting at what the defunct operation consisted of. Except for the hoisting system, visitors to historic prospect shaft sites will find a striking parallel between the remains at a shaft and the facilities that served prospect adits.

Each type of hoisting system had the potential to leave unique forms of physical evidence. While horse whims were transitory and intentionally impermanent, they left characteristic clues such as the cable reel foundation and turntable, the animal track, and the cable trench. Usually, though, the prospect operation uprooted the foundation along with the whim, leaving a depression where the reel had been situated. Deep, spacious depressions, generally around seven to ten feet in diameter and two to three feet deep, strongly suggest that the whim was a malacate, especially when the pit has been lined with planks or dry-laid rock walls. Small, shallow pits, on the other hand, reflect the location of timber foundations for either factory-made horizontal reel or geared whims. In either case the depression is often connected to the shaft by the remains of a cable trench.

A draft animal path always encircles the pit left by a horse whim. The path, usually twenty to twenty-five feet in diameter, manifests as a circular pad that the prospectors had cleared of obstacles, and the cable trench usually bisects it. If the prospectors had to locate the path on uneven ground, they may have erected circular dry-laid rock walls to bolster the loose gravel fill and keep the path level, or they may have excavated the circular track out of the hillside. If the shaft collar remains intact, it may also feature hardware for the whim's brake and clutch levers.

Prospect operations that used horse whims also may have left specific artifacts that, when associated with the above characteristics, can help the visitor to confirm the use of such a hoisting system. The visitor should watch for horse and mule shoes, horseshoe nails, tack buckles and singletree hardware, large strap brakes and control linkages, reel bearings, and even the whim's harness beam. Whim makers often manufactured control linkage sections out of rectangular iron bars perforated with boltholes that the prospecting outfit custom-assembled it to fit the distance between the shaft collar and the whim.[42]

Prospect outfits that used fixed steam hoists usually left behind more lasting evidence than the operations that relied on horse whims. Fixed hoists almost always required a substantial foundation readily identifiable at historic mine sites today. The foundations typically manifest as a set of anchor bolts projecting out of the remains of timber cribbing, out of a concrete pad, or out of masonry. The footprint of the anchor bolts, which usually are at least three-quarters of an inch in diameter, reflects the hoist's size and shape. Foundations for duplex steam models tended to be rectangular, while single cylinder hoists, discussed with production-class plants below, featured an elongated rectangle marking the location of the engine and a small adjacent square denoting the cable drum. The visitor to a mine site will always find the hoist foundation aligned with the orientation of the shaft, and usually situated between approximately twenty and forty feet away. In most cases the visitor will also find the hoist foundation on a cut-and-fill platform that workers graded to accommodate the hoist house and shop.

Visitors to mine sites can usually expect to encounter evidence of the boiler near the hoist foundation. Mining engineers customarily placed the boiler near the hoist to permit the hoistman to fire it, to minimize heat lost through the steam pipes, and so the machines could have been sheltered under one roof. The skids of locomotive,

Pennsylvania, and Scotch marine boilers often left two linear depressions in the ground approximately one-foot wide and up to sixteen-feet long, or they stood on rock alignments of similar dimensions. Upright boilers left circular depressions or raised rectangular pads where their bases rested. No matter the specific type of boiler, the features left in the ground should be oriented at the same angle as the hoist foundation, because mining engineers as a rule used the site's master datum line when installing boilers.[43]

Feeding coal and cordwood to a boiler firebox on a prolonged basis resulted in the accumulation of fuel residue that the mine crew had to dispose of. The fireman periodically shoveled the residue out of the boiler's ash pit and into buckets or a wheelbarrow, and ejected most of the refuse over the edge of the waste rock dump, usually in the vicinity of the boiler. Therefore historic mine sites often feature a deposit of fuel residue around the boiler area, and a concentrated dump nearby.

The fuel residue dumps encountered at prospect operations today consist of either wood ash and charcoal, or *boiler clinker*, which is much like the clinker blacksmiths cleaned out of their forges. The major differences between dumps of boiler clinker and dumps of forge clinker include a lack of forge-cut iron scraps mixed in with the matrix, and the residue itself. Because the dark and scoria-like mass was extracted from boiler fireboxes much less frequently than from shop forges, large clasts and nodules, often greater than one-and-one-half inches in diameter, had time to form among the glowing coals. While shop coal had to be pure, mining companies attempted to minimize costs and consequently purchased low-grade bituminous coal laden with impurities, especially slate, to feed their boilers. Boiler clinker dumps usually manifest today as a mixture of large clinker nodules, fragmented, heat-altered slate, and unburned bituminous coal heavy with impurities. The ash dumps indicative of the use of cord wood fuel, on the other hand, are relatively rare at prospect shaft sites because the frequent and heavy winds of the West, as well as rain and snowmelt, easily mobilized and blew away the light ash.

The use of steam hoisting systems often left specific types of artifacts that today's visitor may observe in conjunction with the remains discussed above. Pipes, valves, and pipefittings may lie near the hoist foundation, in addition to machine parts, nuts and bolts, and cast-iron bedplate fragments. Heavy iron rivets, pipefittings, and boiler water level sight tube fragments tinted aqua from exposure to sunlight may be scattered around the boiler location, or mixed in with the clinker dump. Occasionally the visitor to a site that had steam power may also encounter fractured remains of boilers, including legs for the skids, fire doors, and smokestack sections.

Identification of a donkey hoisting system can present the visitor to a prospect shaft with a challenge because these machines left little direct evidence. Based on a combination of factors, the visitor to a mine site may conclude that an operation used a donkey hoist. First, these machines were self-contained and they required only a bare, earthen platform situated fifteen to forty feet from the shaft. Because donkey hoists relied on steam, they were likely to leave behind a deposit of either boiler clinker or a substantial ash deposit, and an artifact assemblage including machine parts, rivets, and water level sight tube fragments. Further, the hoist platform may feature guy posts that anchored the rear of the donkey hoist's frame. Last, the association of the above site characteristics with a shaft too deep for a windlass strongly suggests that the operation used a donkey hoist. The relative depth of a shaft can be gauged by the quantity of waste rock present at the site. Lack of clinker or ash leaves the question open as to whether the hoist was powered by a wood-burning boiler or by a small petroleum engine.

When prospect operations installed petroleum hoists, they almost always followed the convention established for steam equipment in terms of the layout of the hoisting system components and the use of construction material. However, petroleum hoist foundations were distinct and different in shape from those for steam hoists. Petroleum hoist foundations, like those discussed for gas compressors, tended to be long, narrow, and relatively tall, although the mine construction crew may have back-filled soil around them. The average size tends to range from two-and-one-half by five feet in area for a light-duty sinking hoist to six-and-one-half by thirteen feet for a deep sinking and light production hoist. Petroleum hoists were directly aligned with the shaft and installed on the hoist house platform.

Mining operations that used petroleum hoists left behind an artifact assemblage that differed subtly from steam hoists. First, petroleum hoists generated no ash or boiler clinker. Instead, they consumed petroleum, which oil companies typically shipped in tall square five-gallon cans that featured heavy wire handles and pour spouts. To prevent the volatile gasoline from evaporating out of the can's seams, the can makers sealed the joints with solder. After the mine crew emptied the cans, they threw them out onto the waste rock dump, or converted them into pails for other uses. In addition, the prolonged operation of petroleum hoists generated small machine parts such as engine valves, spark plugs, springs, piston rings, and electrical wiring.[44]

The visitor to a historic prospect shaft may also encounter the remains of other surface plant components associated with the hoisting system. When the prospect operation was abandoned and the headframe taken down, the demolition crew may have left behind the structure's timber footers embedded in waste rock. The footers should closely flank the shaft and extend toward the hoist house platform. Often they feature mortise-and-tenon joint sockets or notches where the headframe's vertical posts were fastened. The foundations for four-post derrick headframes feature four sockets at each corner of the shaft, while two-post gallows structures left one socket on each side of the shaft. The visitor might be able to identify the spot where the headframe's backbraces had been anchored to the timber footers. Construction crews tied the backbraces in with mortise-and-tenon sockets or notches, and timber bolts at a location between the shaft and the hoist house platform. In many cases, however, demolition crews managed to win the timber headframe footers from the waste rock, leaving nothing more than parallel depressions.

The prospect shaft site may also feature the traces of the hoist house, which includes a cut-and-fill platform, and structural and industrial materials. In addition, prospect shafts also featured a simple rail system which generated specific artifacts, and remote operations may have had associated habitation structures.

Production-Class Surface Plants for Shafts

Mining companies that had proven the existence of ore underneath their claims were uncommon in the West. Further, the companies that had struck major ore deposits, enough to excite experienced mining men the likes T.A. Rickard and Herbert C. Hoover, were a true rarity. The investors, promoters, and directors comprising such fortunate outfits understood that they were not able to profitably extract their precious ore without the proper machinery, experienced miners, and an engineer to oversee efficient operation of the mine. Mining companies often mandated the extraction of maximum tonnages of payrock for minimal expense. In response, engineers erected production class surface plants, also known as *permanent plants*, which were large, capital-intensive, and efficient. The machinery that the engineer selected proved to be a major factor in profitable ore production.

No two mines were identical, each having individual requirements, and in response mine supply houses offered a spectrum of production-class machines and products to fit the economic and performance needs of the wide variety of western operations. For example, mine machinery makers offered small production-class hoists for mines with limited capital, but sold enormous hoists to wealthy operations seeking around-the-clock production of huge volumes of ore. Generally, production-class plants were capital-intensive across-the-board, physically demanding to transport and install, and meant for long-term operation.

Many western mining companies gutted their ore bodies in surprisingly short periods of time. In so doing, they wasted substantial sums of capital on machines designed to deliver savings only over decades, that ironically were used for only a short time. Of course, the West played host to many fabulously wealthy mines that produced ore for over a half-century, justifying the replacement of lasting machinery, made obsolete by time and by wear.

A production-class surface plant was either a pre-planned wholesale installation, or a slow development over an extended operating period. Professionally-trained engineers with access to substantial quantities of capital, eagerly supplied by profit-hungry investors, were more likely to install instant production-class surface plants than were self-made engineers. To the extreme, the Woods brothers, capitalists and promoters of the burgeoning townsite of Victor in the Cripple Creek District, discovered gold when grading a hotel site, and rather than undertake cautious development of their property, they hired an engineer who built a substantial production-class surface plant at the outset of operations. When the wooden shaft house burned within several years, their engineers replaced the plant with one of the district's most extravagant mine complexes, which included a stained-glass appointed brick shaft house that enclosed a massive double reel flat cable hoist, a huge duplex air compressor, full machine shops, and other facilities.[45]

Self-taught engineers often took the conservative approach of limiting expenses by upgrading the surface plant only as the mine proved itself worthy. Even so, the upgrades rarely included big, highly efficient machines. Frank Crampton, one such engineer, exercised this approach with a number of properties he had been charged with developing in the Southwest, such as the Copper Belt Mine, the Wren claim, and the Boss Mine.[46] W.S. Stratton followed a similar path with his wealthy Independence Mine in Cripple Creek. Ultimately the mine received a large production-class plant only after an English company purchased the property from him. Many of the West's small operations did not prove worthy of full-scale upgrade and as a result they were abandoned while their surface plants were relatively simple. Highly profitable mines continued operations with plants consisting of a combination of sinking and production-class equipment at the sacrifice of long-term operating efficiency.

$$\times\hspace{-0.8em}\times$$

Production-Class Hoisting Systems

The components of production-class hoisting systems conformed to a few fundamental patterns of physical arrangement and composition. Specific combinations of hoists, power sources, hoisting vehicles, and headframes worked well together. The size and depth of a shaft directly influenced the type of hoist an engineer selected. The type of hoist in turn influenced the power source an engineer arranged for, and the headframe he erected affected his choice of hoisting vehicle. As a result, when a mine began production, the engineer had to upgrade portions of the plant simultaneously.

One of the engineers' first considerations when bringing a mine into production was whether the existing shaft was adequate in size. The enormous costs of shaft sinking and the speculative aspect of mining discouraged most operations from sinking three compartment shafts for balanced hoisting at the outset of claim development. Most mining companies sank double compartment shafts because they cost less. As a result, many mining outfits that sought to maximize the tonnages of ore while realizing energy savings through balanced hoisting were forced to drill and blast a third compartment. The expenses of work stoppages and the dangerous undertaking of enlarging a two-compartment shaft ensured that most western mining companies would continue to use their constricted two-compartment shafts.

In rare cases profitable mining companies in the hands of skilled engineers had their miners add a third compartment to their shafts. Several of Cripple Creek's noteworthy mines that relied on balanced hoisting, including Stratton's American Eagle, the Vindicator, and the Hull City began with two-compartment shafts, which the companies later expanded. When adding a third compartment, standard practice dictated that miners build a strong timber bulkhead deep in the shaft to stop falling material, and begin driving a third compartment from the bottom up, using gravity to advantage. In a few cases, however, miners worked from the top down. The Delmonico Mine, in Cripple Creek, had been documented by the Colorado State Mine Inspector in 1916 as having a three-compartment shaft from the collar down to the 600 foot level, where the shaft suddenly constricted into the standard two compartments.[47]

After the engineer made a decision regarding the shaft size, he had to select a compatible hoist. The engineer considered the number and sizes of the hoisting compartments, the depth of the shaft, and the type of hoisting vehicle, but available capital proved to be the most significant factor. Mining engineers almost always installed *single-drum hoists* to serve two-compartment shafts, and usually installed *double-drum hoists* for balanced hoisting in three-compartment shafts. In general, few mines possessed the capital or production levels to purchase a double-drum hoist.

During the early years of western mining, technology was severely limited and little machinery was available. As mining progressed into the twentieth century the variety of power sources, hoists, and other components increased greatly. During the 1850s and 1860s the mining world recognized the horse whim as conventional production-class hoisting technology. The clumsy, slow, and light-duty nature of the horse whim limited ore production, but nothing better existed until practical steam hoisting arrived. The first steam hoist was an experimental affair in which a Cornish mining engineer adapted a massive beam engine at the Wheal Maid Mine in Cornwall to turn a rope drum encased in heavy timber framing. The early hoist was slow, cumbersome, and could not be readily reversed, but it proved successful and led to further development. By the early 1800s other well-financed Cornish mines had followed suit and their engineers erected similar but more efficient and faster models, still slow by late nineteenth century standards. In Britain, steam hoisting remained a luxury that was limited to large mining companies until the 1850s. At that time mechanical and mining engineers reduced the sizes and costs of the machines, improved performance, and their availability increased.

By this time the mining industry in the United States had caught up to the British in developing power hoisting. Cornish mining engineer Joseph Rawlins installed one of North America's first steam hoists at the Cliff Mine on Michigan's Upper Peninsula in 1851, and other large American interests imitated Rawlins. Nevada's wealthy Comstock, with its deep mines and high volumes of rich silver ore, provided an ideal setting for one of the West's first experimental steam hoists. In 1860 the Ophir Mining Company harnessed steam power for hoisting by lashing a rope to the drive-shaft of its pumping engine, pulling a small skip up an incline. However, the Ophir made a tacit statement regarding the state of steam hoisting during this time when it quit using the pumping engine and returned to the traditional, reliable, slow horse whim.[48]

The experiments in developing power hoisting in the West did not stop with the Ophir's failure. By the late 1860s, mining engineers, undoubtedly backed by the deep pockets of large companies, devised a mechanized hoist that revolutionized western mining. They had adapted the common horizontal steam engine to turn a cable drum through a set of reduction gears. Mining engineers anchored the engine to a brick or rock masonry foundation and fitted its flywheel shaft with a pinion gear, which turned a large bull-gear mounted to a timber frame. The large gear was connected by a clutch to the cable drum, which was mounted adjacent to the engine. While this hoist mechanism was new and practicable, the basic principle remained unchanged from the cumbersome contraption first employed at the Wheal Maid Mine across the Atlantic. Through the 1870s only well-capitalized mining companies were able to afford these innovative hoisting systems. No convention existed for engineers to follow when building hoists, and each system was virtually a custom affair.[49]

The first steam hoists used for hardrock mining in the West consisted of a cable drum geared to a horizontal single cylinder industrial steam engine. The photograph illustrates one of the few surviving single cylinder hoists in the West. In keeping with steam hoist engineering of the day, the cable drum is mounted onto a timber frame and connected to the engine's flywheel via a pinion gear. Unfortunately, soil creep has buried the horizontal steam engine, which is located along the right side of the hoist behind the large flywheel. The mine is located in Nevada's Jackrabbit District.

Author.

A far cry from horse whims, single cylinder hoists were still inefficient and they presented mining companies with many problems. The greatest drawback was that the engine had to run continuously, for fear of it stopping on top-dead-center. In the occurrence of such an undesirable event, the hoistman had to disengage the cable drum and pull the brake, then go for help while the miners on the cage waited in the shaft below. The hoistman and at least one other mine worker had to wrestle with the heavy flywheel and turn the engine over to reset the piston, which proved to be a Herculean task. Because the engine had to run in one direction, the hoistman had to disengage the cable drum and lower the hoisting vehicle with the handbrake. Also, single cylinder hoists did not permit balanced hoisting. Another drawback to the single-cylinder hoist was that the engine, when constantly running, consumed large quantities of fuel.

While the engine rhythmically chuffed and hissed, and its large heavy-spoked flywheel spun around, the hoistman worked the brake lever and lowered the cage to a shaft station deep underground. He watched the level indicator, similar to a large clock face with a hand showing the position of the hoisting vehicle in the shaft, until it registered that the cage was perfectly at the station. There, a trammer loaded a full ore car onto the cage, rang the bell, and the hoistman engaged the clutch. The engine momentarily slowed then resumed a quick tempo until the cage had arrived at the shaft landing. The hoistman disconnected the clutch, the engine sped up slightly, and he lowered the cage onto a *landing chair* engaged by the topman who switched the full car with an empty one.

Only the best hoistmen intimate with the mines where they worked were able to accurately stop the cage at shaft stations, or at the collar, so that the rails on the floor aligned with the rails of the mine. To remedy the disparity in track alignment, mining engineers had retractable iron bars installed at each shaft station, and at the collar, on which the cage could rest while trammers switched ore cars. Topmen manually operated landing chairs by levers, which were used as early as the late 1860s.[50]

Mining on the Comstock was in full swing during the 1870s and the companies there had plenty of resources to increase production at ever-greater depths. Engineers overseeing the operation of particularly large Comstock mines, such as the Yellow Jacket and the California & Consolidated Virginia, recognized the limitations posed by single cylinder hoists. In their attempts to increase tonnages of ore and improve the technology, engineers contracted with a few San Francisco foundries, such as the Pacific Iron Works, to design and build double-drum hoists. Most of the new hoists were of mammoth proportions, featuring dual engines placed over twenty feet apart that flanked cable reels over ten feet in diameter. The new hoists weighed dozens of tons and delivered 1,000 to 2,000 horsepower. Known as the *duplex pattern*, these titanic hoists with opposing cylinders became the convention that the rest of the mining West would follow for fifty years. By 1880 the duplex engine pattern had become standard for all sizes of both single and double-drum hoists, and within less than ten years single cylinder hoists had become a thing of the past.[51]

Steam hoisting technology maintained mechanical supremacy among the western mines from the 1880s until gasoline and electric power superceded it by the early 1920s. During this time period mining machinery makers such as Allis-Chalmers, the Lidgerwood Manufacturing Company, the Lambert Hoisting Engine Company, Hendrie & Bolthoff, the Union Iron Works, and the Ottumwa Iron Works offered steam hoists in a wide array of sizes. These manufacturers also offered hoists equipped with either *first-motion* or *second motion* drive trains. First-motion drive, also known among mining engineers as *direct-drive*, meant that the steam engine drive rods were coupled directly onto the cable drum shaft, much like the way the drive rods were directly pinned onto a steam locomotive's wheels. Second motion drive, also commonly known as a *geared-drive*, consisted of reduction gearing like the sinking-class hoists discussed earlier.

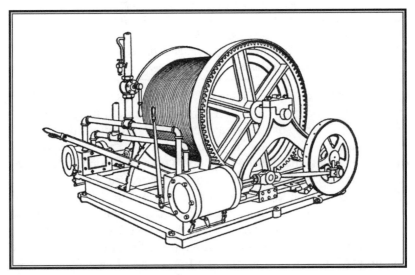

"Made by the mile and cut off by the yard", mining companies throughout the West favored the single drum duplex geared steam hoist above all other varieties from around 1880 until electric power and petroleum fuel replaced steam in the early 1920s. These hoists were relatively inexpensive and easy to install because all of the components were assembled onto a common bedplate. However the machines were slow and fuel inefficient. The rear of the hoist, where the hoistman operated the controls, is located at left, and the front where the drive shaft rotated is at right.

International Textbook Company International Library of Technology: Hoisting, Haulage, Mine Drainage International Textbook Company, Scranton, PA, 1906 A50 p8.

During the 1880s and early 1890s some mining companies employed a variety of single drum geared steam hoist in which the engine, reduction gears, and cable drum were mounted onto an elongated bedplate, as shown. While the footprint of these hoists mimicked that of the first-motion hoist illustrated below, the foundations for these old-fashioned geared models tended to be smaller, encompassed by anchor bolts, and had a flat top-surface.

Ihlseng, Magnus A *Manual of Mining* John Wiley & Sons, New York, NY, 1901 p79.

The difference in the driving mechanisms was significant in both performance and cost, and each served a distinct function in western mining. Gearing offered great mechanical advantage that permitted the use of relatively small steam cylinders. The arrangement of the gear shafting and the small cylinders on a common bedplate permitted the hoist's footprint to be compact because the steam cylinders flanked the cable drum and shaft bearings. First-motion hoists, on the other hand, required that the cable drum be mounted at the end of large dual steam cylinders so that the drive rods could gain leverage. Where the footprint of geared hoists was almost square, the footprint of first-motion hoists was that of an elongated rectangle. First-motion hoists were intended by manufacturers to serve as high-quality production-class machines designed to save money only over protracted and constant use, while geared hoists were intended to be inexpensive and meet the short term needs of small, modestly capitalized mines. First-motion hoists were stronger, faster, and more fuel-efficient than geared models. Their large sizes, the necessity of using high-quality steel to withstand tremendous forces, and the fine engines made the purchase prices of first-motion hoists three to four times that of geared hoists. The latter cost between $1,000 and $3,000 for light to heavy production-class models. First-motion hoists had a speed of 1,500 to 3,000 feet per minute, compared with 500 to 700 feet per minute for geared hoists. These hoisting speeds reflect the ability of first-motion hoists to work in shafts with depths well into the thousands of feet. Geared hoists usually relied on old-fashioned but durable slide valves to admit steam into and release exhaust from the cylinders, while first-motion hoists usually were equipped with corliss valves for the engine, which were initially more expensive but consumed half the fuel. In short, mine supply houses sold first-motion hoists to handle numerous tons of payrock in deep shafts while consuming little energy.[52]

First-motion steam hoists consisted of two powerful steam engines coupled directly to the drum shaft. The direct-drive motion enabled these mighty machines to raise loads much quicker while using less fuel than the geared hoists illustrated above, suiting them for heavy ore production in deep shafts. Because first-motion hoists were very costly, generally only well-capitalized mining outfits installed them. Note the level gauge at the upper left which displayed where the hoisting vehicle was in the shaft. The drum in the illustration has been wound with two cables for balanced hoisting.

International Textbook Company International Library of Technology: Hoisting, Haulage, Mine Drainage International Textbook Company, Scranton, PA, 1906 A50 p16.

If the costs of purchasing first-motion hoists were not high enough, the expenses associated with their installation were exorbitant, discouraging operations from using them for anything but advanced production. Because geared hoists were self-contained on a common bedplate, they required less time to set up, and installation was much easier. The surface crew at a mine merely had to build a comparatively small foundation with anchor bolts projecting from a flat surface, and drop the hoist into place. First-motion hoists, on the other hand, required raised masonry pylons for the steam cylinders, pylons for the cable drum bearings, a well for the drum, and anchor bolts in masonry between the pylons for the brake posts. The hoist pieces then had to be brought over, maneuvered into place, and simultaneously assembled.

Mining engineers chose specific hoists based on the power delivered by the engine, which had a proportional relationship with the hoist's overall size. Geared hoists smaller than six-by-six feet were usually made for deep exploration and delivered less than fifty horsepower. Hoists between seven-by-seven feet and nine-by-nine feet were for minor ore production and offered seventy-five to 100 horsepower. Hoists ten-by-ten feet to eleven-by-eleven feet were for moderate to heavy production and generated up to 150 horsepower, and larger units were exclusively for heavy production. Mining engineers rarely installed geared hoists larger than twelve-by-twelve feet, because for a little more money they could have obtained an efficient first-motion hoist.[53]

Double drum geared steam hoists fit a special niche among highly productive mines. Mining companies that sought to achieve balanced hoisting while limiting their expenses on equipment installed these compact machines, which afforded them a great ore output. However, these geared hoists were slower and less efficient than first-motion models.

Ingersoll Rock Drill Company Catalog No.7: Rock Drills, Air Compressors and Air Receivers
Ingersoll Rock Drill Company, New York, NY, [1887] p58.

Engineers felt that if they were to spend large sums of capital on installing an efficient production-class plant, double-drum hoists offered greater economical performance for the dollars spent, while maximizing tonnages of rock extracted. As a result, few engineers recommended that their companies install first-motion steam hoists featuring only one drum. Like single-drum hoists, double-drum units came with geared or first-motion drives, which were either self-contained on a bedplate or consisted of components that had to be anchored to masonry foundation piers.

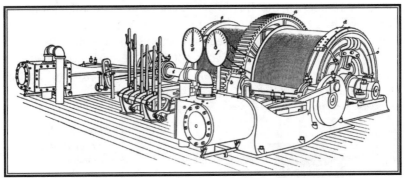

The line drawing depicts a large and powerful double drum geared steam hoist. Workers usually laid plank flooring around the hoist and they routed the steam pipes underneath, as illustrated. The plethora of levers at the hoistman's station control a clutch, brake, throttle, and reverse link. A portion of the reverse link is visible immediately left of and behind the stand of levers. Note the two level gauges.

International Textbook Company International Library of Technology: Hoisting, Haulage, Mine Drainage International Textbook Company, Scranton, PA, 1906 A50 p9.

The primary purpose for spending a small fortune on such an impressive apparatus was to realize maximum production through balanced hoisting. Double-drum geared hoists, ranging in size from between seven-by-twelve feet to twelve-by-seventeen feet, were slower, weaker, and noisier than their direct-drive brethren, and they cost much less to purchase and install. Like single-drum geared hoists, double-drum geared models had weight, speed, and depth limitations mining engineers with high expectations, such as T.A. Rickard and Herbert C. Hoover, would not tolerate. The ultimate answer for raising the maximum quantity of ore in minimal time was the installation of a double-drum first-motion hoist. The extreme difficulty and exorbitant costs of transporting and installing these massive machines relegated them to heavily capitalized mining companies with highly productive operations in well-developed districts. Not only did these types of hoists permit mining companies to maximize profits, but also they served as a statement to the mining world of a company's financial status, levels of productivity, and quality of engineering. Mines equipped with such extravagant machinery often caught the attention of notable mining men.

Double-drum first-motion hoists ranged in size from approximately eighteen by twenty-five feet to over thirty-by-forty feet in area, and their performance kept pace with their sizes. Small models, large by comparison to other types of hoists, generated tremendous 600 horsepower while large units created over 1,200 horsepower. During the 1880s some of the highly profitable and deep mines on the Comstock had hoists that delivered an amazing 2,000 horsepower. The massive steam pistons' sheer strength permitted these hoists to raise a payload over five tons at a speed of at least 3,000 feet per minute, making geared hoists seem like toys. Their working depths were well over 3,000 feet.[54]

MINING AND SCIENTIFIC PRESS

When the Moon-Anchor Mining Company made arrangements to install a massive double drum steam hoist around 1900, the boom times in Cripple Creek were in full-swing. The photo depicts a newly competed eight-post derrick headframe standing over a three-compartment shaft, and a pile of specially cut dressed stone blocks which will be used for the hoist foundation.
Mining & Scientific Press
Feb. 21, 1903 p1.

The installation of such immense hoists required exacting engineering, highly skilled labor, and significant site preparation. While mine machinery makers shipped the smaller geared hoists either intact or in several large components which were easily assembled, the large first-motion units came in many pieces that required special anchor bolts built into an elaborate masonry foundation. The steam assist cylinders that powered the clutches and brakes, as well as the brake posts, control linkages, and the main hoist parts all had to be perfectly mounted onto bolts set in masonry pylons placed at exact heights and locations. Because they were highly specialized and their installation required precision work, manufacturers such as Webster, Camp & Lane, Wellman-Seaver-Morgan Company, Allis-Chalmers Company, and the Stearns-Roger Manufacturing Company dispatched mechanical engineers to assist in site preparation and final assembly.

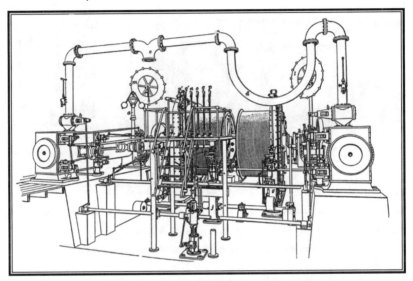

Double drum first-motion steam hoists represented the culmination of efficient and costly production-class hoisting systems. The line drawing provides a rear view of the complex machinery comprising the hoist, which includes two massive steam cylinders flanking the hoistman's platform, the cable drums, and small steam cylinders that activated the clutches and brakes, located in the pit. Few mining companies were productive enough and possessed sufficient capital to install these titanic behemoths. For scale, the span between the cylinders is approximately 18 feet.

International Textbook Company International Library of Technology: Hoisting, Haulage, Mine Drainage International Textbook Company, Scranton, PA, 1906 A50 p18.

During the 1860s and 1870s, progressive engineers made improvements in hoisting cables, which proved to be a significant technological development. During the early 1860s when western mining companies were experimenting with power hoisting systems, engineers experimented with heavy *hemp rope* and *wire rope*, commonly known as cable, for hanging the hoisting vehicles. Hemp rope had a short life in wet mines, and mining companies preferred steel cable, but it was costly and difficult to obtain, A.S. Hallidie in San Francisco being the only manufacturer in the West. Cable-making technology was nascent at that time, and its performance was limited. In 1863 Hallidie and the Comstock's Ophir Mine attempted an experiment intended to test an alternative to the conventional round cable. Hallidie had devised the now-famous flat hoist cable, which proved superior to round cable in terms of tensile strength, wear, and resistance to spinning when unwound off the hoist drum.[55]

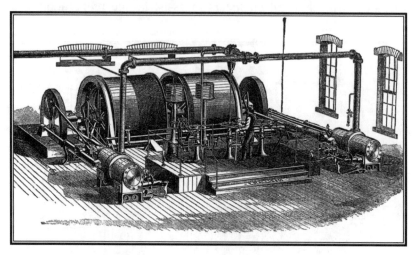

The illustration provides an aerial view of a double drum first-motion steam hoist. Mining companies used these mighty machines from the 1870s into the 1920s.

Ihlseng, Magnus A Manual of Mining John Wiley & Sons, New York, NY, 1901 p76.

While Hallidie developed flat cable as a more lasting and stronger replacement for round cable, mining engineers had stumbled upon unintended consequences of its application. When mining engineers oversaw the installation of hoists, they had the workers wind round-wire rope onto cylindrical cable drums that were much like enlarged thread spools. When the hoist operated, the cable either paid out or wound itself up on the drum in an orderly fashion. No so with flat cable; it had to be wound onto a large-diameter reel much like those used on cinema projectors, because it would have become hopelessly tangled on a flat drum. Mining engineers took note that when a flat cable had been fully paid out, the hoisting engine had an easy time re-winding it because the diameter of the empty reel was relatively small, offering the engine mechanical advantage. As increasing amounts of cable gathered onto the reel, the diameter of the reel increased, which gradually put a load on the engine. Ordinarily, hoisting engines had to exert the greatest force, therefore consuming the most energy and fuel, when the cable had been paid out down the shaft. Not only did the engine have to begin turning the drum, but the combination of the cable in the shaft and the hoisting vehicle constituted great weight. An empty flat cable reel, with its small diameter, offered the straining engine leverage at the critical moment. The result was that the hoist vehicle's ascent rate increased while the load on the hoist remained the same. The ultimate effect, to many engineers' delight, was a great saving in energy.[56]

By the 1890s steel-making and wire rope technology advanced greatly. Machinery manufacturers offered wire ropes with a performance superior to flat cable for less money. The new hoist cables lasted much longer and did not require periodic unspooling and reweaving, as did flat cable. Yet, some of the highly productive western mines with esteemed engineers, such as Herbert C. Hoover and T.A. Rickard, continued to use flat cable into the 1920s because of the energy savings it offered. However, the high costs associated with flat cable, and the replacement of steam with inexpensive electricity, proved to be the demise of the fifty-year-old technology.

Flat cable hoists bore great resemblance to the conventional double drum first-motion machines illustrated above. Flat cable was expensive to maintain, but many western mining engineers felt that it contributed great savings in energy, in addition to making for an impressive hoist that aroused the interest of other engineers.

Mining & Scientific Press Oct. 21, 1899 p470.

During the 1890s mining engineers devised a first-motion steam hoist that featured a conical cable drum intended to permit the use of less costly wire rope while offering the economy of flat cable. These special drums let the cable off the broadest portion of the cone first and progressed toward the narrow portion as the hoist paid cable down the shaft. When the hoist rewound the cable, it benefited from starting the critical moment with the cable on the small portion of the drum. Professionally-educated mining engineers admired the clever use of physics and geometry in the design of these interesting machines, but their high costs and unconventionality ensured that only a few well-financed and wealthy mines in the West would ever install them. Conical hoists incorporated an expensive grooved drum to wind the cable. They were inefficient for working multiple shaft levels, and the cable had a propensity to gather up onto the narrow portion of the drum. Mining machinery makers sold both single and double conical drum hoists between the 1890s and 1910s, but the outrageous costs and replacement of steam by electricity relegated them to economic obsolescence by the 1920s.[57]

A few profitable and progressive mining companies in the West attempted to make use of conical drum hoists in hopes of combining the advanced application of physics, calculation, and round cable to affect savings in energy expenditures and maintenance. Conical drum hoists were innovative and expensive, and they never saw great popularity. The illustration depicts a single drum version, which rarely saw use in the West.

Ihlseng, Magnus A Manual of Mining John Wiley & Sons, New York, NY, 1901 p84.

Mining engineers were as concerned with saving power costs at the generation source, the boiler, as they were with the efficiency of the hoist. Nearly all mechanized western mines operating during the Gilded Age were run by steam. Adhering to the objectives of maximizing production and minimizing operating costs, professionally-trained mining engineers used calculation and mechanical specification in their attempts to meet the power needs of a mine. The mining engineer had to add up the steam demands of all machines, usually measured in *boiler horsepower,* to calculate the size, type, and number of boiler units he would need. While this task may have been a major undertaking for some self-educated engineers, professionally-trained engineers were taught this type of problem solving, which they also applied to calculating compressed air needs.

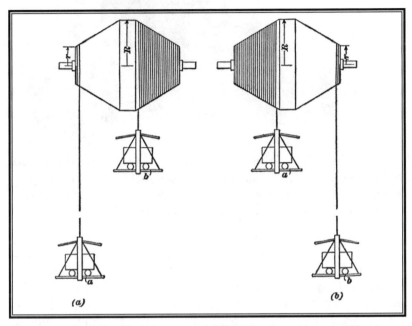

The concept behind conical drums was brilliant. As the cable was paid out down the shaft, the diameter of the conical drum continually decreased, as shown in the illustration. The small diameter gave the engines great mechanical advantage in beginning the movement of lifting heavy weight, and as the cable wound onto increasingly larger portions of the conical drum, the hoist's engine came under gradual strain while the hoisting vehicle's rate of ascent increased. The net result proved to be a significant savings in energy.

International Textbook Company International Library of Technology: Hoisting, Haulage, Mine Drainage International Textbook Company, Scranton, PA, 1906 A52 p12.

Between the 1880s and 1910s an engineer may have included in his calculations the hoist, an air compressor, a de-watering pump, and a tiny pump to feed water to the boilers. Small mines may not have had a compressor, while large operations may have included a donkey engine to drive a ventilation fan or mechanized shop appliances. Engineers almost never relied on portable boilers to supply the plant with steam because of their inefficiency. Rather, the masters of mine mechanics predominantly used production-class *return-tube boilers* in masonry settings, or they erected *water-tube boilers*, which offered the ultimate fuel economy. In a few rare cases, engineers working deep in the backcountry were forced to make do with Pennsylvania and locomotive boilers.

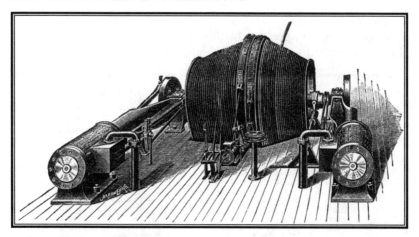

The installation of a first-motion hoist with double conical drums was a news-worthy event in the western mining industry because these machines spoke of capital, heavy production, and a progressive company with a flair for modernity and extravagance.

Ihlseng, Magnus A Manual of Mining John Wiley & Sons, New York, NY, 1901 p78.

Manufacturers designed most steam machinery to work under pressures between 100 and 150 pounds-per-square-inch. At this pressure a return-tube boiler five feet in diameter and sixteen feet long provided enough steam to run a hoist, a Cameron sinking pump, heating pipes in the shaft house, and another engine such as a compressor, all totaling approximately eighty boiler-horsepower. Mine plants that included production-class single and double-drum hoists, duplex compressors, and additional machinery usually required the steam generated by at least two to three return-tube boilers totaling over 200 boiler-horsepower. Mining companies with a progressive engineer and plenty of capital often installed an extra boiler so that if one of the others had to be cooled off for servicing, the mine could have continued operating.[58]

The use of steam as a power source for mining dates back to the early 1700s. The Cornish originally used it for running their pumps, and for some hoisting by the 1800s. Early steam boilers were similar to giant teakettles, and they were not able to withstand the high pressures that best powered mining machinery. During the British Industrial Revolution, boilermakers introduced cylindrical vessels better able to withstand steam pressure, but attempting to heat these huge tanks of water consumed vast quantities of coal. The high fuel costs provided the Cornish with the stimulus to improve boiler efficiency. In 1800 Oliver Evans, an American engineer, and in 1802 Richard Trevithick, a Cornish mining engineer, developed varieties of the *Cornish boiler*, also known as the *flue boiler* and the *Lancashire boiler*. The Cornish boiler consisted of a plate-iron shell six to eight feet in diameter, thirty to forty feet long, perforated lengthwise by one or two tubes one to two feet in diameter.[59]

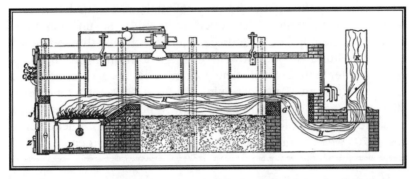

The plain cylindrical boiler, developed in Britain during the early nineteenth century, consisted of a water-filled boilerplate iron shell 5 to 6 feet in diameter and over 30 feet long. As the profile illustrates, the shell stood over a brick setting that featured a firebox F at one end and a smokestack K at the other end. The superheated flue gases H traveled across the belly of the boiler, imparting heat to the water within. Brick bridge walls G were built to force the flue gases in close contact with the boiler shell. A fireman fed fuel through fire doors J, and after combustion the cinders dropped through an iron grate E and accumulated in the ash pit D. The fireman periodically cleaned the ashes out through the ash pit door Z. A few mining companies in the United States used plain cylindrical boilers, but the Cornish boiler quickly rose to supremacy due to its greater efficiency.

International Textbook Company A Textbook on Metal Mining: Steam and Steam-Boilers, Steam Engines, Air and Air Compression, Hydromechanics and Pumping, Mine Haulage, Hoisting and Hoisting Appliances, Percussive and Rotary Boring International Textbook Company, Scranton, PA, 1899 A18 p21.

The concept behind the Cornish boiler was brilliant. The boiler shell, part of a complex structure, was suspended from iron legs known as *buckstaves,* so named because they prevented the associated masonry walls from *bucking* outward. Brick walls enclosed the area underneath the boiler shell, and a heavy iron façade shrouded the front. A *firebox* lay behind the façade underneath the boiler shell. Under the firebox lay an *ash pit,* and both were sealed off by heavy cast-iron doors. Superheated flue gases traveled along the belly of the boiler shell from the firebox, rose into a *smoke chamber* at the rear of the structure, reversed direction and traveled toward the front through the large *flue tubes,* and then exited the shell through the *smoke manifold.* Unlike the previous water tank-like boilers, the path under and then back through the boiler shell offered the flue gases every opportunity to transfer energy to the water within and convert it into steam. This design not only dramatically boosted energy efficiency, but also set precedent for boiler systems that remained a convention until electricity superceded steam in the 1920s.

When properly built with heavy boiler-plate iron and hot-riveted joints, Cornish boilers developed unheard-of steam pressures, upward of one-hundred pounds-per-square-inch, and they generated up to one-hundred horsepower, which gave mechanical engineers much greater latitude for efficiency. Cornish boilers became standard in both British and American mining by the 1830s, at a time when they were used mostly to run Cornish pumps. Cornish boilers remained unrivaled for powering mining machinery in America until the 1870s when return-tube boilers began to replace them.

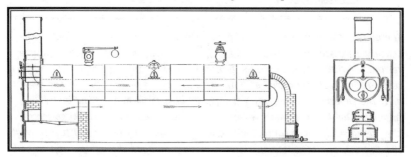

The Cornish boiler, also known as the flue boiler, was the true predecessor to the return tube boiler that powered the western mining industry during the Gilded Age. Flue boilers were highly popular among productive western mines during the 1860s and much of the 1870s, and they had fallen out of favor by around 1880. The setting, portions of which may still be encountered at historic western mine sites today, consisted of brick over a stone footing, and they were often over 10 feet wide and greater than 30 feet long. The arrows in the illustration depict the path that the flue gases traveled; from the firebox, along the shell's belly, then back through the two flue tubes and out the smokestack.

Rand Drill Company Illustrated Catalog of the Rand Drill Company, New York, U.S.A. Rand Drill Company, New York, NY, 1886 p49.

By the 1880s mining engineers had achieved major advances in both boiler shells and in an overall understanding of the mechanics of steam power, and they recognized the return-tube boiler as the production-class power source of choice. Return-tube boilers were merely an improvement on the concept of the Cornish boiler. Instead of one or two large flue tubes, several dozen tubes two to four inches in diameter penetrated the boiler shell, immensely increasing the surface area of the water. The greater surface area afforded by the numerous tubes permitted boiler manufacturers to reduce the shell's length by half. These return-tube boilers proved so efficient, durable, and affordable that they remained the principal source of power for the western mining industry into the 1920s. Mining engineers sought the generation of what they termed *superheated steam*, which lacked the visible vapor contained in inferior *wet steam*. Superheated steam was a true gas and carried more energy to run the pistons of mine machinery. Usually the engineer set the working pressure of boilers between 100 and 150 pounds-per-square-inch, which they determined to be most efficient for driving hoists and compressors. They had established the fact that the draft feeding oxygen to the fire influenced boiler efficiency. Too little air allowed the fire to suffocate, while too much air flushed the system of the flue gases before they transferred their energy to the water. As a result, engineers offered specifications detailing the relationship between the diameter and height of the steel smokestack, the number of boilers one smokestack should have served, and how to control the dampers in hopes of balancing back pressure against incoming air. While the combination of these factors may have had a notable economic benefit for large mines with many boiler units, they were usually not taken into account by the everyday engineer when he was arranging a power plant at a small or medium-sized mine.[60]

The profile clearly depicts the fundamental features comprising the settings of the return tube boilers. These power plants were common among western mines, as well as other industrial facilities.
Croft, Terrell Steam Boilers McGraw-Hill Book Co., New York, NY, 1921 p252.

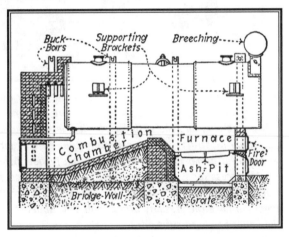

The masonry enclosures surrounding return-tube boilers changed little from their Cornish boiler ancestry. Return-tube boilers used throughout the West were supported by four to six buckstaves. Thick brick walls enclosed the firebox and smoke chamber underneath, and the flue gases traveled the same zigzag path as Cornish boilers. Mining engineers and their labor crews undertook a major project when they attempted to install such complex power sources. First, the engineer had to allocate a source of water and have a work crew lay a pipeline from it to a holding tank at the mine. Developed mining districts often had the benefit of a local water company which took care of this arduous logistical task. Engineers working in the backcountry attempted to secure a spring at a high elevation so that gravity would pressurize the system, but when that proved impossible a small pump adjacent to the boiler raised the water. Boiler feed pumps were only approximately two foot long, eight inches wide, and mounted onto a heavy timber block.[61]

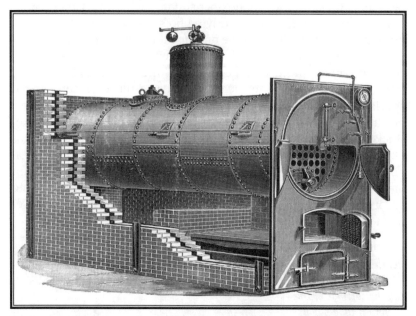

The return tube boiler was a great improvement over the Cornish boiler in terms of size, weight, and fuel efficiency, but the general structure of the masonry setting and the path for the flue gases remained the same. The lithograph illustrates the typical return tube boiler that many western mines relied on for power into the 1920s. The full façade includes double doors that provide access for swabbing soot out of the flue tubes. Note the removable iron grates lining the floor of the firebox which is accessed by the middle door, and note the masonry bridge wall that abuts the rear of the firebox. The boiler has been suspended from riveted brackets that rest on the masonry walls, which engineers only permitted when the masons had used cement mortar instead of sand.

Rand Drill Company Illustrated Catalog of the Rand Drill Company, New York, U.S.A. Rand Drill Company, New York, NY, 1886 p44.

Engineers experienced with western mining discovered that water had a significant economic impact in addition to the cost of piping it to the boiler. Hard water, which plagued the Great Basin and Southwest, caused mineral deposits to form inside the boiler shell. This greatly reduced the efficiency of the boiler, and it was almost impossible to completely remove. In regions lacking fresh water, mining companies used the water pumped from mine shafts, which often proved highly caustic, corroding boilers and shortening their lives. For these reasons, and for general maintenance, boiler shells featured ovoid ports for cleaning the interior. Mineral scale was impossible to remove from between the numerous flue tubes, and as a result, boiler life was so short in some regions that mining engineers advised companies there to return to the old-fashioned and less fuel-efficient Cornish boilers, which were easy to clean.[62]

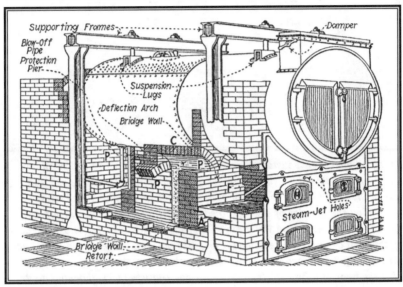

The line drawing illustrates a return tube boiler with a half façade. The shell has been suspended from iron buckstaves, and boiler grates A are visible lining the firebox. The smoke manifold, on which stood the smokestack, appears as a rectangular box on the top right portion of the boiler shell, and the filler pipe and blow-off tubes are visible at the lower left. The brick setting, including the arched bridge walls P, is more elaborate than the structures typically erected at western mines.

Croft, Terrell Steam Boilers McGraw-Hill Book Co., New York, NY, 1921 p253.

Installation of a return-tube boiler also meant that the engineer had to allocate a source of inexpensive fuel. Engineers agreed that cordwood, usually one-dollar-and-fifty-cents to two-dollars per cord, was the most economically appropriate fuel for remote districts in the mountain states and northern portions of the Great Basin. But professionally-educated engineers advised that coal was the most efficient boiler fuel and should be used when possible, because a half-ton of bituminous coal equaled the energy output of one cord of wood. A well equipped mine could consume up to five cords per boiler per day. For example, during the 1890s the Ophir Mine on Nevada's Comstock consumed approximately sixty cords of wood per day when its twelve return-tube boilers were at full steam. Despite the advantages presented by coal, the fossil fuel only proved economical when the district was either served by or was near a railroad. Mining companies in the West generally used coal when economically possible.[63]

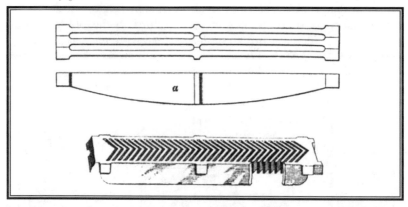

Firing a boiler over a sustained period of time created a variety of worn and broken parts, among which were boiler grates thrown out with the ash and clinker. Today's visitor can determine the type of fuel a mine relied on by examining the remains of the discarded grates. The herringbone pattern at bottom was for coal, and the grate at top was heavily used for cordwood. Coal grates also came with a rectangular grid-like pattern, and each individual unit was usually 3 feet long and 6 inches wide, numerous units being required to line the floor of a firebox. The grate for wood consisted of individual bars that could have been spaced with any desired gap. Because wood generated large quantities of smothering, fine ash, the air gaps had to be wider to allow the ash to drop through while providing the fire with enough oxygen.

International Textbook Company International Library of Technology: Mine Surveying, Metal Mine Surveying, Mineral-Land Surveying, Steam and Steam Boilers, Steam Engines, Air Compression International Textbook Company, Scranton, PA, 1924 A23 p54.

The engineer's selection of fuel directly influenced the construction of the boiler setting. Wood required that masons build a firebox that either featured a V-shaped profile, or provide a grate with broad air gaps. As the wood segments burned and disintegrated, they worked their way toward the bottom of the V where the ashes dropped through a grate into the ash pit below. Engineers claimed that the V-shape also funneled oxygen upward. In actuality, many western mining engineers, probably self-educated, did not build V-shaped fireboxes, but they did ensure that the boiler grate offered wide gaps to permit a free-flow of air through the bed of fine ash. Boilers designed to burn coal featured fireboxes with a broad, flat floor of boiler grates. Each grate unit featured small gaps to discourage unburned coal from falling through. Boiler grates were subjected to wildly fluctuating temperatures, and as a result they cracked and had to be replaced. Most boilermakers offered interchangeable grates designed to drop into place after the boiler tender extracted the broken ones. By the 1910s a very few highly productive western mines attempted to reduce costs by firing their boilers with fuel oil, which was finally being produced in economic volumes.[64]

Erecting a return-tube boiler was no small feat. After arranging the logistics, the engineer sent wagons to fetch the boiler, plumbing, fittings, and bricks, and he set a crew to work building a foundation near the hoist. Under the engineer's direction, the crew excavated a pit for the foundation, and masons usually constructed it with local fieldstones. By around 1900 concrete became popular because of its quick and easy handling.[65]

The photograph shows grates for wood fuel in-situ lining the remains of a boiler firebox at Colorado's Mary Murphy Mine. The boiler shell and the backside of the façade are visible at top and at left, respectively.

Author.

After the labor crew laid the foundation, they erected buckstaves to support the boiler and ensure that it would stand independent of the masonry setting. Mining engineers used one of two methods to stand boiler shells on buckstaves. In one method the boiler shell featured attachments on top so it could hang from steel rods, and in the second, the boiler shell featured brackets that rested on the tops of the buckstaves. Boilermakers sold factory-made buckstaves, but many mining engineers used salvaged lengths of railroad rail or four-by-four timber posts. Regardless of which suspension method employed, they rarely let their workers simply place the boiler shell on the masonry walls. The structure would have collapsed and the boiler may have exploded within a very sort time due to expansion, contraction, and exposure of portions of the shell to fire.[66]

Next, laborers with plumbing skills hooked up the pipes before the masons constructed the brick setting. They usually screwed in a one-inch *water feed pipe* into the bottom rear of the shell, an *overflow pipe* at the shell's three-quarter mark, and a two to six inch *steam pipe* and *blow-off pipe* onto the *steam dome* at top. As a final touch, they added the water-level glass sight tube and a pressure gauge to the front where the fireman could keep an eye on them.

After construction crews erected the boiler on the buckstaves and assembled the plumbing, masons began building the setting's walls and floor. Prior to raising the walls, they built a brick baffle known as a *bridge wall* that stood between the firebox and smoke chamber to force the hot flue gases to flow up against the boiler's belly. The front face of the baffle was vertical, and the structure tapered off at an angle toward the rear of the boiler. The masons then began laying bricks for the setting's side and rear walls. To allow for expansion and contraction of the boiler shell against the masonry setting, the workers left an inch gap between the walls and the iron shell, and they sealed the chinks with bricks. The mining engineer instructed the masons to arch bricks across the boiler's back for insulation. In most cases the masons also created ports at ground level in the rear of the setting to allow an unlucky miner to clean the soot out of the inside following protracted use of the boiler.

Masons building proper production-class boilers lined the firebox with firebrick because the intense heat destroyed other materials, and they used red brick for the remainder of the setting. In remote regions some outfits substituted local stone for red bricks in an effort to save money. When laying bricks or rocks, masons used well-sorted dampened sand as mortar, which not only saved the company the cost of purchasing and hauling thousands of pounds of cement to the site, but also eased the task of replacing damaged firebricks. However, engineers preferred to use cement mortar because it resisted the elements better, and permitted workers to build a setting strong enough to help support the boiler shell. As the quality of concrete improved through the 1890s, mining engineers began experimenting with it as a cheap substitute for the red brick masonry traditionally used for the setting.[67]

While building the front footing and sidewalls, the workers embedded anchor bolts in the brickwork to fasten a heavy cast-iron façade to shroud the firebox and underlying ash pit. Nearly all boilers at western mines featured a half-façade, but mining companies with substantial capital paid extra money to have a heavier full façade bolted onto the setting.[68]

Return-tube boilers were workhorses that withstood the harsh treatment and neglect endemic to western mines. Boiler tenders and firemen had to ensure that they carried out at least some maintenance to avoid disastrous explosions and ruptures. First, they had to keep the boiler at least two-thirds full of water. Second, the fireman had to clean the ashes out of the ash pit regularly to ensure that the fire did not suffocate. Shoveling ashes was a foul and dirty job that no one enjoyed. Usually the fireman shoveled the unwanted refuse into a wheelbarrow and trundled it out to a crook in the waste rock dump where the crew regularly dumped other trash. Pity the unfortunate worker who had to undertake such an unpleasant task on a gusty day! Third, the fireman ensured that the valves in the steam lines were operational, and that the pressure did not exceed the critical point. Last, the fireman had to feed the fire. Skilled firemen were able to throw on just enough fuel in an even distribution so that the fire kept a fairly constant glow. To ensure that firemen and boiler tenders had easy access to plenty of coal, the mining engineer usually placed a coal bin facing the firebox doors. In other circumstances cordwood may have been stacked in the bin's place.

Professionally-trained mining engineers with access to plenty of capital, employed additional devices designed to improve the energy efficiency and performance of their return-tube boilers. They may have elected to install up to three feed water holding tanks to allow sediment and mineralization to settle out. Some engineers working in the West installed feed water heaters, which were small heat exchanging tanks that used some of the boiler's hot water or steam to preheat the fresh feed water. These had been proven to moderate the shock of temperature changes to the boiler, prolonging the vessel's life, and increasing fuel efficiency. A few engineers working at the largest mines attempted to mechanize the input of coal into the fireboxes of heavily used boilers with mechanical stokers. While they were costly, mechanical stokers did a better job than laborers. Engineers also fitted heavily-stoked boilers with rocking or shaking grates that sifted the ashes downward, promoting better combustion of the fuel. Last, many engineers had mine workers wrap the heater, the steam pipes, and exposed parts of the boiler with horsehair or asbestos plaster as an insulation. Only a few large western mining companies employed these accessories because of the expense involved.[69]

In 1856 while boiler technology was young and alternatives to the long and heavy Cornish units were being developed, an American inventor named Wilcox devised a boiler radically different and much more efficient than the best return-tube models. Wilcox's system consisted of a large brick vault capped with several horizontal iron water tanks. The vault contained a firebox, an ash pit, and a smoke chamber, all underneath fifty to sixty water-filled iron tubes. The tubes drew water from one end of the tanks and sent the resultant steam to the other end. By 1870 the design, known as the *water-tube boiler*, had been commercialized and was being manufactured by the firm Babcock & Wilcox.[70]

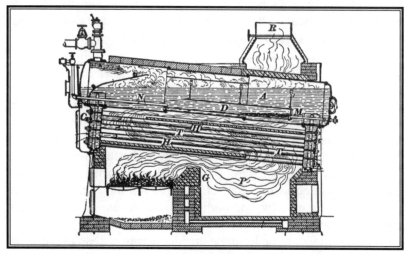

The profile illustrates the interior workings and flue gas path of a Heine water tube boiler. The flue gases P flowed over a bridge wall G, then wafted through a cluster of water tubes H and T, they flowed around the overhead drum D, and then rose out of the smokestack R. This path gave the flue gases every opportunity to transfer heat. When the water in the tubes turned to steam, the gas worked its way left, it entered a manifold C, bubbled up into the overhead drum, and escaped through the main pipe K. The Heine boiler was second in popularity only to the Babcock & Wilcox model.

International Textbook Company A Textbook on Metal Mining: Steam and Steam-Boilers, Steam Engines, Air and Air Compression, Hydromechanics and Pumping, Mine Haulage, Hoisting and Hoisting Appliances, Percussive and Rotary Boring International Textbook Company, Scranton, PA, 1899 A18 p38.

After Babcock & Wilcox's water-tube boiler had proven itself in a number of industrial applications, mining engineers began to take an interest. The fact that the water ran through the tubes, not around them, increased the liquid's heating area, which resulted in much greater efficiency than return-tube boilers. In addition, the threat of a catastrophic explosion was minimal. By the 1890s a number of manufacturers produced other water-tube boilers, such as the Heine, the Sterling, the Wickes, the Hazelton, and the Harrisburg-Starr.

All of the above models required much more attention than the rugged return-tube boilers. They were significantly more costly, and they were beyond the understanding and field skills of average mining engineers. As a result, water-tube boilers saw use only at large, well capitalized mines under the supervision of talented, professionally-trained engineers. As the prices of water-tube boilers fell during the 1900s and capital became more abundant as the mining industry recovered from the Silver Crash of 1893, the popularity of the efficient steam generators grew. Increased interest in practical electricity around 1900, however, prevented the widespread adoption of water-tube boilers.[71]

Mining engineers working in the West began experimenting with electricity as early as 1881 when the fabulous Alice Mine & Mill in Butte, Montana attempted to illuminate its passages and buildings with Edison's new lightbulbs. At that time electric technology was brand new and its practical application evaded not only mining engineers, but many industrial engineers as well. During the 1880s visionary inventors demonstrated that electricity worked, enticing mining engineers who dreamt of sending power to hoists, compressors, pumps, and other machinery through slender wires rather than through cumbersome and expensive steam pipes.[72]

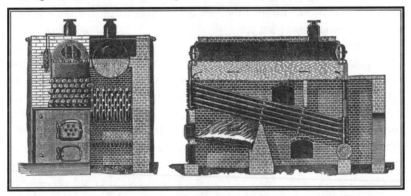

The illustration depicts the type of brickwork typically used for water tube boiler settings. This unit, the classic design manufactured by Babcock & Wilcox, was the most popular water tube boiler in the West.

Greeley, Horace; Case, Leon; Howland, Edward; Gough, John B.; Ripley, Philip; Perkins, E.B.; Lyman, J.B.; Brisbane, Albert; Hall, E.E. 'Babcock and Wilcox Boiler' The Great Industries of the United States J.B. Burr, Hartford, CT, 1872.

During the late 1880s and into the 1890s, a few mining engineers working for profitable and well-capitalized western mines attempted to turn their dream into reality. They made their first attempts to run machinery in locations that featured a combination of water and topographical relief suitable for the generation of hydro-power. In 1888 the Big Bend Mine on California's Feather River experimented with electricity, and the Aspen Mining & Smelting Company, in Colorado, used electricity to run a custom-made electric hoist that served a winze underground. A few years later progressive mining companies in Silverton, Telluride, and Creede attempted to adapt electricity to run machinery and illuminate the mines.

These early electrical applications were experimental and while they offered mixed results, they forwarded the technology. One mining engineer writing about electricity during the early 1890s asserted that:

> "Today the installation of a [electric] plant ceases to be a novelty; and its utility as an illuminator, and a power capable of long-distance transmission, is unquestioned. It is true, it has not yet realized all the hopes and anticipations of its zealous advocates. Serious objections have been raised against it, and many plants have proven failures, yet it has demonstrated its merits that, with better understanding, it cannot fail to work an entire revolution in the industries." [73]

Electric plants continued to be a rarity in the West until the late 1890s when more mining districts attempted to utilize the curious and promising power source.

Several factors excited interest in electrification during this time. The nation's economy and the mining West were recovering from the severe economic depression associated with the Silver Crash of 1893, and mining companies once again had capital. Electrical and mining engineers had made great strides in harnessing electricity for the unique work of mining. The earliest electrical circuits were energized with Direct Current (DC) which had a unidirectional flow. But during the late 1890s mining engineers began experimenting with Alternating Current (AC), which oscillated. Neither power source was particularly well suited for western mining. AC current could be transmitted over a dozen miles with little energy loss, but motors wired to it were incapable of starting or stopping under load. Therefore AC was worthless for running hoists, large shop appliances, and other machines that experienced sudden drag, or that required variable speed. AC electricity was effective, however, for running small air compressors and ventilation fans because they were constant-rotation machines that offered little resistance. DC electricity, on the other hand, had the capacity to start and stop machinery under load as the Aspen Mining & Smelting Company had successfully demonstrated, but the electric current could not be transmitted more than a few miles without suffering debilitating power loss. Therefore DC current had to be used near the point of generation. In addition, DC motors were incapable of running the massive production-class machines mining companies had come to rely on for profitable ore extraction.[74]

In general, electrical technology as it existed during the 1890s offered mining companies little incentive to replace even their small pieces of sinking-class steam equipment. However, enough progressive industrialists and engineers saw the benefits that electricity offered to keep the movement going. In the mid and late 1890s, a few capitalists formed electric companies that wired the accessible and heavily-developed mining districts such as Cripple Creek and Central City, Colorado, Mercur, Utah, and several portions of California's Mother Lode. More electric companies formed in similar districts during and shortly after 1900. These mining districts shared the characteristics of being limited in geographic area, lending themselves to DC power distribution, and they encompassed numerous deep, large, and profitable mines, which constituted a potentially significant consumer base. To further the demand for power in these districts, the electric companies leased motor-driven hoists and compressors to mining operations at discount rates. As a result small operations with little capital installed electric hoists and compressors amid their surface plants. Large mining companies used electric hoists underground to serve winzes, and they equipped their shops with motor-driven power appliances.[75]

Few remote mining districts saw electrification at this time as they lacked the necessary density of potential consumers, and resident mining companies often lacked the capital and will to experiment with electric machinery. A few mine supply houses interested in promoting electricity attempted to sell DC generators, which cost hundreds of dollars, to remote operations. In most cases mining engineers realized that the benefits offered by electricity did not offset a system's price, especially if the steam engine required a boiler which could be used to run conventional mining machinery.

Around 1900 electrical appliance manufacturers made several breakthroughs which accelerated the mining industry's acceptance of the modern power source. Electricians developed the three-phase AC motor, which could start and stop under load while using a current that could be transmitted long distances. The other major breakthrough was the development of practical DC/AC converters, which permitted the use of DC motors on the distribution end of an AC electric line. Thus electricity became an attractive power source to a broad range of electric consumers.

Still, most western mining companies were not yet willing to relinquish steam technology because even the new three-phase AC motors could only drive sinking-class hoists and small compressors. In addition, voltage, amperage, and current had not yet been standardized among machinery manufacturers or the various power grids in the West, discouraging the use of motors for critical mine plant components. Many pragmatic, professionally-educated mining engineers, including T.A. Rickard and Herbert C. Hoover, felt that electricity during the 1900s and 1910s was not ready to replace steam power.[76]

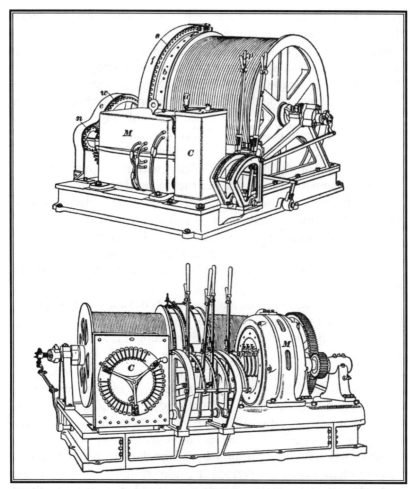

The western mining industry saw the introduction of practical single drum electric hoists in the late 1890s. Similar in form to the geared single drum steam hoist, early electric models were not able to match the performance of steam. While the electric hoists were adequate for sinking, productive mining companies shunned them until the late 1910s when the technology had improved. Both images depict single drum hoists with DC current motors. The box labeled C on the left unit is the controller, which acted like a throttle, and the box labeled M is a motor sealed against dust.

International Textbook Company International Library of Technology: Hoisting, Haulage, Mine Drainage International Textbook Company, Scranton, PA, 1906 A50 p39.

The rigors of mine hoisting proved to be one of the greatest obstacles to the acceptance of electricity. But by the 1900s mining machinery manufacturers had developed a variety of small AC and DC models that were reasonably reliable. The early electric hoists were similar in design to sinking-class geared steam hoists, and were manufactured by mining machinery makers with motors purchased from electrical appliance companies such as General Electric. Most of the hoists consisted of a cable drum, reduction gear shafts, and a motor fixed onto a rectangular bedplate. In many cases the *controller,* which served the same function as a throttle on a steam hoist, was also mounted onto the bedplate. A second popular electric hoist configuration consisted of a cable drum and a gear shaft fastened onto a main bedplate, with the motor bolted onto an extension projecting outward from the side. While electric hoists used bedplates similar to geared steam hoists, their performance rating was less.

A company of lessees at the Los Angeles Mine in Cripple Creek employed an electric hoist around 1900 in hopes of avoiding having to lay down the enormous capital required to install a steam plant. Lessees quickly caught on and began to favor electric equipment for this reason. The man at right is probably the superintendent. The motor is an early AC unit.

Engineering & Mining Journal March 1, 1902.

Though the electric hoists made during the 1900s were able to start and stop under load, they were very slow and had a limited payload capacity. Most of them featured motors rated at seventy-five horsepower or less. The early electric hoists had speeds under 600 feet per minute, payloads less than three tons including the weight of a hoisting vehicle, cable in the shaft, and ore, and their working depths were rated to around 2,500 feet.[77]

By the 1910s, professionally-educated mining engineers could not deny the potential savings from the use of electricity, and the performance of electrical machinery was rapidly approaching that of production-class steam-powered machines. As steam machinery wore out, mining outfits replaced it with electric models. Conversion to electric power occurred quickest in the well-developed mining districts of the Great Basin and Southwest such as Tonopah and Goldfield in Nevada, and Eureka and Park City in Utah, where fuel had to be shipped from distant sources at great expense. One engineer asserted that in well-developed mining districts, a steam-driven compressor cost up to one-hundred dollars per horsepower per year to run while an electric model cost only fifty dollars. The cost savings were probably even greater for hoisting.[78] Electrical converts wisely maintained steam boilers in an operable condition in the event the motor, or the entire electrical grid, failed. Electrical engineers had standardized electric grids and motors in the West's mining districts for either 220 or 440 volts and sixty cycle AC current; other voltages and DC current had fallen out of favor by this time.[79]

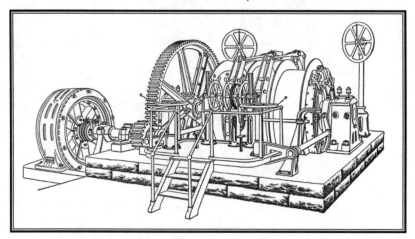

By the late 1910s electric hoists had come a long way in terms of performance and efficiency, and machinery makers began offering viable double drum models, as shown. The arrangement of the components of the illustrated hoist made by Wellman-Seaver-Morgan changed little into the 1930s, when steam had become obsolete. Note the stone masonry foundation for the hoist and the separate motor mount.

International Textbook Company International Library of Technology: Hoisting, Haulage, Mine Drainage International Textbook Company, Scranton, PA, 1906 A50 p40.

Mining machinery makers made the greatest advances with electric hoists during the 1910s. Not only did electrical engineers and machinery makers improve the performance and reliability of single-drum electric hoists, they also introduced effective double-drum units for productive mines interested in achieving economies of scale through balanced hoisting. Engineers began installing these new machines at large mines in well-developed districts including Cripple Creek, Colorado, and Nevada's Goldfield and Comstock. By the mid-1910s electric hoisting had the full attention of professionally-trained mining engineers, who devoted time and energy to improve the technology. Within ten more years, except for remote and poorly capitalized operations, most of the mining West completely adopted electric power for hoisting, as well as for running other types of mining machinery.[80]

During the 1910s, mining engineers developed two basic electric circuits for production-class operations. They could wire their machinery directly to an electrical substation connected to a power grid, or they could first run the hoisting circuit through a rotary converter that had the potential to save electricity and moderate the demand on the system. Large hoists came under great load when beginning movement with a cage, and siphoned a tremendous amount of power from other plant components, resulting in a brownout. In response, electrical engineers introduced rotary converters that played a dual role in the hoist's circuitry.

The converter fed electricity to the hoist when needed, but it allowed the hoist motor to act as a generator when the hoistman shut off the power and used the motor's mechanical drag to lower the hoisting vehicle down the shaft. The electricity generated by the hoist motor went to the converter and powered a motor there that set in motion a large iron flywheel. When the hoistman powered up his machine and raised another load from the depths of the mine, the hoist again drew full current, but the motor in the converter, kept in motion by the flywheel, reversed its role and became a generator that supplemented the power drawn by the hoist.[81]

Rotary converters such as the unit illustrated were designed to moderate the wildly fluctuating drains on the electric circuits serving large mines. However, few western mining outfits installed them because the devices, which occupied an area 6 by 12 feet, were both costly and beneficial only for long-term applications.

Mining & Scientific Press August 13, 1910 p203.

The mining industry used three basic systems of rotary converters. These included the *Lahmeyer,* the *Siemans-Ilgner,* and the *Westinghouse,* the latter of which was by far the most popular. Rotary converters were capital intensive, and because electricity by nature was inexpensive, few mining companies saw the necessity of installing such machinery. However, a few heavily electrified mining companies operating their own generators found converters to be economical, and as one converter was able to serve several mines at once, some electric companies also found them to be cost effective to wire into their grids. But in general, rotary converters saw little application in the West.

While hoisting machinery underwent significant transformation from the 1870s to the 1920s, the headframe changed little in form and function during this time. The types of timber headframes designed and erected by professionally-trained engineers were truly captivating industrial structures. Tall, well-braced, and made with heavy timbers, headframes had a look of strength, longevity, and permanence, which was an image many mining companies tacitly strove to convey.

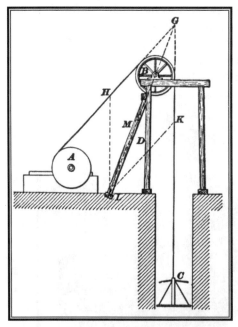

The diagram illustrates the geometry many mining engineers attempted to apply when designing production-class headframes. By locating the backbraces M between the shaft C and hoist A, the sum of horizontal and vertical forces were distributed equally among all of the headframe's members. Note that angles H and K, and G and L, are the same.

International Textbook Company A Textbook on Metal Mining: Steam and Steam-Boilers, Steam Engines, Air and Air Compression, Hydromechanics and Pumping, Mine Haulage, Hoisting and Hoisting Appliances, Percussive and Rotary Boring International Textbook Company, Scranton, PA, 1899 A23 p105.

Designing a production-class headframe can be viewed as an art, because the mining engineer had to balance significant stresses with a multitude of functions and the structure's interaction with other hoisting system components. Western mining engineers have been criticized for overbuilding their headframes and wasting capital. However, the stresses they had to consider in their designs were many. They had to build a structure capable of withstanding vertical forces including an immense *dead load,* which was the static weight of the structure, *live load* which was the weight of a loaded hoisting vehicle and cable paid into the shaft, and *braking load* generated when the hoistman brought to a halt the descent of heavy machines and supplies sent underground. Engineers had to calculate horizontal forces including the powerful pull of the hoist when raising a heavy load, and *wind-shear,* which could not have been underestimated in the rugged West. Last, mining engineers had to plan for *racking* and *swaying* under loads, *vibration,* and *shocks* to the structure.[82]

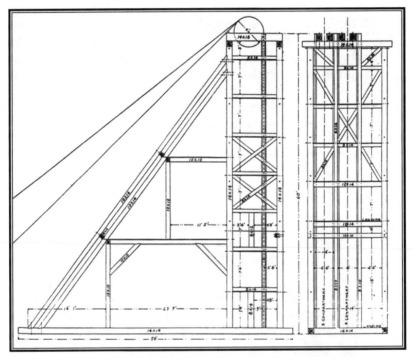

The line drawing depicts a classic production-class four-post derrick headframe. The structure stands 60 feet tall, it is well-braced, and features two sets of sheave wheels and guide rails for balanced hoisting with cages. This headframe belongs to the Portland Shaft No.2 and can be seen looming over Victor, Colorado today.

Forsyth, Alexander 'The Headframes of Shafts at Cripple Creek' Engineering & Mining Journal
March 7, 1903 p366.

Yet, building a headframe that could withstand the sum of the above forces was still not enough for service at a producing mine. Engineers had to forecast how they thought the headframe would interact with the mine's production goals, and how it would interface with the rest of the hoisting system. The depth of the shaft, the speed of the hoist, and the rail system at the mine directly influenced the height of the structure. Deep shafts served by fast hoists, such as direct-drive steam units, required tall headframes, usually higher than fifty feet, to allow the hoistman plenty of room to stop the hoisting vehicle before it slammed into the sheave at top. Highly productive mining operations often utilized vertical space on their claims, which required multiple shaft landings. For example, the Joe Dandy, Nichol, and Vindicator mines in Cripple Creek all featured landings at the shaft collar to permit topmen to push ore cars laden with waste rock out to the dump. The resident engineers designed second landings in the headframe where topmen pushed cars loaded with payrock off the cage and across a high trestle into ore sorting houses. Some mines using skips as hoist vehicles had rock pockets built into the headframe, and this also required height. The headframe had to be tall enough to permit the hoistman to raise a skip to a point well above the rock pocket where a special guide track upset the vehicle, emptying the rock into the bin.[83]

While most engineers had the skills to erect structures that proved lasting and functional under the conditions of western min-

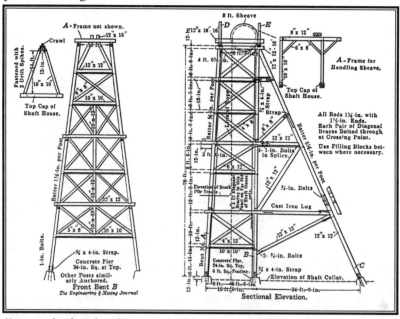

Many production-class four-post derrick headframes were built with legs set at a batter, meaning they leaned slightly inward, to counter the sum of horizontal and vertical forces. According to the drawing, the headframe, which stood over the Hiawatha Mine in Michigan, was 76 feet high and enclosed in a massive shaft house. Note the small gallows frames at top that workers used to remove the sheaves for servicing.

Botsford, H.L. 'A Timber Headframe' Engineering & Mining Journal June 10, 1911 p1148.

ing, professionally-trained engineers created headframes that possessed the greatest economical value and grace. Four basic designs met the rigors of heavy ore production. These included the *four-post derrick,* the *six-post derrick,* an *A-frame* known also as the *California frame,* and a heavily-braced two-post structure known also as the *Montana type.* As the names suggest, engineers working in specific regions in the West favored certain headframe designs. While the above structures were intended to serve vertical shafts, two-post gallows headframes and a variety of A-frame up to thirty-five feet high were also erected to serve inclined shafts.[84]

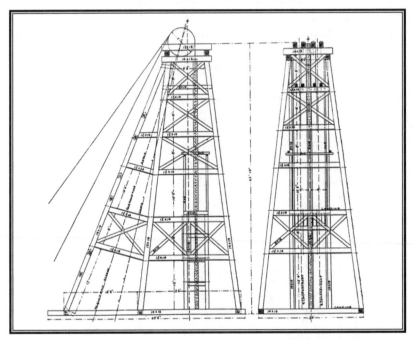

The illustration depicts another form of four-post derrick in which the posts have been set at an exaggerated batter. This structure stood over the fabulous Gold Coin Shaft in Cripple Creek.

Forsyth, Alexander 'The Headframes of Shafts at Cripple Creek' Engineering & Mining Journal
March 7, 1903 p366.

To meet the combination of horizontal and vertical forces and the performance needs of ore-producing mines, nearly all mining engineers in the West built their headframes with heavy timber beams assembled with mortise-and-tenon joints, timber bolts, and iron tie rods. In general they used ten-by-ten inch posts for headframes up to around forty feet high, twelve-by-twelve to eighteen-by-eighteen inch posts for headframes up to sixty feet high, and up to twelve-by-twenty-four inch timbers for large two-post headframes. Mining engineers attempted to allocate full-length uncut timbers for the posts and back-braces because of the solidity they offered. Skilled carpenters assembled the materials into towers that featured cross members and diagonal bracing spaced every six to ten feet. Solid, relatively clear sixteen-by-sixteen inch timbers sixty to seventy feet long are almost unimaginable commodities today, yet they were standard materials. All four and six-post headframes featured stout backbraces anchored between the shaft and the hoist, and the entire structure stood on foundation footers straddling the shaft. The posts on A-frames were set at an exaggerated batter, meaning they splayed out to absorb all of the vertical and horizontal stresses, and as a result A-frames rarely had backbraces. Four and six-post headframes were much more common in the West than A-frames, though more costly to build, because these vertical structures were within the technical means of most engineers. A-frames, by contrast, required a greater knowledge of mechanics and physics, and were harder to build.[85]

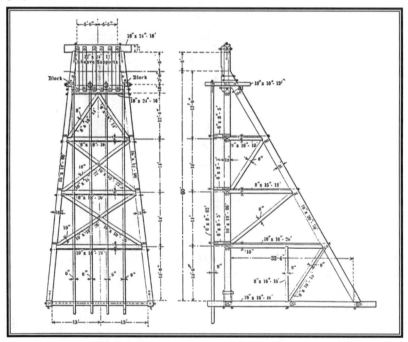

The Montana type headframe can be described as being a large heavily braced version of the two-post gallows headframe. The structure illustrated here belonged to the Goldfield Consolidated Mines Company in Goldfield, Nevada, and it stood 55 feet high. The Montana design, which incorporated less timber than other forms of headframes, was well-suited for productive mines in remote districts where the cost of materials was high.

Barbour, Percy E. 'Details of a Wooden Headframe' Engineering & Mining Journal Aug. 19, 1911 p344.

Mining engineers determined that production-class headframes, which weighed dozens of tons, required a sound and substantial foundation in order to remain stable. When an engineer erected a production-class surface plant from scratch he simply put a crew to work clearing soil to bedrock around the proposed shaft, on which the crew built a timber framework for the headframe. The engineer who inherited a semi-developed prospect shaft had significant and expensive work ahead of him, because the previous operation may have left a large, unconsolidated waste rock dump that workers had to clear away to expose bedrock.

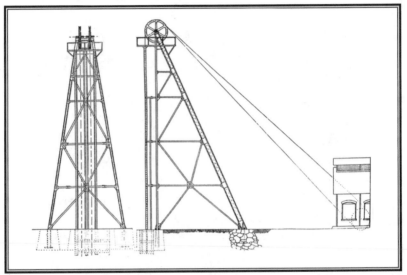

The superior strength of steel permitted engineers to design headframes that incorporated less bracing. Most steel headframes were patterned after the A-frame, or after the Montana design, such as the structure in the drawing. Because of their great weight, steel headframes required strong foundations, as shown.

Engineers used one of three foundations to support production-class headframes. The first consisted of a squat timber cube featuring bottom sills, timber posts, and caps bolted over the posts. Construction workers stuffed logs, timber blocks, and boulders under the bottom sills to level them when the foundation was built. Other foundations included a group of hewn log cribbing cells assembled with saddle-notches fastened with forged iron spikes, and a hewn log or timber latticework consisting of open cubes between four and six feet high, capped with dimension timbers. When any of the three foundation systems were built, workers sided the shaft walls with plank lagging and shaft-set timbering, and filled the surrounding foundation framework with waste rock as quickly as the miners generated it. The last two foundation systems were the most popular in the West because the raw materials were plentiful, and they did not require exact engineering. The problem with all of the above systems was that the wood rotted when covered with waste rock, especially when the rock was highly mineralized. A few progressive engineers attempted to substitute concrete for wood to create a lasting foundation, but only well-financed companies anticipating lengthy operation erected such foundations.[86]

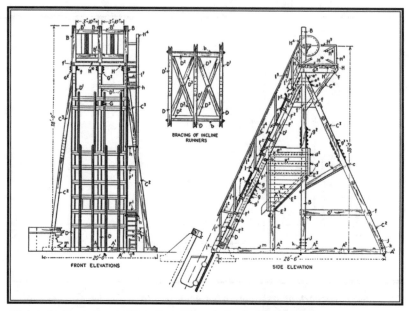

Mining engineers favored erecting A-frames for inclined shafts. Because inclined shafts were favorable for the use of a skip, A-frames often featured rock pockets to accept the loads disgorged by the vehicles when emptied.
Ketchum, Milo S. C.E. The Design of Mine Structures McGraw-Hill Book Co., New York, NY, 1912 p227.

All headframes were crowned with enormous pulleys known as sheaves that guided the hoist cable down into the shaft. Mining engineers took very seriously the type, size, and placement of the sheave because it was the focal point where the tremendous vertical forces met the horizontal force of the hoist's powerful pull. Both the sheave and its timber braces had to be well built and properly designed not only to resist the forces put upon it, but also to work with the other components of the hoisting system. In the first several decades of hardrock mining in the West, companies had to seek out reputable foundries willing to custom make these specialized pieces of mine hardware. But in later decades mine supply houses offered what amounted to factory-made kits that included the sheave, a high-quality heavy steel axle, and lead-lined bearings.

The four-post derrick head-frame at Cripple Creek's Joe Dandy Mine featured a second landing where miners pulled loaded ore cars off the cage, pushed them over a trestle, and emptied the cars in the ore sorting house at left. This arrangement was common among productive mines in the West.
Author.

Much calculation went into determining what kind of sheave to place atop a headframe. Mining engineers had to consider the diameter of the hoist cable, the distance from the shaft to the hoist, the type of hoist installed at the mine, the height of the headframe, and the weight of the load. First they had to select the proper diameter for the wheel. Professionally-trained engineers chose diameters of three, four, six, and eight feet depending on whether the hoist cable was three-quarter, one, or one-and-one-half inches in diameter, respectively. The reason for large-diameter sheaves with thick hoist cables was that when the cable arced over the wheel, its top strands came under considerable tension and those closest to the wheel were highly compressed. This tension-compression relationship wore out hoist cables, causing the wire strands to fatigue and break. In addition, when a hoist had been anchored close to the shaft, the cable had to wrap around much of the sheave, straining a significant portion of the cable. Large-diameter sheaves minimized the stresses of arcing and prolonged the lives of hoist cables. Self-taught mining engineers who might have been less sensitive to the physics and mechanics involved may have opted for a small diameter sheave in hopes of saving money.[87]

The photograph depicts, in profile, the initial stages of constructing a headframe foundation. The headframe will stand atop the framework, and the diagonal braces at right countered the pull of the hoist, which is out of view. The shaft is located underneath the foundation's left portion. Eventually the foundation will be buried by waste rock.

Mining & Scientific Press Feb. 6, 1904 cover.

Engineers could purchase a cast-iron sheave or one with numerous steel spokes that had been assembled like a bicycle wheel. Because of its greater strength and light weight, most engineers used the latter form of sheave in conjunction with large cables. The groove in the sheave had to be deep to prevent the cable from jumping out, but it had to be wide enough to permit the slight side-to-side motion of the cable. This movement, known as *fleet angle,* occurred when the hoist wound the cable onto the drum, where it crossed from one side to the other and back again. Sheaves for flat hoist cable were broad, and they did not have to make allowances for fleet angle since the cable reel on the hoist did not permit side-to-side motion.

After much use, the lead linings in the heavy sheave bearings wore out and required replacement. This meant that a small crew of workers had to scale a ladder to the top of the headframe, pull the sheave out of the bearings, and reline them. To accommodate this dangerous undertaking, the engineer usually designed a platform atop the structure to provide the workers with footing, and a small A-frame beam to attach a block-and-tackle for lifting the sheave, which could have weighed over a thousand pounds.

In the 1890s professionally-trained mining engineers working for the West's wealthiest and largest mining companies began experimenting with steel girders for headframes. In their view, steel was the ultimate building material for production-class operations. One turn-of-the-century mining engineering text summed up this sentiment:

> *"Wherever permanency of headframes is required, if steel is obtainable at a price at all comparable with wood, steel structures are being used, as timber frames rot."* [88]

The foundation for the two-post gallows headframe standing over the Locan Shaft in Eureka, Nevada consisted of four elegant pylons of faced breccia blocks. Even though the headframe was originally covered by a shaft house, the foundation made a statement to the mining industry that the mining company possessed capital and ore.

Author.

Professionally-educated engineers felt that steel was superior to wood because it did not decay, it was much stronger, it was non flammable, and it facilitated the erection of taller headframes. However, we can infer from the above quote that steel was significantly more expensive than timber, especially in the West where the distances from steel manufacturing centers were vast. The engineer at the Gwin Mine, located in California's Mother Lode, had a one hundred foot high steel A-frame put up around 1900 which cost approximately $13,300. In 1911 the Tonopah-Belmont Mining Company erected a steel two-post gallows frame for $4,210. By comparison a thirty-five foot high four-post derrick made of timbers cost $900, and a timber A-frame of similar height cost around $675. As a result of the marked price differences, only heavily capitalized and highly productive mines in the West put up steel structures. Well capitalized mining companies, engineers, and promoters appreciated steel headframes for their superior performance and the statement of wealth and permanence they conveyed to the mining world.[89]

Carpenters pose after completing a massive two-post gallows headframe in the eastern portion of the Cripple Creek district around 1900. The headframe stands at one end of a timber foundation, which extends to the left where the hoist has been anchored. Their work is not finished, for the piles of lumber indicate they have yet to put up a large wooden shaft house to ward off the Rocky Mountain winters.

Mining & Scientific Press
Feb. 23, 1903 cover.

The design and erection of steel headframes were beyond the means of most mining engineers during the Gilded Age. Mine supply companies, including Hendrie & Bolthoff, Wellman-Seaver-Morgan Company, and Stearns-Roger, usually contracted with the interested mining company to design, prefabricate, ship, and erect the impressive structures. Most designs built during this time were based on the A-frame, while a few others included variations of the two-post Montana type. Girders usually consisted of two steel channels laced together with iron rods, and the structure had to be riveted together. The use of steel allowed headframes to soar to dizzying heights of between 100 to 180 feet. Undoubtedly mine workers dreaded having to climb a ladder up such a high structure to service the sheave bearings at top.[90]

Steel headframes saw increasing popularity through the 1910s and 1920s. Yet, because of their high cost and the overriding conventionality of using wood, they never attained a commonplace status in any mining district. In fact, steel headframes were a novelty, a feat of engineering, and a statement of wealth that commanded attention whenever they were built. Ironically they greatly outlived their designers, their builders, and the mines over which they loomed by over a century.

By nature production-class hoisting systems consisted of substantial components intended for long-term use. Like most of the facilities of historic mine surface plants, the machines and structures comprising hoisting systems succumbed to predation by neighboring mining outfits, creditors, and war-related scrap drives. Yet, they left lasting evidence for today's visitors to investigate.

Hoist foundations are perhaps the most elemental of the remains available to visitors, and they are ubiquitous among productive shaft mine sites. The visitor can analyze foundation shapes, dimensions, and methods of construction to ascertain the type, size, and age of a hoist. The large single cylinder steam hoists possessed a distinct footprint. These machines consisted of a large horizontal steam engine connected to a cable drum that rotated inside of a heavy timber frame. The mining engineer usually specified that the steam engine had to be bolted onto a brick or rock masonry foundation for stability. The engine foundations for medium-sized hoists tended to be from eight to twelve feet long, three feet wide, and oriented so that the long-axis pointed toward the shaft. On medium-sized hoists, the associated drum frame was bolted onto a timber foundation situated immediately adjacent to one end of the engine mount. On large versions of this type of hoist, the drum and bull-gear bearings were often bolted onto heavy stone pylons instead of timbers.

The single-drum duplex geared hoists, common after 1880, sat on foundations that were simple and can be easily identified today. The foundation possessed a square or slightly rectangular footprint, and it was studded around the perimeter with anchor bolts. Small and medium-sized single-drum-geared hoists usually were affixed onto a totally flat foundation, while the foundations for large models, nine-by-nine feet and greater, consisted of a raised rim that encompassed a well for the cable drum. Prior to the 1890s mining engineers usually had workers construct foundations of rock or brick masonry, and afterward they had their crews use concrete. As with air compressor foundations, hoist foundations predating the late 1910s tended to consist of natural concrete, while later hoist foundations were made of portland concrete.

   The physical constitution and the characteristic footprint for single-drum first-motion hoists are different than those for single-drum geared hoists. These machines, rarely larger than fourteen-by-nineteen feet, consisted of two long horizontal steam engines flanking a big cable drum. The overall foundation plan was that of a rectangle oriented with the long axis pointing toward the shaft. Power-assist steam cylinders and control linkages were located between the engines, with the hoistman's platform at the rear. By nature these machines required complex foundations which generally consisted of two parallel raised masonry blocks for each engine flanking a well for the cable drum. The floor of the drum well may have featured small anchor bolts to fasten heavy-duty brake posts, and the rear of the foundation may have featured steam pipes projecting out of the ground. Because of the great stresses the hoist endured, mining engineers usually had construction workers build the foundations out of stone masonry because it was stronger than concrete.

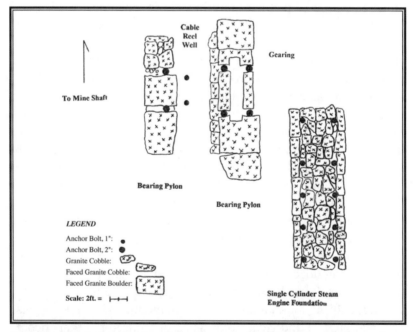

The plan view depicts the foundation for a single cylinder steam hoist at the Lander Shaft in Austin, Nevada. The foundation block at right anchored a large horizontal steam engine, and the other two blocks anchored drum bearings. The narrow space between the bearing pylons suggests that the hoist featured a reel for flat cable. Smaller single cylinder foundations were similar in footprint; only mining engineers substituted timbers for the rock masonry pylons.

Author.

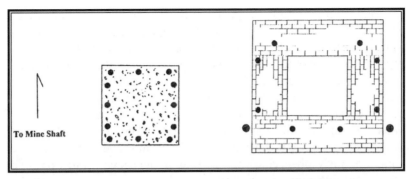

The illustration depicts the footprints typical of single drum geared hoists. The left foundation anchored a hoist at the Graphic Mine in New Mexico and the right foundation anchored a hoist at the Tornado Shaft in Cripple Creek. The well in the center of the right foundation provided clearance for a large cable drum.

Author.

The geometry, great weight, and thrust of first-motion single drum hoists necessitated that engineers design substantial foundations that firmly supported the steam cylinders and the cable drum. Engineers preferred that the foundations consist of a rock masonry footer and two masonry pylons capped with dressed stone blocks, as shown in the illustration. Compare this foundation, recorded at the Yankee Girl Mine in Colorado's Red Mountain Mining District with the hoist example several pages back.

Author.

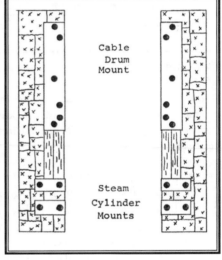

Like the above types of single-drum steam hoists, the foundations for double-drum steam hoists also were distinct in shape and size, and they are usually readily apparent to today's visitors. Geared double-drum hoists required foundations that had either totally flat-top surfaces, or raised rims encompassing wells for the drums. The sizes of the foundations for these relatively compact hoists typically ranged from around six-by-twelve feet to ten-by-sixteen feet. Because the drums had been mounted onto the bedplate side-by-side with each other, the foundations were oriented with the long axis ninety degrees to the shaft. Mining engineers had their work crews built such foundations with masonry, although they also used concrete in some cases.

First-motion double-drum steam hoists were complex and powerful machines that had to be assembled onto special foundations that are unmistakable at today's historic mine sites. These behemoths were powered by massive steam engines that flanked dual cable drums located at the front of the machine. The pit between the steam cylinders near the rear of the foundation enclosed small power-assist steam cylinders for the clutch and brake, while the elevated hoistman's platform stood behind. From this position, the hoistman could view the workings of the massive machine while simultaneously observing the hoisting vehicle level indicators and the action at the headframe.

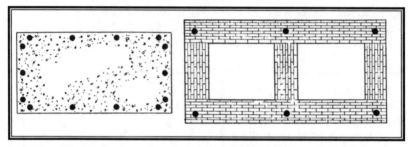

The illustration depicts two foundations for double drum geared hoists. The left foundation anchored a relatively small hoist, while the right foundation featured wells that provided clearance for large-diameter cable drums. In some cases mine workers used dressed stone blocks instead of brick or concrete. Compare these foundations with the illustration of the double drum geared hoist above.

Author.

The foundation was an enlarged version of the masonry that supported direct-drive single-drum hoists. The massive steam engines and drive rods were anchored onto two long masonry footings, and the drums rotated over depressed wells located between the engines. These hoist foundations also featured a pylon erected between the cable drums to support the drum axle's center. Masons built rock or brick footings and capped them with dressed sandstone or granite blocks intended to bear the weight of the machine's primary components. Mining engineers very rarely used concrete for the primary portions of the foundation because it was not lasting. The drum wells for hoists equipped with round cable were at least five feet wide while wells that accommodated flat cable reels were three to four feet wide, and deep. The overall sizes of direct-drive double-drum hoist foundations were at least twenty-by-twenty feet in area.[91]

Today's visitors may encounter foundations for electric hoists at mines active during the Gilded Age. The foundations for these machines will appear similar in shape, size, and construction to those for single-drum geared steam hoists, presenting the visitor with the potential dilemma of sure identification. The visitor may be able to infer that a mine used an electric hoist based on the absence of a boiler foundation, the lack of fuel residue, and the presence of tell-tale electrical artifacts. Often when a salvage crew removed electrical machinery and dismantled the enclosing structure, they left behind porcelain insulators, switch panel fragments, electrical wiring, and motor parts. Construction crews rigging electrical wiring often utilized *knob insulators* to pin exposed wires down to walls. *Tube insulators,* which were hollow, provided a safe passage for wiring through walls and framing studs. Many electrical companies used *pony* and

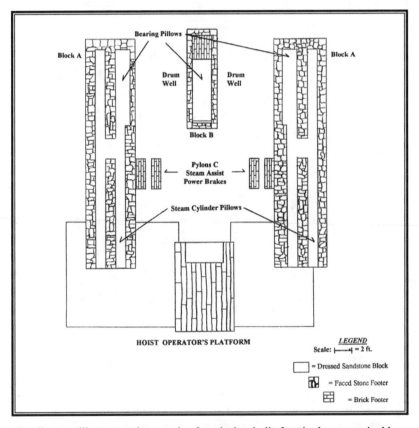

The diagram illustrates the complex foundation built for the immense double drum first-motion hoist at W.S. Stratton's American Eagle Mine in Cripple Creek. The foundation's composition, reflective of similar foundations elsewhere in the West, is complex and consists of a heavy rock footer supporting faced rock masonry pylons. The pylons have been capped with huge dressed sandstone blocks known as pillows, which bore the brunt of the hoist's weight and horizontal forces.

Author.

two-piece transposition insulators, typically made of glass and screwed to telephone poles, for minor power distribution circuits to mines. The wiring for the electrical circuit converged at the switch panel, which usually was a sheet of marble, slate, or Bakelite. Switch panels were often at least one-by-one feet in area, and they featured screw holes for porcelain switches, insulators, and fuses. The motor parts that mine workers may have left include flat copper or steel disks with an appearance similar to a circular comb, and motor brushes which were graphite blocks with attached wires.[92]

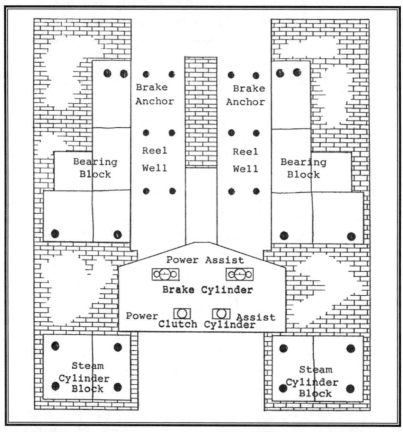

The foundations for first-motion flat cable hoists differed little from the foundations for conventional double drum hoists. Compare this foundation, recorded at Cripple Creek's fabulous Gold Coin Mine, with the flat cable hoist illustrated several pages above. The foundation consisted of a broad rock masonry footer capped with brick masonry blocks. The brickwork supported large dressed sandstone pillows held in place by 2 inch diameter anchor bolts. The power-assist clutch and brake steam cylinders were bolted onto a cast iron plate laid over the masonry. The hoistman operated the mighty machine from a raised cast iron platform.

Author.

Visitors to historic mine sites may encounter piles of red bricks and firebricks denoting the location of where the mine boilers had been. While the setting walls may no longer stand, the footprint of the brick structure often manifests as truncated brick walls amid the chaos of loose bricks. The setting remnant may feature four to six buckstaves, and the bridge wall, constructed of brick, may be apparent bisecting the setting interior. The end that featured the firebox is typically open and lined with slag-encrusted firebricks and may have horizontal anchor bolts for the cast-iron façade. Remote mining operations occasionally substituted local stone for some of the bricks in the setting.

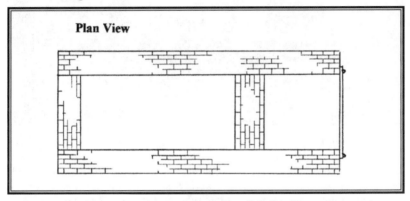

Even though salvage operations may have demolished boiler settings and removed most of the iron, often the telltale footprint of the setting walls and bridgewall are visible today, as shown in the illustration.

Author.

By analyzing the remains of boiler settings, today's visitors can determine the number and sizes of the return-tube units once employed by a mining company. Convention dictated that the construction crew build the setting walls eighteen inches to two feet thick. Therefore, the void remaining between the setting walls roughly corresponds to the diameter and length of the boiler shell.

At some historic mine sites the masonry settings for boilers have been disturbed to the point where the interior space is unmeasurable, leaving the visitor to determine the boiler size based on the setting's exterior dimensions. When the total thickness of the setting walls are subtracted from the total dimensions, the resultant area should reveal the boiler shell's approximate size. Where masons erected settings for two or more return-tube boilers, they usually built partition walls eighteen inches thick with an additional air gap up to four inches wide. By comparing the boiler shell size to the table provided in the appendix, the visitor can determine the horsepower for a single unit. By adding up the total number of boilers at a mine site, the visitor can determine total boiler horsepower, from which they can infer the type of machinery the mine featured.

Salvage operations often left behind boiler hardware that today's visitor may encounter at historic mine sites. The photo illustrates the smokestack manifold that had been affixed onto the end of a return tube boiler. The smokestack rested on the round flange currently pointing right, and the hatch provided access for swabbing soot out of the flue tubes in the shell. The tape measure has been extended three feet for scale.

Author.

Water-tube boilers at historic mines rarely escaped at least partial destruction, yet their remains are often distinct from the crumbled settings of return-tube boilers. The brickwork for water-tube boilers usually consisted of four brick walls mortared with cement. The brick structure was often at least twenty-by-twenty feet in area and surrounded by a girder framework that supported the heavy tanks and water-tubes. The visitor may be able to identify the firebox based on the existence of a slag-encrusted firebrick masonry lining, and the presence of ash pit doorways at the foot of the setting front. In addition, a masonry bridge wall, built to force the flue gases to rise through the water tubes, may be visible immediately behind the firebox. The visitor to a historic mine site may note other qualities that distinguish a water tube boiler setting from other brick structures. The front, sides, and rear of the setting may have been breached by large arched cleaning ports with cast-iron doors, and heavy iron pipes for heating water may be visible inside the interior. Last, the brick walls required support in the form of masonry bolts and washers, which may have been left amid the pile of rubble following demolition.[93]

The large tube union joined the manifolds of two return tube boilers at the Yankee Girl Mine so they could have shared one common smokestack. The tape measure has been extended three feet for scale.

Author.

The last significant component of a hoisting system that visitors are likely to encounter today consists of the headframe. Mining engineers had so successfully designed production-class headframes and their carpenters so skillfully carried out construction that examples of these ponderous structures unique to mining have persevered in sound condition into modern times. Today, however, standing headframes are relatively rare because most of them have succumbed to fire, demolition, and salvage efforts.

Today the visitor to a historic mine site can encounter any one of a variety of manifestations left by a headframe. Aside from a standing structure, the next best circumstance a visitor can hope for at a mine site would be the wreckage of a collapsed headframe. By analyzing the jumble of timbers and the foundation footers the visitor can determine the size and type of headframe. The foundation footers will usually have mortise-and-tenon sockets or square-notched joints denoting where the posts stood. If the foundation exhibits joints at all four corners of the shaft, then the structure probably was a four-post derrick. Because the legs for A-frames were set at an exaggerated batter, the footers they stood on were spaced some distance from the shaft. In keeping with mining engineers' conventions regarding headframe construction, the visitor might examine the midpoint between the shaft and hoist foundation for the locations of diagonal braces.

Water tube boilers were irresistible to salvage operations interested in scrap iron and bricks, and as a result most of these complex steam generators have been reduced to ruins, such as the remains of the boiler at the Locan Shaft in Eureka, Nevada. The remains portrayed in the photograph are fairly typical of what today's visitor may encounter, including a portion of the brick setting and members of the steel framework.

Author.

The visitor should inspect either the standing headframe, if the mine is so blessed, or the structural remnants for hardware that can lend greater detail to interpretation. The existence of four-by-four inch hardwood guide rails for the cage or skip hoisting vehicle constitute a primary feature of production-class headframes. The guide rails should feature countersunk lag bolts fastening them onto timber cross-members, and they may feature grease stains and wear from the hoisting vehicle's steel guides. If the headframe features a small ore bin known as a *rock pocket,* and curved guide rails built into the framing, then the mining company employed a skip. Otherwise the outfit used the traditional cage for moving the materials of mining.

Wealthy mining companies concerned for the safety of their miners, a rarity during the Gilded Age, occasionally installed a special safety catch on the end of the hoist cable that was designed to release the cage in the event an inattentive hoistman accidentally ran the vehicle up into the sheave. The *Akron Detaching Hook* was the most common variety of safety catch, and it is easily distinguishable amid a headframe's structural elements. The device consisted of a socket much like a truncated bell held fast by heavy cross members high up in the headframe. In the event the cage had been run into the headframe, the hook clutching the cage entered the bell, which caused it to release the cage. If the cage had been properly maintained, the sudden release of the cable would have activated the safety dogs, which were supposed to grip the guide rails and hold the cage, loaded with terrified miners, in place.

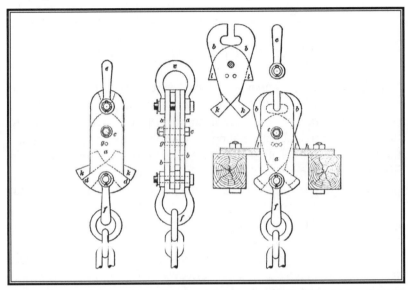

Well-capitalized mines run by competent engineers occasionally equipped their headframes with detaching hooks designed to release the cage in the event the hoistman ran the vehicle up into the headframe. In the diagram at left the hook a clutches both the cage's ring f and the hoist cable's shackle e, and it features two ears k at bottom. In the diagram at right, the hook has been accidentally forced into a special iron socket in the headframe h, and the action forced the ears to squeeze together, opening up the hook and releasing the hoist cable's shackle, which it had been gripping in the diagram at left.

International Textbook Company International Library of Technology: Hoisting, Haulage, Mine Drainage International Textbook Company, Scranton, PA, 1906 A53 p46.

The visitor inspecting headframes might take note of whether the structure includes several landings for removing ore cars at different levels, and whether the headframe had mounts for ventilation ducting and compressed air plumbing. Further, production-class headframes often feature knob insulators for electric power, phone lines, and electric signal bells wired down into the mine. Mining companies usually maintained a manual signal bell communication system which incorporated unique pieces of hardware fastened onto the headframe. The signal bell linkage is the most likely remnant to have been left, and it typically consisted of a sheet iron bracket with two legs that form a 90° angle, forged by the mine blacksmith. Each leg of the bracket featured holes for attaching pull cords, one of which went down the shaft and the other into the hoist house. When a miner down in the shaft pulled the cord, it rang a small bell screwed onto the headframe, and transferred the motion to the cord attached to another bell in the hoist house.

Western Mining Architecture

O nce a mining company had proven the existence of ore, the investors, who often had influence over management policy, fully expected the operation to perform throughout the year, until the ore had been exhausted. Attempting to comply with company wishes, mining engineers responded by erecting structures that sheltered important components of the surface plant against the blistering summer sun and arctic winter winds. Engineers understood that the buildings served two purposes: providing a tolerable work environment, and sheltering sensitive plant components. The engineer and the mining company had a tacit understanding that the buildings could also inspire investors and prominent figures in the mining industry. Large, well-built, and stately structures conveyed a feeling of permanence, wealth, and industrial might while small and poorly constructed buildings aroused little interest from investors and promoters.

Building materials, architectural styles, and structural layouts for mine buildings in the West changed between the 1870s and the 1920s. Perhaps small mining outfits in remote areas realized the greatest gain from changes in conventional construction practices as the expanding network of roads and railroads in the mining West reduced the costs of building materials. Regardless of a mine's location, the buildings erected by well-financed, profitable, and large mining companies tended to be substantial, while the buildings belonging to poorly funded and limited mining companies were crude, small, and rough.

Professionally-trained mining engineers considered four basic costs that influenced the type of buildings they chose to erect. Time had to be spent designing the structure. Basic construction materials had to be purchased and some items fabricated. The materials had to be hauled to the site, and the mining company had to pay a crew to build the structure. Between the 1870s and around 1900 nearly all mining engineers in the West attempted to meet the above considerations by directing their carpentry crews to build wood frame structures and side them with dimension lumber. In a few cases small and poorly funded operations working deep in the mountains substituted hewn logs, but they understood that the log structures were intended to be impermanent, either to be replaced by dimension lumber should the mine prove a bonanza or totally abandoned should the mine go bust.[94]

The introduction of steel and iron building materials to the western mining industry in the 1890s changed the structures erected by mining companies. A number of steel makers began selling iron siding for general commercial and residential construction nationwide in the 1890s. Much of the siding was decorative, and a few varieties were designed with industrial applications in mind. One of these types, corrugated sheet iron, found favor with the western mining industry and its use spread quickly. Engineers increasingly made use of the material through the 1900s, and by the 1910s sided all types of mine and many commercial buildings in the West. The advantages of corrugated sheet iron were that it cost little, its light weight made it inexpensive to ship, it covered a substantial area of an unfinished wall, the corrugations gave the sheet rigidity, and it was easy to work with. These qualities made corrugated sheet iron an ideal building material, especially in the Great Basin and Southwest, where remoteness rendered lumber a costly commodity.[95]

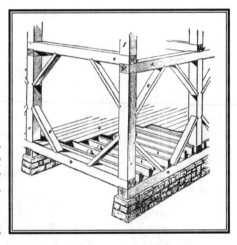

Engineers and architects used square-set timber skeletons like the unit shown here to support large and spacious shaft houses, compressor houses, and mill buildings. The square-set cells ranged from 10 by 10 by 10 feet to twice that size.

International Textbook Company A Textbook on Metal Mining: Preliminary Operations at Metal Mines, Metal Mining, Surface Arrangements at Metal Mines, Ore Dressing and Milling International Textbook Company, Scranton, PA, 1899 A42 p46.

The other significant use of steel in mine buildings occurred during the 1890s when a few prominent western mines began to experiment with the use of girders for large buildings such as shaft houses, paralleling the rise in the construction of steel headframes. Architects began using steel framing to support commercial and industrial brick and stone masonry buildings as early as the mid-1880s, but western mining companies found that wood framing met their needs as well and for less money. By the 1890s architectural steel-work improved, and steel makers offered light-weight beams which a few mining engineers adapted to the framing of huge shaft houses. However, taking shelter in a steel building during violent lightning storms undoubtedly caused a stoic concern among otherwise hardened miners.[96]

Even though steel framing became slightly more common during the 1900s, its high costs rendered it beyond the means of all but large, wealthy mines located in well-developed districts. For

A construction crew stands for the camera while erecting a grand wooden shaft house at Leadville's Wolftone Mine around 1900. The carpenters have clearly built a square-set support structure for the building, and the crew appears to be standing on what will become a second shaft landing. The headframe is a four-post derrick, and the hoist is located to the right and out of sight.

Mining & Scientific Press Feb. 23, 1903 cover.

example, among the hundreds of well-capitalized mines in the Cripple Creek Mining District, in 1900 the Hull City Mine alone featured a steel-framed shaft house. In 1909 the Vindicator Mine became the second operation to erect a steel shaft house, followed by the famous Golden Cycle Mine. In fact, steel mine buildings were such a novelty that Alexander Forsyth granted the structure enclosing the Hull City shaft special recognition in his article "The Headframes of Shafts at Cripple Creek" printed in the *Engineering & Mining Journal* in 1903. In general, during the Gilded Age shaft houses and other structures at most western mines rarely attained the sizes necessary to make the use of steel framing cost effective.[97]

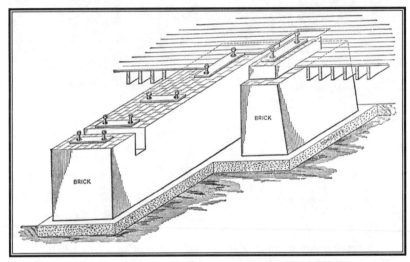

The block diagram illustrates how carpenters built flooring flush with and around machine foundations. Well-built production-class shaft houses featured such flooring, while the buildings at poorly capitalized operations featured earthen floors liberally sprinkled with crushed waste rock and forge clinker.

Hawkins, N. New Catechism of the Steam Engine: With Chapters on Gas, Oil, and Hot Air Engines Theo. Audel & Co. Publishers, New York, NY, 1900 p32.

The general forms, types, and layouts of the structures at shaft operations followed a few general patterns, regardless of the building materials the mining engineer used for construction. Between the 1870s and 1890s most mining engineers enclosed the primary surface plant components, usually clustered around the shaft, in an all-encompassing *shaft house*. These buildings contained the hoisting system, boilers, the shop, the shaft collar, and a workspace all under one roof. The buildings therefore tended to be unmistakable edifices in a mining district. Relatively small shaft houses in the West were constructed of stout post-and-girt frame walls, gabled rafter roofs, and informal or no foundations. Particularly spacious shaft houses required a square-set timber skeleton capable of supporting the roof independent of the walls. Regardless of the type of frame, carpenters clad the walls with board-and-batten siding or several layers of boards, and they used shakes for roofing material. Curiously, during the 1870s and 1880s, mining companies built mill buildings, as large as typical shaft houses, out of local stone or red brick, yet they rarely used masonry for shaft houses. During the time spanning the 1870s to the 1890s, electric lighting was virtually unheard of, and mining engineers instead had carpenters install large multi-pane windows at regular intervals in the walls for natural light.

The lithograph portrays the interior of a well-equipped hoist house in the 1880s. The layout of machinery is applicable to medium-sized and large mining operations spanning from the 1880s into the 1920s. Usually the boiler was walled off from the rest of the machinery to contain soot and dust. The fireman at left shovels coal, the boiler tender near center is operating the boiler feed pump, and the hoistman stands over an unusual two cylinder double drum hoist. Like many well-capitalized operations, the floor of the hoist house has been lined with planking.

Ingersoll Rock Drill Company Catalog No.7: Rock Drills, Air Compressors and Air Receivers
Ingersoll Rock Drill Company, New York, NY, [1887] p24.

Most shaft houses built in the West conformed to a few standard footprints influenced by the arrangement of mine machinery. Overall, the structures tended to be elongated to encompass the hoist and shaft, and they featured lateral extensions that accommodated the shop, a water tank, the boilers, and either coal or cord wood storage. Professionally-educated mining engineers recommended that at least the boiler, and ideally the shop, be partitioned in separate rooms because they generated soot and dust which took a toll on lubricated machinery such as compressors and hoists.[98]

The interiors of the shaft houses typical of small productive western mines were rough, noisy, and cramped work environments rarely captured on film for public viewing. This photograph depicts the claustrophobic inside of the White Crow shaft house high in the Rockies above Boulder, Colorado around 1900. At left is a return tube boiler with a half façade, at right a hoistman waits at the ready behind a sinking-class steam hoist, and the vacuum bulb of a pump is visible at bottom right. The dark interior of the shaft house was the last scene many miners saw before going on shift underground and it was the first thing that greeted them after a long day in the wet underground.

Courtesy of the Carnegie Branch Library for Local History, Boulder Historical Society Collection

The roof profile typical of most western shaft houses featured a louvered cupola enclosing the headframe's crown, and a sloped extension descending toward the hoist that accommodated the hoist cable and the headframe's backbraces. Tall iron boiler smokestacks pierced the roof near the hoist. The stovepipe for the forge extended through the roof near the shaft collar, and the shaft house may have also featured other stovepipes for the stoves that heated the hoistman's platform and the carpentry shop. The tall smokestacks and stovepipes usually had to be guyed with baling wire to prevent them being blown over by strong winds.

The mining engineer working at high elevations often floored the shaft house with planks to improve heating. In some cases the shop and boiler areas, where workers dropped smoldering embers and hot pieces of metal, were surprisingly also floored with planking, which presented an enormous fire hazard. In mining districts with less adverse climates, the mining engineers felt that wall-to-wall flooring was not necessary, and they elected to leave the floors earthen, except for the areas around the hoist and compressor, which were decked. Customarily, the mining engineer designed the flooring to be flush with the top surfaces of the machine foundations, permitting the steam, air, and water pipes to be routed underneath and out of the way.

The Tamborine Mine, also located in the Rockies above Boulder, Colorado, had enough capital to build a small production-class surface plant enclosed in a frame shaft house. At right is a two-stage straight-line steam compressor anchored to a brick foundation. The compressor's receiving tank is visible behind the headframe legs in photo-center. Left of center stands a cage loaded with an ore car that has been let down onto a landing chair at the shaft collar. The topman at left poses by the levers that engage the landing chair.

Courtesy of the Carnegie Branch Library for Local History, Boulder Historical Society Collection

Particularly well-funded mining companies took extra steps to improve the aesthetics of their buildings by painting them, adding trim, and in some cases using the broad sides as billboards advertising their names. In Cripple Creek, standard practice among productive mines dictated not only painting the mine's name in huge letters, but garnishing the advertisements with elaborate images of horses, women, and eagles. Such immense monuments of wealth stood in stark contrast to the low, dark, and unpainted structures enclosing the small surface plants of less-successful mines.

Shaft houses big enough to cover a bank of boilers, a large hoist, an air compressor, and a shop were extremely costly to build and they required expensive upkeep. In addition, the heat generated by the shop forge, boilers, and a few woodstoves proved no match for the frigid drafts of winter. In response to the economic drain posed by such large shaft houses, during the 1900s and 1910s many mining companies began sheltering key surface plant components in individual buildings. During this time the appearance of the surface plants of

The floor plan of the Peruvian Mine Shaft House in the Peru Creek District, Colorado, is representative of many small productive western mines. A board-and-batten shaft house covered the wooden floor and the boiler. If the Peruvian Mining Company had more capital, it might have installed a small straight-line compressor adjacent to the hoist.

Author.

Shaft

Plank Floor

Blacksmith Forge

Steam Hoist

Water Tank

Firing Floor

Boiler in Brick-lined Rock Setting

Stark, dirty, dark, and barren, the shaft house of the Moose Mine in Butte, Montana serves as an excellent example of types of structures erected by profitable, moderately capitalized mining companies throughout the West between the 1860s and 1890s. The structure consists of a square-set frame covered with board-and-batten siding, and the walls feature a few multipane windows. The headframe stands under the louvered cupola at center and the smokestacks at left mark the location of two boilers and the hoist. The blacksmith shop appears to be located in the small extension featuring a large smokestack left of center. The building at far right probable houses an air compressor.

Courtesy of World Museum of Mining, Butte, MT.

many mines changed to consist of a cluster of moderate-sized buildings surrounding the exposed headframe, instead of one or two.

The Wild Horse Mine proved to be one of Cripple Creek's richest producers, and it boasted a well-appointed surface plant, including a shaft house decorated with attractive images reflective of the mine name. The surface plant provides a good example of the types of structures well-capitalized mining companies erected.

Mining & Scientific Press July 16, 1904 p42.

Shaft houses also began disappearing during this time for another reason. In 1898 three miners were at work drilling blastholes deep in the Londonderry Mine situated in the heart of the Cripple Creek District. As had happened at other mines, somehow the shaft house caught fire and the surface crew was not able to contain the conflagration. The miners underground realized that their only avenue of escape had been cut off by fire, and in a short time smoke began pouring down the Londonderry shaft, filling the workings, and ultimately suffocating all life underground. Due to accidents such as this, western states began outlawing shaft houses during the 1910s and 1920s. The mining industry responded by tearing down the ponderous industrial structures, except in the rare cases where a shaft had been enclosed in a nonflammable building.[99]

Instead of a shaft house, at particularly large and well-equipped mines the engineers had carpenters enclose the hoist and boilers in a *hoist house,* the compressor in a *compressor house,* and the shop in its own building. The mine plant may also have featured a miner's *change house* also known as a *dry,* a storage building, a stable, a carpentry shop, and an electrical substation. Small and medium-sized mines often combined the hoist, boiler, compressor, and shop in one large hoist house, while the headframe and shaft collar remained exposed to the weather. It is important to note that sheltering the surface plant components in a single hoist house was standard practice for poorly-capitalized mines and prospect operations in the Great Basin and Southwest from the 1870s into the 1930s.

The hoist house at the Gleason Shaft in Cripple Creek presents another example of the types of buildings erected by well-capitalized western mining operations. The smokestacks denote the locations of the boilers, specifically the large stack was probably shared by two units and the thin stack was for a standby unit, and they appear to be walled off from the rest of the hoist house. The hoist and a compressor were enclosed inside the main building, which features broad windows for illumination. The greenhouse window afforded the hoistman an unobstructed view of the shaft collar and headframe, and it features two slits for the hoist cables. The shaft and headframe are out of view to the right.

Mining & Scientific Press July 16, 1904 p42.

Wood frame and steel buildings consisted of structural materials that other mining operations, creditors, and district residents prized, and they were quick to remove a building for the lumber following a mine's abandonment. Further, federal tax laws levied assessments on the owners of property with improvements such as structures, and as a result private parties demolished mine buildings in hopes of avoiding payment. As a result, only a tiny fraction of the mine buildings that dotted mining landscapes across the West prior to the 1930s have endured until today. Visitors seeking to identify the size, type, and shape of buildings at mine sites must turn to archaeological remains in the forms of foundations, footprints, and artifacts.

Mining had reached an imposing scale in Butte early in the century. Many of the large operations outgrew their shaft houses, such as the Buffalo Mine in the foreground. The Buffalo presents an excellent example of the nature of large surface plants erected by well-capitalized and profitable mining operations between the 1890s and early 1920s. Left to right stands the hoist and compressor house, a Montana-style headframe, the shop behind, and possibly another shop at far right. This photo brings together several details discussed early in the text, such as the large air receiving tank by the shaft collar, the plain industrial architecture, the rail spurs extending to different portions of the plant, and the fact that the structures share a common alignment. In addition, working mines generated much rubbish, which has been dumped in the foreground.

Courtesy of World Museum of Mining, Butte, MT.

In a few rare cases, mining engineers instructed their crews to lay either concrete or masonry wall footers, which can help the visitor determine the footprint of a structure. But, due to lack of funding and the impermanence associated with western mining, workers rarely built lasting foundations. Instead, they built informal foundations consisting of dry-laid rock alignments or heavy timbers set in the ground. The visitor to historic mine sites may be able to identify the footprints of buildings by abrupt changes in soil character and linear depressions where building walls stood. Heavy foot and animal traffic, grease and oil deposits, forge clinker, dark soil with a high organic content, and differential soil weathering had the potential to result in the ground underlying a structure appearing different from the surrounding soil or waste rock. To further aid visitors puzzling over historic mine sites, the soil differences affect revegetation patterns, which may outline a structure's footprint.

The assemblage of artifacts may also help the visitor to identify the types and locations of mine buildings. Heavy concentrations of nails, window glass, lag bolts, stovepipes, and stovepipe flashings often were left after a wood frame building had been disassembled. Usually the artifact assemblages at mine sites active after around 1900 also include at least a few pieces of corrugated siding. Some mining companies with limited funding used a variety of other forms

Today operations such as the Nettie Mine have left expansive and complex sites packed with numerous features and artifacts representing long-term activity, much capital, and several occupational episodes. The Nettie Mine in Butte, photographed around 1915, presents the scale and scope of the surface plants typically erected by large, productive, and well-financed mines. Left to right, the shop buildings, a Montana-style headframe fitted with rock pockets and screens for the use of skips, a large boiler house behind, the hoist house, and either a storage facility or another shop in the right foreground. Like many large mining companies, the Nettie operation purchased a forest of timbers and planking that timbermen will trim to size and send underground in an unending procession for support timbering and to build structures. The power poles erected over the rail lines traversing the mountainous waste rock dump in the left and center backgrounds indicate that the mining company utilized electric locomotives for moving materials around the surface plant. A pile of discarded carbide drums and other industrial refuse is accumulating against the building at right. Note the wagon and team by the shops at left.

Courtesy of World Museum of Mining, Butte, MT.

of sheet iron siding to cover holes and gaps in mine buildings. Last, the concentration of lumber fragments and light industrial artifacts such as small machine parts, electrical insulators, and nuts and bolts is often greater within the boundaries and immediately surrounding the location of a mine building than at short distances away.[100]

Before leaving this chapter, we should consider several additional surface plant manifestations which visitors to historic mine sites might encounter. Large mining operations that were bent on turning the earth inside-out for profit employed dozens of mining crews underground to drill and blast. In the process of bringing down ore in the quantities that made investors happy, the crews of miners consumed hundreds of pounds of dynamite per day. The mining company had to store enough dynamite to last several weeks spanning freight deliveries, and all too often they stacked fifty pound boxes, the standard shipping container, in shaft houses, compressor houses, storage sheds, and vacant areas underground. Worse, during cold months, which spanned much of the year at high altitude, mine superintendents had boxes of dynamite stored near boilers, in blacksmith shops, and near hoists where it remained in a thawed and ready state. Such storage practices were extremely dangerous. In response, mining engineers had construction crews build explosives magazines where storage was more controlled and orderly.

Well-built magazines came in a variety of shapes and sizes, but they all shared the common goal of concentrating and sheltering the mine's supply of explosives. Academically-trained mining engineers felt that magazines should be bulletproof, fireproof, dry, and well ventilated. They also felt that magazines should be constructed of brick or concrete, and if of frame construction, the walls needed to be sand-filled and sheathed with iron. These structural features not only protected the explosives from physical threats, they also regulated the internal environment which was important, especially in summer. Extreme temperature fluctuations and pervasive moisture had been proven to damage fuse, caps, blasting powder, and most forms of dynamite. This in turn directly impacted the miners' work environment, because degraded explosives created foul and poisonous gas byproducts that vitiated mine atmospheres. In extreme cases, degraded caps, fuse, and dynamite misfired when in the drill hole, meaning they failed to explode, until a miner attempted to extract the compacted mess a little too vigorously with a drilling spoon. Dynamite exploding in the faces of miners was a leading cause of death and injury in the mining West.[101]

Regardless of obvious safety hazards and degradation of the explosives, many small and medium-sized mining companies stored their explosives in very crude and even dangerous facilities.

Engineers, often self-educated, had crews erect sheds sided only with corrugated sheet-iron that offered minimal protection from fluctuations in temperature and moisture. In other cases small capital-poor operations took even less precaution and stored their explosives in sheet-iron boxes similar in appearance to doghouses, in earthen pits roofed with sheets of corrugated iron, or they used abandoned prospect adits. Lack of funding appears to have been a poor excuse for improper storage practices, because most operations had the ability to erect fairly safe, inexpensive vernacular dugout magazines. Large mining operations, on the other hand, found it within their means to build proper magazines.

Progressive and well-capitalized mining companies erected explosives magazines to regulate the storage of explosives and concentrate the hazardous materials away from the plant core. Mining outfits erected a variety of designs, and they favored an arched roof structure made of either brick, stone, or concrete. This structure stands at the Lillie Mine in Cripple Creek.

Author.

Proper magazines manifest as stout masonry or concrete buildings around twelve-by-twenty feet in area with heavy arched roofs and iron doors. Usually these magazines were erected a distance away from the main portion of the mine's surface plant. Well-built vernacular magazines often appeared similar to root cellars. Generally they took form as a chamber that workers excavated out of a hillside, often eight-by-ten feet in area, and roofed with earth, rubble, and rocks.

The last principle surface plant component pertains to a practice known as *highgrading*. Before Frank Crampton had learned enough of the mining trade to deem himself a mining engineer, he had worked in

Goldfield, Nevada as a miner. Highgrading was rampant in Goldfield as it was in many other districts throughout the West, and miners considered pilfering unusually rich samples of ore to be their God-given right. As the miners spread their wealth through the community, local leaders were naturally sympathetic toward the miners when officials cracked down. Illustrating the degree of support miners generally found within the community, Crampton had attended a church sermon in Goldfield during which the local pastor had railed against theft and stealing, but he threw in the clause: "But gold belongs to him wot finds it first" for the benefit of the miners present.[102]

Mining companies, of course, saw highgrading as outright thievery, and by the 1890s they began to employ a number of measures to at least prevent it if not recover the lost ore. One response that companies undertook was the installation of *change houses* near the collars of their shafts. Mining companies claimed that they erected change houses for humanitarian purposes, but all persons involved knew that companies actually built change houses in hopes of discouraging miners from highgrading. There, miners were expected to remove their filthy work clothes, wash if they felt so motivated, and put on clean clothes, often under the supervision of a company official hoping to catch potential thieves. The more persistent the mining companies became the more clever miners had to be. They hid ore in lunch buckets, false-crowned hats, boot soles, and about their bodies. Mining companies never were totally able to eliminate the illicit flow of ore from their mines, but the institution of change houses certainly reduced the volume.[103]

The fact that change houses served a sanitary function could not have been denied. The warm and dry buildings provided miners with a badly needed place to change as they came off shift, and as the years progressed into the twentieth century, miners and mining companies began to emphasize this function. The change houses erected by mining companies during the Gilded Age were variations of the other types of mine buildings. The structures usually consisted of a wood frame sided with either boards or corrugated sheet iron, a gabled roof, and a wood plank floor. Some mining companies furnished the interiors of their change houses with no more than a large stove, benches, a wash counter, and clothing baskets suspended from the ceiling. Well-capitalized, well-meaning outfits, on the other hand, installed shower areas heated by steam pipes. Born out of the disparity between management and labor, change houses became an institution invaluable to the western miner.

CHAPTER 5

IN THE SHADOW OF THE FORTUNE SEEKERS: MINING DURING THE GREAT DEPRESSION

During the 1930s the combination of thousands of destitute workers and an increase in the value of gold spelled a revival of mining in the West after a ten year long decline. Suffering from an acute lack of capital, technical skill, and high-grade ore, most Depression era mining companies were forced to work in the shadows of once prosperous mining operations. During the Depression the Hull City Mine boasted no more than a flimsy hoist house, a sinking-class electric hoist, a headframe, and an ore sorting house where there had once been a steel headframe broad enough to encompass the structures in the photo. Foundations and ruins of titanic machines encompassed the small 1930s plant.

Author.

After sixty long and productive years the glory days of hardrock mining in the West were drawing to a close. Between the 1860s and 1900s parties of prospectors backed by dreamers of wealth combed the West and discovered most of the rich mines. By the 1910s, the close of the Gilded Age, many prominent mining men began to realize that the easily mined ore was nearly gone and that prospectors were making few new discoveries. During the waning years of the Gilded Age, mining companies in established districts were forced to chase ore of declining values ever deeper in efforts to remain solvent. World War I created a glimmer of hope in the darkness of the growing pessimism because it stimulated need for industrial metals and a surge in the price of silver. The demand gave many mining companies renewed vigor, but the high prices they enjoyed collapsed following the Armistice, and western hardrock mining fell into a torpid state. By the early 1920s the grand days of highgrade bonanza ore, self-made mining men, and rags-to-riches dreams came to a close, and many of the marginal mines in the West that persevered through the 1910s shut down. Only a few large companies capable of achieving production through economies of scale kept open what had once been the richest mines in the West's best districts.[1]

Ironically, under the presidency of one of the world's greatest mining engineers, Herbert C. Hoover, the Crash of 1929 brought the nation to its knees, destroying what little was left of metal mining in the West. The demand for industrial metals and gold vanished; capital to upgrade and streamline hardrock mining and milling processes evaporated, and the business community turned away from mining and directed its energy toward a search for meager profits in other areas. The few large mining companies that had been operating during the 1920s either dissolved, or suspended operations.

All that remained of the once-proud Gilded Age operation at the Gemini Mine in Eureka, Utah when the Depression began was the massive headframe. In the 1930s a mining company with limited capital rehabilitated the property and put up a small, inexpensive, and rough hoist house that provides an interesting contrast against the extravagant production-class headframe.

Author.

In 1932 the victory of Franklin Delano Roosevelt over Hoover for the office of United States president set in motion a chain of events that would spell a revival of mining in the West. In an effort to devalue the U.S. dollar, in October of 1933 Roosevelt enacted a plan in which the federal government bought gold at relatively high prices. Price declines began to interfere which this scheme, and in response Roosevelt and Congress passed the Gold Reserve Act and the Silver Purchase Act early in 1934, which set the minimum price for gold at $35.00 per ounce and artificially valued silver.[2] The combination of relatively high values for the precious metals and the hordes of destitute families and jobless men, created a return to the West's mining districts. However, unlike the optimistic gold and silver seekers of the 19th century, the Depression changed miners' priorities and they no longer dreamt of bonanza ore. They were happy to glean enough gold and silver merely to make a living.

The Great Depression changed the way in which outfits mined the West's ores. Because capital was scarce and the available payrock was low in grade, mining companies had to exercise fiscal caution in preparing a mine property for production, and they needed to minimize operating costs. This combination of factors set the stage for a deep division in the western mining industry. On one hand, under the direction of professionally-educated mining engineers, large corporations invested in capital rehabilitating mines that had sound histories of ore production. To profit from the low-grade ore left by defunct Gilded Age operations, the large companies built modern and efficient infrastructures designed to minimize operating expenses and produce ore in economic tonnages.

On the other hand, small capital-starved shoestring outfits set their sights on cobbling together mine plants at absolutely minimal initial cost. They made great use of salvaged materials and machinery, and relied heavily on hand labor. The trade-off these companies made was that they were capable of producing only limited tonnages of ore, but they were able to turn a profit because their costs were exceedingly low, and they carried little debt. Both large corporations and shoestring operations usually worked literally in the shadows of the structural and mechanical remains left by formerly glorious mining companies, each attempting to make profitable once again mines that had been long abandoned.

Regardless of the size and type of mining organization that reworked old ground during the Depression, nearly all operations continued the traditional hardrock mining practices that had become an institution in the West. While some types of mining machinery had changed since the peak days of mining, the work underground and the need for a surface plant at top remained the same.

The Depression Era Mine Shop

Like the mines of decades past, shaft and tunnel operations required the services of a shop to sharpen drill-steels, to manufacture hardware, and maintain other pieces of equipment. A blacksmith who worked in the mines between the 1890s and 1910s would have felt at home in the shops of small Depression-era operations. Not only did these companies and partnerships continue to utilize facilities left over from the Gilded Age, but, out of financial necessity, they sharpened drill-steels and manufactured hardware using the same labor-intensive methods of decades past. Further, while the use of rockdrills had become commonplace by the 1930s, small companies revived the art of hand-drilling for some underground work because it was inexpensive. As a result, blacksmiths in the primitive shops of small mining companies sharpened machine drill-steels and hand-steels with coal-fired forges, hand-held swages, and hammers. The work was often crude and the drill-steels did not remain sharp for long, but it was enough to get by.[3]

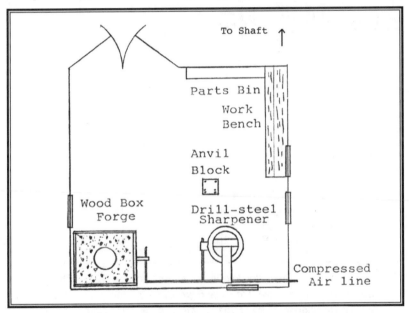

The plan view depicts a blacksmith shop typical of medium-sized mines during the Great Depression. The shop consisted of a grout-capped wood box forge, an Ingersoll-Rand drill-steel sharpener, an anvil, a quenching tank, a workbench, and parts bin. The engineer adapted a compressed air line to blow air into the forge, instead of using a forge blower. Compare this shop, erected in the 1930s at the Valentine Mine in Goodsprings, Nevada, with that of the Delmonico Mine illustrated in Chapter 3.

Author.

In marked contrast, the shops at mines run by well-financed companies tended to be mechanized to a greater degree than their Gilded Age counterparts. Used and new metal and wood shop appliances were readily available for little money. In addition, a greater number of mining districts had been wired with electric grids than in the past, and petroleum-powered generators had become affordable, making the installation of motor-driven shop machines attractive. There was a great variety of used and new air compressors available and their costs had plummeted. Many mining companies operating during the Great Depression installed them not just to run rockdrills, but also to power shop appliances such as drill-steel sharpeners.

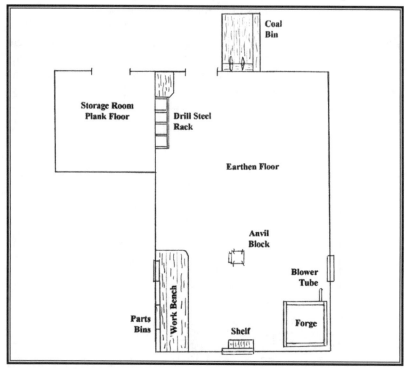

The company that operated the Joe Dandy Mine in Cripple Creek installed a shop typical of medium-sized and large Depression-era mines. The shop included a vernacular wood box forge, an anvil, a quenching tank, a work bench, shelves, a drill-steel sharpener, a drill-steel rack, and coal storage. This shop and the facility at the Valentine Mine bore great resemblance to the shops at mines of similar size in the late years of the Gilded Age.

Author.

Large, more solvent and productive mining companies were able to install additional shop appliances rarely seen in the past. Sets of oil and coal forges, multiple drill-steel sharpeners, drill-rod straighteners, bit grinders, and continuous-flow quenching tanks all became common in the spacious and stately shops of the large mining companies. Some mines even employed machines designed to forge collars on the hexagonal steels used in sinker drills, and to form the lugs on steels used in large hammer drills. During this time machinery makers introduced the hot miller, which heated and sharpened drill-steels in less time than the older drill-steel sharpeners, but the machine's high cost limited it only to the largest mining companies. Power metal and wood saws, drill-presses, pipe threaders, and lathes also became common. Last, the shop buildings at large and well-run mines were spacious and afforded a greater work area than in the past.[4]

✗

Ventilation Systems

M iners' need for fresh air underground was one factor in hardrock work that remained unchanged. The nature of the ventilation systems installed during the 1930s bore great resemblance to those used by mining companies in decades past. One difference was that more outfits employed fans for blowing air into dead-end workings than in times past. Mining companies and partnerships attempted to make-do with natural ventilation when at all possible, because inducing natural air currents cost little. The greater willingness to set up mechanized ventilation systems during the Depression may be attributed to the same factors that led to the increase in the use of automated shop appliances. The combination of a wide variety of inexpensive new and used machines and hardware, and the widespread adoption of electricity made installation of ventilation fans attractive.

To the relief of mining companies rehabilitating abandoned mines in developed mining districts, the nineteenth century operations already made the necessary underground connections with neighboring properties, which had the effect of creating flushing air currents. All that a Depression-era operation needed to do to maintain natural air circulation was to ensure that the drifts and crosscuts interlinking neighboring mines remained open.

Mining engineers continued the conventional practice of installing fans immediately outside of the adit portal or shaft collar, in alignment with the surface plant's master datum line. Many of the fans used in conjunction with tunnels were belt-driven by a motor and were anchored to portland concrete or timber foundations with four bolts. The 1930s saw the rise of an interesting type of squirrel cage fan that was designed to be mounted on its side. Because large horizontal fans were expensive and required engineering to install, they remained within the means of only large and well-financed mining companies. During this time mining engineers found that using portable fans placed inside of mine workings also afforded ventilation at little cost.[5]

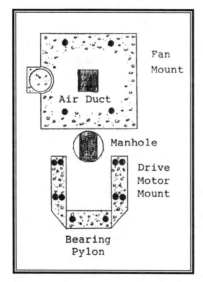

During the Depression ventilation fans saw greater popularity than in times past. Well-financed mining outfits installed high-capacity fans, including an odd unit designed to be mounted on its side. The illustrated foundation, featured at the Enterprise Mine in Nevada's Gold Mountain District, anchored a horizontal squirrel cage fan driven by an electric motor. The fan drew fresh air up through the duct in the foundation's center, and forced it into the mine. The duct extended underneath the hoist house floor and out the building's side where fresh air could have been collected. The manhole provided access to the duct.

Author.

The method of sending the fresh air into the mine through heavy galvanized ducting had changed little from the Gilded Age practices. Depression-era miners usually hung the tube work with baling wires fastened onto wedges driven into the tunnel ceiling, or lashed onto shaft timbers. Mining outfits with access to only modest capital found it economical to salvage ducting from neighboring abandoned mines. Miners attempted to hammer dents out of the tubing and seal holes and joints with metal sleeves, burlap, canvas, and tar. As a result, the mechanized ventilation systems installed by moderately sized Depression-era mining outfits appeared rough and were not very efficient, but they performed their necessary duty.

⚒
Compressed Air Systems

While the ventilation systems and mine shops used by mining companies during the Great Depression had remained virtually static in technology and form, compressed air technology had undergone dynamic changes since the close of the Gilded Age. Between 1870 and 1920, mechanical engineers introduced a variety of air compressors that were smaller, faster, and more efficient than past models, but most were based on the original forms of either the straight-line compressor or the duplex compressor. Mechanical engineers began to experiment with unconventional designs beginning in the 1900s, and during the 1910s several of these models experienced commercial production. By the 1930s, a few western mining companies with substantial capital became interested in installing some of the modern designs in hopes of maximizing efficiency. But, well-financed mining companies requiring large volumes of air at high pressures continued to favor belt-driven duplex compressors, while companies with reduced air needs continued to use relatively inexpensive single-stage belt-drive straight-line compressors.

The popularization of automobile engines gave rise to the invention of several compressors similar in form, and they saw use during the Depression. By the 1910s an *upright two-cylinder compressor,* with valves and a crankshaft like an automobile engine, became somewhat popular. Used on an experimental basis as early as the 1900s by prospect operations, these units were inexpensive, adaptable to any form of power, and weighed little. Further, mining machinery makers had mounted them onto four-wheel trailers or simple wood frames for mobility. As a result, impoverished western mining outfits embraced them because they required no engineering and were ready to use. During the 1930s western mining companies hauled these machines to mine sites where they bolted them to simple timber frames and coupled the drive shaft to salvaged automobile engines, single cylinder gas engines, or motors for power.[6]

Upright two-cylinder compressor on a trailer at the Minietta Mine in the Lookout District, California.

Author.

The *angle-compound compressor*, developed during the 1910s, was a major breakaway from traditional Industrial Revolution compressor designs. This machine consisted of two large compression cylinders oriented ninety degrees from each other, one lying horizontal and the other extending vertically upward. The piston rods for both cylinders were bolted onto a crankshaft in an engine case, much like the piston arrangement for automobile engines. A large belt pulley that also served as a flywheel turned the crankshaft. One of the cylinders had been designed for low-compression and the other for high compression, and the air passed through an intercooler between them.[7]

A V-cylinder compressor sits amid the ruins of Cripple Creek's Gold King Mine. The compressor consists of cylinders arranged like an automobile engine, a radiator, and a belted motor all bolted onto a common steel frame. The frame was bolted onto concrete foundation.

Author.

The operating principles behind the angle compound-compressor were the same as those for compound duplex compressors, and mining and mechanical engineers claimed that the new machines were far more efficient when driven by an external source such as a motor. These innovative compressors were able to deliver a volume of air up to 900 cfm at high pressures for less energy and for less floor space than duplex or straight-line units. Despite superior performance of angle-compound compressors, only a few western mining companies experimented with them during the 1910s and 1920s because of high purchase prices and the unconventionality of the design. These factors continued to suppress employment of the compressors into the 1930s, at a time when many mining outfits were forced to be fiscally conservative to maintain profitability. As a result, angle compound compressors never saw great popularity in the West.[8]

Angle-compound compressors were larger and more complex than other compressors used at western mines, and as a result their foundations were asymmetrical and presented uneven, stepped profiles. The foundations were often eight by ten feet in area and their footprints conformed to "L" and "T" shapes. Like large straight-line compressors, high-capacity angle-compound units often featured an independent concrete pylon to support the heavy flywheel's outboard bearing.

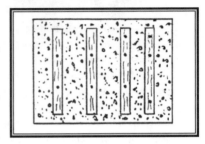

The plan view depicts a timber-studded V-cylinder compressor foundation.

Author.

The last breakaway from traditional compressor form employed during the Great Depression consisted of another design that mimicked the structure of the automobile engine. The Chicago Pneumatic Tool Company and Gardner-Denver both introduced compressors known in the mining industry as *V-cylinder compressors* and as *feather valve compressors*. The machines were virtual adaptations of large-displacement truck engines with between three and eight compression cylinders arranged in a "V" configuration, and the pistons were coupled onto a heavy crankshaft. Further, the new designs no longer relied on circulating water from a storage tank for cooling. Instead they featured grossly enlarged radiators similar to the types automakers installed on the fronts of cars. The compressor makers designed the large machines to be powered by electric motors directly coupled onto the crankshaft. Small machines were belted to a motor. V-cylinder compressors frequently came from the factory mounted onto a heavy steel frame that the mining company bolted onto a concrete foundation.[9]

V-cylinder compressors required concrete foundations that possessed a slightly rectangular footprint, and engineers usually specified that they be capped with timbers, which cushioned the vibrating machines. The large compressor units required long rectangular concrete foundations similar in shape to the footings associated with the antiquated straight-line steam compressors.

In the 1910s mechanical engineers had introduced several innovative air compressors that were more efficient than anything the mining industry had yet seen. The angle compound unit was principal among these, but due to its unconventionality and relatively high cost, it saw little popularity in the West until the 1930s, when mining companies began to rehabilitate old workings. The angle compound compressor required an external motive source, either being belt-driven via a motor, or directly driven by a motor, as shown in the illustration. Note the complex foundation, which may be evident at historic mine sites today.

Explosives Engineer June, 1925.

The variety of new compressors sold by mining machinery makers were of little consequence to small operations with limited money, because they were economically forced to employ salvaged and used equipment. Impoverished companies combined any type of compressor they could get their hands on with a suitable drive motor, creating odd, mismatched sets of machinery. Some poorly funded operations that worked old claims deep in the backcountry, where electric power did not exist, employed old-fashioned straight-line gasoline compressors. A few impoverished outfits even revived archaic steam technology.

During the Depression, some impoverished mining companies lacking mechanical knowledge installed air compressors in manners that made professionally-trained engineers shudder. Some outfits not only attempted to reuse ancient machinery, but they even tried to adapt the machines to foundations originally built by preceding mining operations for other pieces of equipment. Miners used hand-steels to bore holes into old foundations, they inserted custom-made anchor bolts, and placed machines over them. In other cases, mine workers retro-fitted timber footers onto old foundations, and bolted compressors onto the timbers.[10]

Another aspect of compressed air systems that changed during the 1930s was the use of water with modern drills to allay silicosis-causing rock dust. Most mines supplied their miners with portable tanks that were pressurized by air taken from the mine's main lines. These tanks were connected to a drill via a separate, small hose, and when the miner turned on a valve the water surged into the drill and squirted out through the hollow drill-steel. Poorly financed mining companies used no water, and silicosis remained a hazard to their miners.

Transportation: Ore Cars and Steel Rails

Between the Gilded Age and the Great Depression the mining industry found it difficult to improve on the internal transportation system used for over eighty years. Steel ore cars of one ton capacity on steel rails spiked eighteen-inches on-center proved to be the most economical and effective means of moving the materials of mining in the tortuous drifts and passages within metal mines. A severe lack of capital and limited profitability during the long Great Depression ensured that this relatively inexpensive system would persevere for at least another decade.

Well-financed mining companies relied on mechanical locomotives in hopes of producing ore in the high volumes necessary to make a profit, while the outfits working small mines continued to employ mules to pull small trains. For mining companies with capital, electric trolley locomotives remained popular while compressed air and battery-powered locomotives saw increased use. Echoing the arguments of previous mining engineers, during the Depression, western mining companies felt that compressed air locomotives had the advantage of running on track constructed with heavy rail spiked to eighteen-inch gauge, while electric trolley locomotives required an expensive broad gauge and heavier rail. Mining machinery makers reduced the costs and physical sizes of battery-powered locomotives, permitting them to also run on 18 inch gauge track.

The West's small mines did not have to worry about such issues, because locomotives and the necessary improvements to the rail lines were well beyond their financial means. These companies constructed rail lines from whatever they were able to salvage. They straightened bent rails, they used large nails instead of proper rail spikes, and they fashioned ties from pieces of lumber. To save materials, impoverished mining outfits spaced the ties far apart, they spliced rails of varying lengths and weights into the line, and they broke connector plates in two to make them go twice as far. The trestles for dumping waste rock were equally ragged, consisting of uncut lengths of dimension lumber and hewn logs. While these rail lines were wavy, bumpy, and gave the impression that they would fall apart at any minute, they helped miners get by while spending little capital.

$\times$

Hoisting Systems

The overarching template for hoisting systems that Depression-era mining companies employed for work in shafts remained virtually unchanged from those developed during the Gilded Age. Depression-era outfits, well-capitalized and poorly-funded alike, continued to rely on a combination of hoist, headframe, hoisting cable, and hoisting vehicle to move materials in and out of shafts. Some of the individual components comprising hoisting systems, however, evolved between the twilight years of the Gilded Age and the revival of mining during the Great Depression. A mining engineer who had worked during the Gilded Age could have immediately recognized a Depression-era hoisting system. However, he would have noticed the absence of a steam boiler, he might have marveled at the wonderful and clean application of electricity, and would have found that the definition of sinking-class versus production-class equipment still applied.

The 1920s saw America's mining industry, including the companies operating in the West, switch from steam power to both electric motors and gasoline engines for running hoists. During the 1910s and 1920s mining engineers had devised a variety of electric hoists that were able to compete with all but the largest steam-powered double-drum behemoths. Well-capitalized mining companies had the means to purchase and install modern, factory-made, production-class electric hoists during the Depression, while the outfits that struggled to make a profit used either small gasoline or electric hoists, many of which had been assembled from salvaged pieces of equipment.

Professionally-educated mining engineers celebrated the industry's embrace of the electric hoist for most types of shaft work. Machinery makers ironed out wrinkles in the technology during the 1900s and 1910s, and by the 1930s they were producing a variety of single and double-drum models for shaft sinking and for heavy ore production. Like the steam hoists of old, electric models came in four basic varieties. Geared single and double-drum units, and direct-drive single and double-drum units. The geared electric hoists were built much like their steam predecessors in that the motor turned a set of reduction gears connected to the cable drum, and the components came from the manufacturer assembled onto a heavy bedplate. The gearing permitted hoist manufacturers to install small and inexpensive motors ranging from thirty to 300 horsepower. Direct-drive electric hoists, on the other hand, had huge motors rated up to 2,000 horsepower attached to the same shaft that the cable drums had been mounted on. These hoists, considerable in size, had to be assembled as components onto special foundations, as did the old direct-drive steam hoists.[11]

Any twentieth century mining engineer felt immense pride when his mine became host to a direct drive electric hoist. These machines were clearly intended for heavy ore production. They had hoisting speeds in the thousands of feet-per-minute, their payload capacity was over ten tons, and they were able to work at great depths. But because they were very costly, only highly profitable and heavily capitalized mining companies could justify installing such machines. Further, the electrical systems required to operate direct-drive hoists were expensive to install. These large machines typically operated with DC electric current, and as a result they required a substation where the AC current wired to the mine could be converted. In addition, because the massive motors for direct-drive hoists drew heavily from the electrical circuit upon starting, mining companies that installed direct-drive hoists found it best to put in an associated rotary converter to moderate the power drain.[12]

Most productive and well-financed mining companies in the West installed single or double-drum hoists depending on whether the shaft featured one or two hoisting compartments. Like the antiquated steam hoists used during the Gilded Age, during the 1930s mining engineers classified single-drum electric hoists smaller than six-by-six feet in area as meeting the qualifications for sinking-duty. Most of the production-class hoists installed by engineers during this time featured motors rated to at least sixty horsepower for single-drum units and one-hundred horsepower for double-drum units. Even with large motors, these geared hoists had slow hoisting speeds of less than 600 feet-per-minute, their payload capacity was limited, and they were not able to work in the deepest shafts. Yet out of economic necessity many mining companies had to settle for these machines. Less-fortunate mining outfits severely constrained by tight budgets had to settle for smaller, slower, sinking-class hoists. It was not uncommon for these companies to use hoists with motors rated as low as fifteen horsepower.[13]

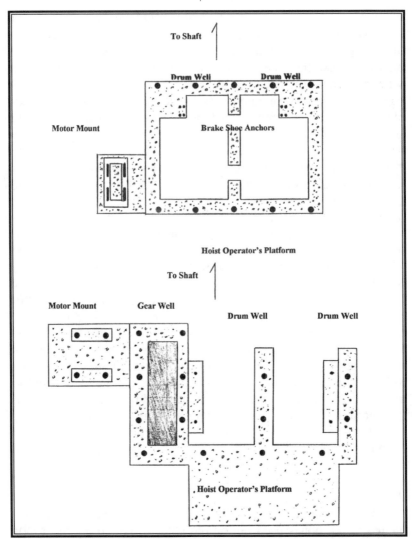

During the 1930s well-capitalized mining companies favored using double drum electric hoists to achieve balanced hoisting. The upper plan view illustrates a typical foundation for a moderate-sized double drum electric hoist. The bottom plan view depicts the foundation for a 150 horsepower double drum electric hoist.

Author.

Some outfits attempting to recondition abandoned mines on shoestring budgets cobbled together hoists from machinery that had been cast off at the close of the Gilded Age. Miners employed creativity and talent in re-using old machinery, and their applications fell into several basic patterns. One method common among operations in the mountain states involved obtaining an old geared steam hoist, stripping it of the steam equipment, and adapting an electric motor to turn the hoist's large bull gear. Miners used whatever type of motor they could secure, and they understood that large motors were most desirable because of their performance. To power the hoist, they had to build a small motor foundation adjacent to the hoist, and they had a machine shop custom make a pinion gear for the motor featuring teeth capable of meshing with the hoist's bull gear. The only other modification that the mining outfit had to make to the hoist was to mount the electric controller on the hoistman's platform. After ensuring that the hoist's clutch and brake worked and that the hoisting cable was sound, the miners were ready to go to work.[14]

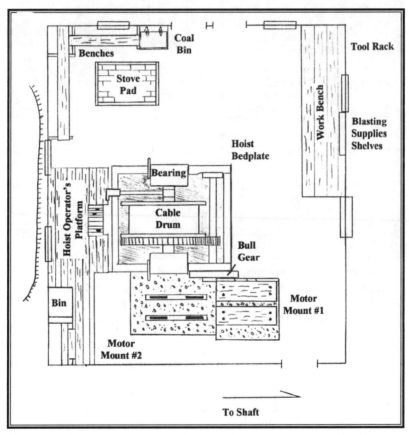

The Joe Dandy Mining Company relied on a steam hoist installed around 1900 that was retrofitted with an electric motor, as illustrated by the foundations in the plan view. The drawing captures the entire hoist house, which was built during the Depression. As the reader may surmise from the illustration, little had changed in terms of the composition and layout of hoist houses built during the Gilded Age from those dating to the 1930s.

Author.

Mining outfits with limited funding practiced another clever means of bringing new life to antiquated steam hoists. Unlike the method described above, the miners left the steam equipment on the hoist intact, and ensured that the pistons, piston rings, and valves were in good condition. They then routed pipes from the hoist to an air compressor to power the hoist. The only drawback to such an innovative use of compressed air was that a large multi-stage compressor had to be installed. In a few cases, impoverished mining operations were fortunate to have as a neighbor a well-funded mine equipped with just such a compressor.[15]

The quarter view shows the hoist at the Joe Dandy Mine. The steam pistons, drive rods, and linkages have been removed, and the mining company fitted the drive shaft at left with a large bull gear which was turned by a motor. Many mining companies low on funding retrofitted steam hoists with electric motors and custom gearing.

Author.

The third practice that impoverished mining companies adhered to when rehabilitating old mines involved assembling mechanical hoists from odd and unlikely pieces of machinery. A system favored by outfits lacking money and an understanding of fine engineering consisted of installing a small friction-drive hoist or geared hoist stripped of everything but the brake and clutch and coupling the gearing to the transmission of a salvaged automobile. Ugly, slow, noisy, and of questionable reliability, these contraptions worked well enough that shoestring mining operations were able to turn a small profit. Lacking the will, money, and possibly the knowledge of how to construct a proper foundation, miners simply bolted the hoist onto a flimsy timber frame that had not necessarily been anchored in the ground, and they mounted the engine to an even less formal timber frame. These small outfits commonly obtained automobiles either wrecked or in disrepair, stripped off the body, cut away the rear portion of the chassis, and left the front portion intact. They aligned the chassis so that they were able to connect the drive shaft to the hoist's gears with custom-made and adapted hardware.[16]

Some small and medium-sized mining outfits installed factory-made gasoline hoists similar to the types used during the 1900s. Mining companies also purchased factory-made donkey hoists offered by machinery suppliers such as Fairbanks-Morse and the Mine & Smelter Supply Company. The donkey hoists manufactured during the 1930s consisted of a small automobile engine that turned a cable drum through reduction gearing. The makers designed the little machines to be portable and affixed all of the components onto a steel frame.

Gasoline donkey hoists experienced mild popularity during the Depression because they required little capital to purchase and operate, they were field-worthy, and they were portable. This unit is preserved at the Central Nevada Museum, Tonopah, Nevada.

Author.

Several conditions played into a small outfit's decision as to which type of hoisting contraption they would install. Mining outfits lacking funding had to resort to alternatives to new equipment. In light of the lack of capital, another influential factor was whether the original hoist remained at the mine. Mining outfits worked with what the original mine plant offered them, and retrofitting steam hoists with motors seems to have been popular.[17]

In the event that a mine under rehabilitation did not possess its original hoist, the mining company was forced to install another apparatus. Outfits with limited capital may have opted to buy a small new or used factory-made electric model, or they may have attempted to retrofit a hoist salvaged from a neighboring defunct mine. For economic reasons, some operations intentionally purchased used steam hoists because they were inexpensive. Due to lack of funding and possibly a lack of training in engineering, small mining companies operating during the Great Depression rarely poured new concrete foundations for their hoists. In most cases they affixed the hoist to a substandard timber foundation, or adapted the new hoist to an old foundation using methods like those for air compressors.

Like mining operations that had been active during the Gilded Age, hoisting systems used by Depression-era mining companies required a headframe to guide the hoist cable into the shaft and to facilitate materials handling. Some mining outfits rehabilitating abandoned mines found that the headframe of the previous operation still stood in good repair. In the arid West some fortunate companies rehabilitating old mines merely had to have a few laborers examine the structure for integrity, oil the sheave bearings, and ensure that the signal bell functioned.

Much to the dismay of Depression-era mining operations, many abandoned shaft mines in the West needed a new headframe before the miners could pick over the ore left by early operations. Large mining companies, under the guidance of a formally-trained engineer, continued the practice of building four and six post derricks and A-frames to meet the rigors of ore production. Mining engineers still considered steel ideal for production-class headframes, although out of financial necessity they usually resorted to timbers. Still, the structures they put up were well built and handsome. Yet, it seems that a certain element of quality in construction, typical of mining prior to the 1910s, had been lost. For example, construction crews no longer took the pains to assemble the structure with intricate mortise-and-tenon joints. Instead, they simply butted the timbers against each other, or created shallow square-notch joints and bolted the frame together.

Impoverished outfits lacked the means to build substantial production-class headframes. When possible they relocated entire headframes from abandoned mines. The headframes they either built or relocated tended to be the old-fashioned sinking-class two-post gallows type because they were simple, inexpensive, and required no formal engineering to erect. In a few cases these small companies built small four-post derricks of the sort that academically-trained engineers would have classified as being sinking-duty.[18]

Many mining companies, rich and poor, tended to use salvaged timbers for building their headframes. Stout timbers were a precious commodity during the Great Depression, and in hopes of saving capital, mining companies reused heavy beams left by abandoned operations.

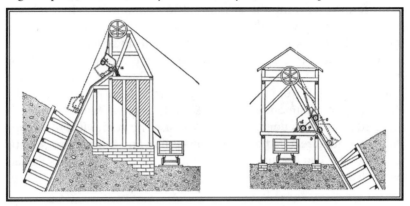

The skip had become the hoisting vehicle of choice during the Great Depression because it permitted more tonnages of rock to be moved in less time and with less labor. However, mining companies interested in employing skips had to make several modifications to a mine's surface plant and underground workings. Surface modifications included retrofitting headframes with special guides designed to force the skip to dump its contents, and either a rock pocket to catch the ore and waste rock shown at left, or a chute to direct it into a waiting ore car illustrated at right.

International Textbook Company International Library of Technology: Hoisting, Haulage, Mine Drainage International Textbook Company, Scranton, PA, 1906 A53 p20.

During the last several decades of the Gilded Age, professionally-educated mining engineers at a few productive mines began to use skips as the principal hoisting vehicle in vertical shafts, instead of the cage and ore car. Skips were nothing new by 1900, having been used in previous decades for extracting ore and waste rock from the depths of inclined shafts. As mining engineers sought ways of increasing ore production while consuming less energy, they came to realize that the skip provided great economy. Because the vehicle consisted of a boilerplate iron box, it had the capacity to contain more rock for less dead weight than the combination of an ore car on a traditional cage. In addition, the skip's open top facilitated rapid loading from an ore chute underground and equally rapid disgorging of its contents once the hoistman had brought it up to daylight. The rapid filling and emptying of the skip increased the tonnage of rock brought out of the mine.[19]

The use of a skip was in some ways easier and safer than relying on a cage. Miners riding up and down the shaft in an enclosed skip were less likely to snag their arms and legs and suffer dismemberment in the mine timbering. The hoistman lowered the skip down the shaft to a station where he stopped it at the mouth of an ore chute. A chute tender or trammer opened the chute gates and let the ore pour forth, then rang the signal bell to communicate that all was ready for the trip up to the surface. The hoistman slowed the ascent of the skip as it approached the shaft collar and raised the vehicle into the headframe. Typical skips consisted of an iron box held by heavy steel hinges in an iron frame that embraced the shaft's guide rails. The box featured small iron wheels or iron pins fixed to the sides. As the hoistman raised the skip into the headframe the wheels entered a special set of curved iron guides. The guides forced the box to pivot until it reached a point of instability and tipped over, spilling its contents. The hoistman reversed the action and the box's wheels were dragged back through the guides, righting the hoisting vehicle.

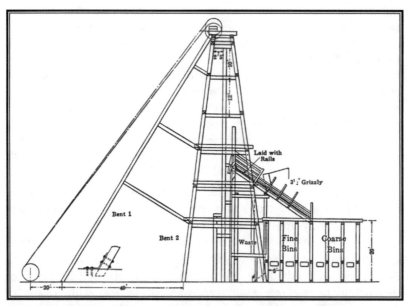

A few mining companies had used skips in vertical shafts as early as the 1890s, but the universality of the practice surged during the Great Depression. The profile depicts a production-class headframe fitted with the necessary structures to accommodate a skip. The hoisting vehicle dumped its contents into an ore chute floored with a screen designed to separate the rich fines from the low-grade cobbles. Both accumulated in individual ore bins built at the headframe's base.

Engineering & Mining Journal August 26, 1911 p392.

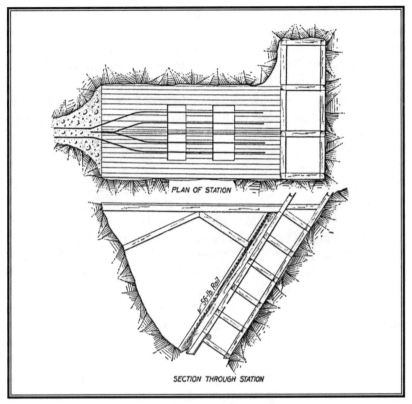

PLAN OF STATION

SECTION THROUGH STATION

The profile and plan views illustrate the modifications Depression-era mining operations had to make to shaft stations underground to facilitate the transfer of rock from an ore car into the skip. Miners had to blast out an underground rock pocket, build a bulkhead with a chute that directed material into the skip, and they had to floor the area with decking.

Engineering & Mining Journal March 17, 1917.

Not only did the employment of the skip require miners to drill and blast rock pockets for ore and waste rock at shaft stations, but the mining engineer had to retrofit the headframe with a rock pocket to accept the skip's load when dumped. In mines simultaneously experiencing underground development and ore production, the engineer had to arrange for two rock pockets, one to catch waste rock and the other for ore. By the 1920s mining engineers recognized the skip as the production-class hoisting vehicle of choice and they designed headframes that incorporated rock pockets into their structural composition.

The alterations mining engineers customarily made to headframes for use with a skip may be apparent to visitors examining the remains of yesteryear's historic mine sites. When a Depression-era mining company retrofitted a headframe with a rock pocket, they were rarely able to match the hue, grain, and cut of the original timbers, which today manifests as a visible contrast. Out of financial motivation many Depression-era mining companies constructed rock pockets with hewn or raw logs, salvaged lumber, and corrugated sheet steel that typically exhibits old nail and bolt holes. Further, elegantly built heavy timberwork seemed not to have been the forte of many construction workers rehabilitating old mines during the Depression, and this manifests today as a notable disparity between the well-built heavy beam work of the headframe and a modestly constructed rock pocket.

Struggling outfits rehabilitating small mines often lacked the capital and resources necessary to install the components for a skip. Instead, shoestring operations employed one of two hoisting vehicles to send miners and equipment down the shaft and to raise out rock. Many of these small companies employed sinking buckets because they were inexpensive, even free when salvaged from abandoned mines, and they required no special timbering underground. However, Depression-era shoestring operations experienced the same problems using free-swinging ore buckets in deep shafts that their Gilded Age ancestors had, so some outfits hung their buckets from crossheads.

Some mining companies employed a hoisting vehicle that was a cross between a sinking bucket and a skip. This vehicle consisted of a large cylindrical body, a bail with a loop for the hoist cable, and heavy iron pins welded onto the lower portion of the body. When hoisted, the bucket skidded upward on wood guide rails, and the pins helped steady the vehicle. When the bucket rose out of the shaft, the pins transferred onto special guides that overturned the bucket, allowing the rock to spill out.

⚒

Mining Architecture During the Great Depression

The architecture encountered by today's visitors at Depression-era mines speaks loudly about the nature of the mining operation that worked a given property. Some mining outfits had the good fortune of rehabilitating sites that possessed the structures erected during the Gilded Age. But many other mining operations had to shelter their plants with buildings constructed from scratch. As was standard, prosperous and well-capitalized mining companies erected sound and well-built structures, while the poorer outfits erected ram-shackle structures capable of fending off only the worst of weather.

The general construction methods and architectural styles of the 1930s changed little from the practices of the late nineteenth century. Like their predecessors, 1930s-era engineers designed stout structures that typically consisted of a dimension lumber frame and rafter roof, which laborers sided with corrugated sheet steel. Engineers continued to take advantage of natural light by designing buildings with multi-pane windows at regular intervals in walls, and they provided broad custom-made doors at important points of entry.

During the 1930s the use of flooring materials for well-built buildings became more common than during the Gilded Age. Engineers either floored principal structures with poured portland concrete, which became less expensive in many areas of the West due to truck transportation, or they stood the buildings on proper foundations and used wood planking. Some operations in mining districts that once bustled with activity often were able to scrounge enough plank lumber to side the exteriors of their mine buildings with boards-and-battens instead of corrugated sheet steel.

Depression-era buildings erected by well-capitalized mining companies shared a few characteristics that separated them from the simple structures typical of lesser mines. Not only were the buildings spacious and lofty, but the companies provided their workers with virgin lumber and sheet iron, and factory-made hardware. The work-ers, often skilled in their trade, built lasting structures with an orderly industrial appearance. In most cases mining engineers emphasized function and cost in their designs and added little ornamentation, unlike the large buildings of the Gilded Age.[20]

Poorly funded outfits were forced to keep construction within a tight budget. These outfits could not afford first-rate materials and tools, they were not able to hire experienced engineers or architects, and they lacked the funding to hire skilled workers. As a result, their buildings tended to be small, low, made with salvaged materials, and poorly constructed. The buildings fabricated by small outfits were personal and unique to each operation, being a true expression of the outfit's nature.

While large and well-financed companies customarily erected multiple buildings at a given mine, small operations tried to save money by enclosing their crucial plant facilities in a single building. At shaft mines this structure usually consisted of the hoist house, or a combination shop and compressor house at adit mines. In a few cases, to save money, impoverished mining companies moved entire buildings from abandoned operations located nearby. The construction of one building, with minor additions and extensions, minimized the cost and time required to erect a structure.

Laborers frequently built such structures with no formal frame. Instead, they preassembled the walls, stood them up, and nailed them together, or established four corner posts, added cross braces, and fastened siding to the boards. The builders may have used a combination of planks and sheet steel for siding, which was often layered to withstand high winds. In temperate climates they cut window frames out of the walls, and neglected to put panes in. Many outfits favored the *shed* structural form, which featured four walls and a roof that slanted from one side of the building to the other, because it was simplest to erect.

The workers at small mining outfits constructed these buildings poorly for several reasons. They sought to minimize cost and effort. Many workers at Depression-era mines were not the jacks-of-all trades that characterized miners during the Gilded Age. Rather, they came from a variety of urban and rural backgrounds, and lacked carpentry skills and proper construction tools. They also built shoddy structures because they wanted to fulfill an immediate need and anticipated abandoning the mine after a brief period of time. The architectural style of the mine buildings erected by such mining companies during the 1930s may truly be termed *Depression-era western mining vernacular.*

CHAPTER 6

RICHES TO RUST: INTERPRETING THE REMAINS OF HISTORIC MINES

As the embers of the California and the Pikes Peak excitement began growing dim in the early 1860s, miners began turning their attention away from the exhausted stream and river channels where they had sluiced placer gold, and embarked on a search for the gold's source. In the course of their wanderings, miners and prospectors discovered that the West abounded with money-producing ores in addition to gold, such as silver, copper, lead, and mercury. In the twentieth century they also began to mine tungsten. Prospectors and miners carried their quest to all corners of the West during the Gilded Age. The frenetic search for and extraction of precious and industrial metals became the foundation for the western mining industry, which was a phenomenon that affected the world. Western mining created extravagant fortunes in America and Europe as capitalists rushed to invest in mining ventures. The industry gave rise to unparalleled advances in technology, engineering, and an understanding of geology that mining experts applied to other parts of the globe.

By nature mineral resources are finite and the metal ores began to play out toward the latter years of the Gilded Age. By the 1910s miners and prospectors had explored most of the West; the easy finds had been discovered and mined, and the rushes that sparked gold fever became rare. A vigorous economy during the 1900s and 1910s and the outbreak of World War I had kept the demand for industrial metals healthy, but the market collapsed in the 1920s, and the mining industry that had been a focus of the western world for over sixty years virtually ceased. The Great Depression stimulated a revival, but the renewed interests consisted of jobless men and families literally picking over the bones left over from the Golden Days.

Today the West possesses a legacy of historic mining districts from this momentous time in American history. Pop culture holds the valid perspective that the West abounds with historic mine sites, but time, the elements, and humankind have all taken their toll, leaving few of these irreplaceable touchstones of our past intact. The damage caused to the legacy left us by Gilded Age mining should not paralyze historians from trying to make sense and meaning of the remains of historical mine sites. In the last three chapters we have taken an in-depth look at the technological dimension of hardrock mining during the Gilded Age and the Great Depression, and we experienced a taste of how we can read the remains of those surface plant components today. This chapter provides a lens to magnify this analysis and to explore in detail the interpretive contexts and patterns through which we can perceive historic mine sites.

⚒

Reading the Mine Site

M ine sites often speak of the nature of the operation that worked the ground. The material evidence can offer clues about when the activity occurred, the longevity of the operation, productivity, the degree of capitalization, and the work environment experienced by the miners. Reading the ruins of mines requires the historian to think like a detective and rely on material evidence to reconstruct the past. The first stage of reading and interpreting the site is the examination of the remnants of the surface plant.

The historian should identify and delineate each surface plant component, determine its function, figure out when it was installed and when it was abandoned, and of particular importance, define its duty classification. The technological materials discussed in the three preceding chapters should help guide the visitor with this difficult task.

By becoming familiar with a mine's written history, the historian may then know at least some of what the surface plant featured, and can therefore better interpret the extant remains. Information can sometimes be found in mining district geological reports, state mine inspectors' reports, Sanborn Insurance Maps, and mining industry trade journals.

To fully read the site, the historian must take into consideration material evidence in the form of artifacts. Relics can often aid in determining accurate dates of a site's occupations. The visitor will encounter three basic categories of artifacts at historic mine sites. These consist of building materials, industrial items, and domestic refuse mostly in the form of food and beverage containers and food preparation utensils. While much of the building materials used for mining were ubiquitous in form throughout the Gilded Age and Depression, a handful of key items evolved over time and today serve as date indicators. The appendix at the end of the book includes a table of artifacts and their date ranges.

Window and bottle glass fragments constitute two of the most common groups of these date indicators. During the Gilded Age glass manufacturers found that the sand they had been using as a source of silica contained impurities that discolored bottles and window panes made with it. To remedy the problem, they found that the addition of minute quantities of purified metals, especially manganese, rendered the finished glass colorless. The principal sources of manganese were European suppliers, and to the horror of

American glassmakers, World War I brought a halt to the import of the metal. After some experimentation, American glass companies discovered that selenium served as an acceptable substitute for manganese because it, too, clarified bottle and window glass.[1]

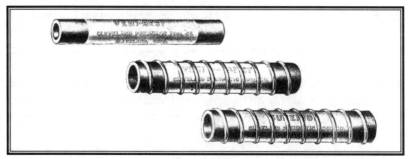

At top is a sample of the rubberized canvas air hose commonly used during the Great Depression, and the lower sample consists of the steel-spiral wrapped air hose universally used during the Gilded Age.

Cleveland Pneumatic Tool Co. Air Hose, Air Hose Couplings, Air Hose Clamps, Air Hose Nipples, Air Hose Menders Cleveland Pneumatic Tool Co., Cleveland, OH, ca. 1905 p1.

At left is a 25 pound carbide drum, and at right is a 25 pound blasting powder keg. The carbide drum, which features an internal-thread widemouth lid, usually reflects activity post-dating 1920, when the carbide lamp had supplanted candlelight underground. Blasting powder kegs, identifiable by a large diameter, a small bung opening, and embossed labeling, are common at hardrock mines that operated prior to the 1890s, when dynamite replaced blasting powder. A few impoverished mining operations continued to use blasting powder into the 1910s because it cost less than dynamite.

Author.

While the consumers of the glass products realized no difference between manganese and selenium, today's visitor to historic sites certainly can appreciate the distinction. After exposure to solar radiation over the course of years, the clarifying metals used prior to World War

I caused window and bottle glass to take on subtle tints of amethyst and aqua discoloration. Selenium caused glass to subtly discolor with hues of golden yellow. During the 1950s glass makers substituted other materials for selenium which remain totally colorless today, even after decades of exposure to sunlight. The visitor should be aware that the transitions between any of the types of glass were not instantaneous. They occurred over the course of years.

Left to right: hole-in-cap can with a lapped side-seam, hole-in-cap can with an inner rolled side-seam, sanitary can, and a vent-hole can.
Author.

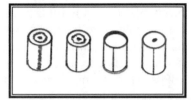

The types of nails that construction workers used to assemble mine buildings also serve as date indicators. During most of the nineteenth century America's builders used *cut nails,* commonly misnomered as "square nails," to assemble wooden structures, but in the late 1880s, hardware manufacturers began to introduce *wire nails,* commonly referred to as "round nails," which we still use today. Cut nails were manufactured by machines that drew out and cut square iron rods into nails. Machines made nails out of inexpensive heavy steel wire in a few simple steps. As the quality of wire improved and inventors developed efficient machines, the cost of making wire nails decreased. As a result, wire nails exploded in popularity during the early 1890s and by around 1892 they had almost totally replaced cut nails.

Industrial artifacts specific to mining fall into a pattern similar to that of building materials. Not only did form and variety of industrial items change over time, but artifacts such as dynamite boxes, blasting powder kegs, blasting cap tins, candlesticks, and carbide lamps offer makers' marks that can enable the visitor to narrow down the time in which they were used.

Food-related items, including cans and bottles serve as excellent date indicators because they were thrown away after a single use. While detailed discussions of makers' marks and of bottle manufacturing technologies are beyond the scope of this work and are covered by other references, cans warrant brief coverage.

During the mid-nineteenth century when hardrock mining in the West became serious business, American food packers sent fruit, vegetables, meat, milk, and tar to the mines in cans. These early vessels, known as in *hole-in-cap cans,* consisted of a body, a top, a bottom, and a filler cap, all of which a worker soldered together by hand. A food packer filled the can through a large opening in the can's bottom, soldered on a cap, then heated the can to sterilize the contents. The pressure escaped out of a pinhole that had been punched into the cap. Immediately following sterilization, another cannery worker sealed the pinhole with a dab of molten solder, pasted on a paper label, and packed the can into a wooden box.[2]

Originally, workers built can bodies by curling a sheet of tin-plated iron around a form and they soldered the overlapping edges to form a *lapped side-seam,* followed by soldering on the can's top and bottom. Beginning in the late 1880s, some can makers began mechanizing their manufacturing processes, and workers used a special machine that bent and locked the side-seam. The new construction method, known as the *inner-rolled side-seam,* was cheaper and faster to produce, and it swept the canning industry by the mid-1890s. Most cannery workers added a solder seal to ensure that the seam did not leak.

Food packers continued to use hole-in-cap cans with inner-rolled side-seams to pack food as late as the 1910s, but in the early 1900s a new can emerged that supplanted the older manufacturing technologies. A problem with the old hole-in-cap cans was that the lead solder had a propensity to contaminate the food inside, especially when the contents were acidic. They were also costly to produce. The public began to see the *sanitary can* as a superior alternative. The sanitary can, still in use today, consisted of a body constructed with an inner-rolled side-seam, and the ends were crimped in an automated and solder-free process. The sanitary can grew in popularity during the 1900s, but it did not replace hole-in-cap cans until around 1920. At about the same time the *vent-hole can* started to become popular for packing milk. The vent-hole can appeared similar to the old hole-in-cap vessels, except it lacked the large cap that a cannery worker had to solder on. Instead, vent-hole cans featured a single tiny hole in the center of the bottom through which a needle injected milk, then the can was sterilized and the hole sealed with solder. The vent-hole can experienced a rise to supremacy for preserving milk safely parallel to that of sanitary cans. Today's visitor is likely to encounter any one of the above types of cans at historic mines, which can help date activity at the site.[3]

⚒

Interpreting the Mine Site

A mine is like an iceberg. The surface plant forms a visible cap that reflects the extent and complexity of the unseen underground workings. The surface plant also serves as the physical expression of mining engineers' and miners' reactions to the interplay of six fundamental influential factors. These include the presence or absence of ore, the company's available capital and investor confidence, the mine's geographic location, the climate, the structural geology, and the operating timeframe. By contemplating the physical remains of the site within the context of these six major influencing factors, we can begin to perceive trends that lend color and dimension to an interpretation of the mine. In this section we will explore how the remains of historic mine sites reflect these factors, and we will examine some common patterns.

The above six factors molded how miners and engineers developed and equipped mines, and while they were all influential, they fall into a hierarchy of significance. The primary factor that influenced how engineers and miners carried out their work was the combination of investor confidence and available capital. Building and running a mine was a function of money, which had to be coaxed from investors. Mining companies that inspired confidence in the investing community were able to provide substantial funding for the development of their properties, including the construction of production-class surface plants. Small mining companies that lacked investors had to spend their money wisely, and they were not able to afford advanced technology. Such outfits had to solve the problems of mining through labor-intensive temporary-class surface plants. In some senses, temporary-class surface plants locked companies into a limited income, because such facilities impeded ore production.

Capitalists responded both subconsciously and consciously to several factors when deciding whether to invest in an operation, and how much money to invest. Investors felt more comfortable financing operations in prominent and well-developed mining districts than those in remote regions. Operations with proven ore were easy to promote, and hence were likely to inspire investor confidence. Small operations situated in remote regions, prospects featuring little or no ore, and companies organized during the early years of hardrock mining were not easy to promote, and did not inspire investors to the same degree.

The second most influential factor that impacted how engineers and miners developed and operated a mine consisted of the presence or absence of ore, and how much payrock had been proven to exist. On one hand, mining companies with large, proven ore reserves developed their properties for production in high volumes. This translated into advanced, mechanized surface plants. On the other hand, mining companies with little or no proven ore usually equipped their operations with temporary-class surface plants because they both struggled to allocate capital and had no need for costly production-class plants.

The time period during which a mining company operated also influenced how miners and engineers set up and ran a mine. The technology employed by mining companies increased in efficiency, complexity, and availability, and decreased in price between the 1860s and 1910s. As a result, mining operations organized toward the end of the Gilded Age tended to be better equipped and more heavily mechanized than the early operations.

Structural geology constitutes the fourth significant factor that influenced how miners and engineers developed a mine. The presence or absence of ore and the structure of the ore body is an aspect of geology. While the mere existence of ore influenced the financial status of a mining company, the physical nature of the ore deposit posed engineering problems that determined the type of surface plant the engineer would organize. When an ore body was gently sloping, miners drove tunnels to tap it, and when the ore body was steep, miners sank a shaft. As we have seen in the preceding chapters, tunnels and shafts required different types of surface plants, and consumed different quantities of capital. Last, deep ore required complex, efficient, and heavily mechanized plants.

Structural geology also affected the topography overlying a mineral body, which impacted plant form. The mining engineer planned to drive a tunnel when the hillsides were steep, and he opted to sink a shaft when the terrain was flat. The topography also influenced transportation from commercial centers to the mine. Engineers planning the movement of large quantities of materials to and from mines located high on the sides of steep mountains favored aerial tramways, while gentle topography was conducive to the grading of roads and railroads. When viewed on a more narrow scope, geology also impacted day-to-day mining operations. The nature of the rock that the mine workings penetrated determined how heavily miners had to timber shafts and tunnels, whether they had to provide extra ventilation to remove natural gases as at Cripple Creek, and whether groundwater threatened to flood low workings.

The physical climate constitutes the last of the six major factors that miners and engineers had to consider when they established a mine. Harsh climates typical of the mountain states and eastern Great Basin rendered the costs of mining high and capital did not go as far. Cold climates had the potential to shut down an operation for at least half the year unless the engineer enclosed key surface plant components in buildings, and spent money maintaining transportation avenues. The cold also reduced the energy efficiency of boilers and made the work environment unpleasant. In some cases mining companies had a difficult time keeping a skilled mining crew on staff because of the climate. The short warm season limited the timeframe the mining company had to complete construction and improvements to its surface plant. On the other hand, mining companies found that operating in warm climates did not require as much development capital and did not generate the same operating expenses because shelter was not as much of an issue, they were able to transport materials all year, and fuel requirements were less.

In light of the six basic influential factors discussed above, mine sites fall into a handful of predictable interpretive patterns, which are reflected by the remains of their surface plants. Primitive temporary surface plants at mine sites in both remote mining districts and well-developed areas generally represent outfits that had poor funding, little or no investor confidence, limited underground workings, and a lack of economic ore. These operations, characterized as typically having only one building, a small waste rock dump, the barest of temporary-class facilities, and sparse deposits of artifacts were usually underground prospects abandoned after a short life.

Simple and small mine sites featuring one to two buildings, temporary-class plant components, and a modest waste rock dump represent the next general interpretive pattern. The surface plants associated with adits tend to possess basic components and modestly sized waste rock dumps. The plants associated with shafts may also feature the remains of a sinking-class mechanized hoisting system. Early operations relied on horse whims while outfits active after around 1880 employed steam power, and those in the Great Basin and Southwest may have installed gasoline hoists. When viewed in the context of well-developed districts where capital went far, these sites represent deep prospecting backed by restricted financing. Generally, such operations failed due to a lack of ore. When such sites are viewed within the framework of the difficult conditions encountered in poorly developed mining districts, and operations active prior to the late 1880s, the interpretation shifts. Such sites suggest the mining company had modest but restricted capital reserves, trepedacious but optimistic investors, and that the company had possibly struck minor ore deposits. Mining companies working in such difficult conditions found that the costs of building and running a mine were high, and therefore were forced to make do with simple surface plants.

The next general interpretive pattern includes historic mine sites that feature a surface plant consisting predominantly of temporary-class components, possibly a few small production-class components, and a substantial waste rock dump. The associated artifact assemblage may be broad in terms of the variety of items, and dense in distribution. Shafts falling into this pattern possess well-built sinking-class hoisting systems in addition to the facilities in common with adits. In the context of remote mining districts far from commercial centers, sites of this magnitude may have represented modest financing, confident investors, the extraction of some ore, and an operation of moderate duration. In terms of well-developed, prominent mining districts, these types of operations may have been merely deep and unsuccessful prospects, or they may have been poorly financed attempts to extract ore. Mine sites with waste rock dumps less than 125 by 125 feet in area and the surface plant core smaller than fifty-by-fifty feet in area generally fall into this pattern.

Mechanized plants featuring a mixture of production-class and temporary components constitute the first interpretive pattern in the production-end of our spectrum of mine sites. These sites typically consist of a well-appointed shop, evidence of machinery, and a substantial waste rock dump. The surface plants associated with shafts may include hoists that were either production-class or temporary-class in nature, and well-built headframes. When the visitor encounters such sites in well-developed mining districts proximal to commercial centers, they may perceive such mines in one of two ways. In one case such sites may have been deep but unsuccessful prospects backed by substantial financing. The associated companies were rife with optimism that their miners would strike ore. In the other case, the mine may have been run by a marginally profitable operation with limited financing. When well equipped mine sites of mixed composition are viewed in the framework of remote and poorly developed mining districts, they represent a higher level of capital and productivity than their counterparts in well-developed districts. The attempt at mechanization reflects an effort at increasing the tonnage of rock hauled out of the mine. If the operation produced ore in economical quantities, then the remains of the surface plant should feature vestiges of an ore bin.

Mine sites that feature the remains of light production-class facilities fall into another interpretive pattern in our spectrum. The plants associated with these mines consisted of components that met production-class specifications, but the machines and other facilities were small, simple, and relatively inexpensive. The mine's shop tended to be well equipped and may have had a vernacular power hammer and drill-steel sharpener, but lacked other power appliances. The mine may have featured a small air compressor and steam boiler, a ventilation system, and an ore bin. The artifact densities at such productive mines tend to be moderate, and the remains usually include boiler clinker and shop refuse dumps, reflecting an operation of moderate duration. When viewed in the setting of a well-developed mining district, the sites of small productive mines confirm the expenditure of capital, modest investor confidence, and minor ore production. Further, the mining company may have employed a professionally-educated mining engineer who built the best plant he could within financial constraints. In terms of remote and little-known mining districts, and mines active prior to the late 1880s, small production-class plants represent sound investor confidence, an earnest attempt at maximizing ore production, and modest profitability. Sites comprised of light production-duty components often feature a waste rock dump around 175 by 175 feet in area and the core portion of the surface plant around fifty-by-seventy-five feet in area.

Well-developed mechanized plants represent an interpretive pattern in the advanced portion of the spectrum of mine sites. Such mine sites are typically medium-sized to large in scale, their surface plants are often exclusively made up of production-class facilities, and their waste rock dumps are huge. These sites commonly feature foundations for machines such as air compressors and boilers, the remains of ore storage facilities, and other capital-intensive components. The artifact densities of mechanized mine sites are typically moderate to high and the assemblage includes a variety of machine-related items, which reflects long-term occupation of the site and intense activity. The visitor may also encounter distinct shop and boiler clinker dumps that also represent long-term operation. In terms of prominent and wealthy mining districts, well-developed mine sites represent sound investor confidence, a heavily financed attempt at maximizing ore production, and an operation that endured for years.

Mechanized medium-sized and large mine sites encountered in remote mining districts represent a greater degree of capital expenditure and investor confidence than mines of similar magnitude in prominent mining districts. Mining operations of this scale were usually among the wealthiest in a given remote area, and the company spent considerable sums of money in efforts to develop large and deep ore bodies. Yet, because the costs of mining were high in districts that lacked well-graded roads or railroads, the mechanized medium-sized and large mines there may not have been as profitable as their brethren in well-developed districts. Mines that fall into the advanced portion of the spectrum of sites generally possess waste rock dumps at least 225 by 225 feet in area and surface plants greater than seventy-five-by-seventy-five feet in area.

Plants equipped for heavy production were rarely seen outside of well-known, industrialized mining districts. The surface plants of such operations consisted of substantial and advanced production-class machines such as first-motion hoists, multiple boilers, large compressors, and mechanized shops. Today, large mines often feature a broad variety of substantive remains left by several occupations. The artifact densities are high, and the sites are expansive. The visitor to such sites may intuitively perceive that the operation produced significant volumes of ore that made the investors wealthy. The outfits running such mines were heavily funded, progressive, well staffed, and had guidance from a prominent professionally-trained engineer. Additionally, the mine may have had the attention of prominent investors and other notable mining men.

Historic mine sites do not always fit neatly into the broad interpretive patterns discussed above. Mining companies engaged in some practices that impacted both the underground workings and the associated surface plant. Three such practices, leasing, contract mining, and stock scams, each impacted the composition of the surface plant, and its relationship with the underground workings, levels of production, and the duration of the operation. Research on the mining district and the mine may enable mining historians and archaeologists to determine whether a site has been the subject of either practice.

Both contract mining and leasing gave miners every incentive to move as much rock as possible while incurring minimal expenditures. Mining engineers and management complained that in their zeal, lessees and contract miners gutted mines of ore in a manner that ruined the underground workings, and left passages and shaft stations in a mess.[4]

Lessees' aversion to spending capital and their avoidance of deadwork manifested in the surface plants of the mines they operated. Lessees and contract miners chronically installed inexpensive and substandard equipment, they performed minimal maintenance and upkeep, and they put little effort into completing proper repairs. For example, lessees anchored compressors and hoists to timber foundations, which mining engineers abhorred, and they erected flimsy buildings. In its early years, the profitable Joe Dandy Mine in Cripple Creek had been leased to a party of miners that installed its own plant. In keeping with the typical behavior of lessees and contract miners, they had slapped together a flimsy wooden shaft house, which completely blew over on a windy day in the spring of 1896. The remains of a mine operated by lessees for much of its life might appear today to have been small, poorly funded, lacking in investor confidence, and marginally productive.[5]

Wealthy scam artists and shady promoters also had the potential to heavily impact the mine as a physical entity. Adhering to the philosophy that one must spend money to make money, crooked promoters had construction crews erect handsome surface plants at worthless properties with the intent of establishing an artificial sense of confidence among investors. Crooked promoters used the confidence they had gained to sell stock in the mine in hopes of reaping large profits. The promoters then made off with the money, leaving the investors to discover that they had been duped.

While mine scams were not common in the West, the physical remains left by such practices can manifest at historic mine sites today. A mine that had been equipped in a manner realistic enough to fool yesteryear's investors will appear today to possess the characteristics of a well-funded and profitable mine that had inspired investor confidence. However, upon close examination, historians may detect a lack of significant work at such mines. The remains may include a lack of waste rock, a substandard road, and an artifact assemblage that is limited in quantity and variety, reflecting a brief occupation. In addition, underground workings are limited and possess no significant stopes. Today's visitor may also find that remaining machinery and structures such as ore bins exhibit little wear.

$$\times\!\!\!\!\times$$

Conclusion

The historic mine sites that dot the western landscape hold a place of great significance today. When we as visitors examine historic mine sites, we walk among the ghosts of one of the world's most incredible and exiting eras. Mine sites constitute the physical remains of the collision of Industrial Revolution technology, westward exploration, the wilderness, and the search for wealth. Each waste rock dump, every adit, and all of the shafts were created by the backbreaking labor of hardrock miners, often aided by the planning of mining engineers. Between the 1860s and World War II thousands of miners called the mine and all of its dangers, discomforts, and benefits their workplace. The western mining landscape also represents a huge outlay of capital by America's middle and upper classes. Every prospect shaft, exploratory adit, and mine required money to develop and operate. Most mining companies went bust, and as a result many investors lost out, while only a few became wealthy. The lust for mineral riches, the men who financed the dream, and the miners who made it a reality, paved the way for the urbanization and industrialization of the West.

Despite the importance of the remains of our mining heritage, many hardrock mine sites face serious threats. Yet these sites are the subject of greater historic preservation, analysis, and evaluation in terms of cultural resource laws than ever before. The public, government agencies, and conscientious elements of private industry are increasingly recognizing the importance of historic mine sites. As sites fall to land-use projects, development, and decay, finding meaning in these irreplaceable cultural touchstones is more important than ever. Mine sites possess the capacity to portray a unique, ground level facet of western history.

TABLES FOR IDENTIFYING AND INTERPRETING SURFACE PLANT MACHINERY AND OTHER FACILITIES

Air Compressor Specifications

Table 1: Compressed Air Consumption of Piston Rockdrills
Air consumption is measured by the piston displacement within the drill's cylinder.

Air Pressure (cfm)	Diameter of Rockdrill Piston							
	2 1/4 in.	2 1/2 in.	2 3/4 in.	3 in.	3 1/4 in.	3 1/2 in.	3 3/4 in.	4 in.
60 psi	60 cfm	68 cfm	82 cfm	90 cfm	100 cfm	108 cfm	115 cfm	123 cfm
80 psi	76 cfm	86 cfm	104 cfm	114 cfm	127 cfm	131 cfm	145 cfm	155 cfm
100 psi	92 cfm	104 cfm	126 cfm	138 cfm	152 cfm	166 cfm	176 cfm	188 cfm

Table 2: Air Compressor Specifications: *Type, Time Frame, and Capital Investment*

Compressor Type	Age Range	Capital Investment
Upright: 2 Cylinders, Belt-driven	1900s-1940s	Low
Upright: 3 to 4 Cylinders, Integral Gasoline Piston	1930s-Present	Moderate
V Cylinder	1930s-Present	Moderate to High
Straight-Line, Single-stage, Gasoline Engine Driven	1900s-1930s	Low
Straight-Line, Single-stage, Steam-driven	1880s-1920s	Moderate
Straight-Line, Two-stage, Steam-driven	1890s-1920s	High
Straight-Line, Three-stage, Steam-driven	1890s-1920s	Very High
Straight-Line, Single-stage, Geared to Electric Motor	1900s-1920s	Moderate
Straight-Line, Various Stages, Geared to Electric Motor	1900s-1920s	High
Straight-Line, Single-stage, Belt-driven by Electric Motor	1900s-1940s	Low
Duplex, Single-stage, Steam-driven	1890s-1920s	Moderate
Duplex, Two-stage, Steam-driven	1890s-1920s	High
Duplex, Three-stage, Steam-driven	1890s-1920s	Very High
Duplex, Two-stage, Belt-driven	1900s-1940s	Moderate
Duplex, Three-stage, Belt-driven	1900s-1940s	Moderate to High

Table 3: Air Compressor Specifications: *Type, Duty, Foundation*

Compressor Form	Power Source	Compressor Duty	Foundation Footprint	Foundation Size	Foundation Material
Upright: 2 Cyclinders	Belt-Driven Petroleum Engine	Temporary	Rectangular Foundation	2x6 ft.	Timber Frame
Upright: 3-4 Cylinders	Integral Petroleum Engine	Temporary & Production	Rectangular	3x7 ft. to 3x8 ft.	Timber or Steel Frame
V-Cylinder: 2-3 Cylinders	Integral Petroleum Engine	Production	Rectangular	3x6 ft. to 3x8 ft.	Concrete
V-Cylinder: 4-8 Cylinders	Integral Petroleum Engine	Production	Rectangular	3x8 ft. to 3x12 ft.	Concrete
Straight-Line Gasoline	Integral Petroleum Engine	Temporary	Rectangular	2x5 ft. to 2x7 ft.	Concrete
Straight-Line Single-stage	Integral Steam Engine	Temporary	Rectangular	2x6 ft. to 2x7 ft.	Timber, Concrete or Masonry
Straight-Line: Single-stage	Integral Steam Engine	Production	Rectangular	3x9 ft. to 3x15 ft.	Concrete or Masonry
Straight-Line: Two-stage	Integral Steam Engine	Production	Rectangular, or Shallow L	4x14 ft. to 6x27 ft.	Concrete or Masonry
Straight-Line: Three Stage	Integral Steam Engine	Production	Rectangular, or Shallow L	5x16 ft. to 6x30 ft.	Concrete or Masonry
Straight-Line	Geared Electric	Production	Rectangular, Extension for Motor	3x9 ft. to 5x15 ft. Motor: 2x3 ft.	Concrete
Straight-Line	Belt-driven	Production	Rectangular with Aligned Motor Mount	2x7 ft. to 5x15 ft. Motor: 2x3 ft.	Concrete
Duplex: Single-stage	Integral Steam Engine	Production	U-Shaped	5x6 ft. to 9x10 ft.	Concrete or Masonry
Duplex: Single-stage	Integral Steam Engine	Production	Parallel Rectangular Pads	7x10 ft. to 10x30 ft.	Concrete or Masonry
Duplex: Two-stage	Integral Steam Engine	Production	U-Shape	5x6 ft. to 15x15 ft.	Concrete or Masonry
Duplex: Two-stage	Integral Steam Engine	Production	Parallel Rectangular Pads	10x15 ft. to 15x35 ft.	Concrete or Masonry
Duplex: Multi Stage	Geared Electric	Production	U with Aligned Motor Mount	5x6 ft. to 15x16 ft.	Concrete
Duplex: Multi Stage	Belt-driven Electric	Production Motor Mount	U with Aligned	5x6 ft. to 15x16 ft.	Concrete

Table 4: Air Compressor Specifications: *Straight-line Single-stage Steam, 1880s-1920s*

Floor Space of Compressor (in feet)		Compression Cylinder Size (in inches)		Boiler Horsepower Required (hp)	Cubic Feet of Air Produced per Minute (cfm)	Number of Rockdrills Powered (At 90-110 pounds per square inch)	
Length	Width	Diameter	Stroke			Piston Drill	Small Hammer Drill
7 ft.	2 ft.	6 in.	8 in.	10 hp	50 cfm	1 small drill	1 drill
9 ft.	3 ft.	8 in.	10 in.	24 hp	70-115 cfm	1 drill	1 drill
9 ft.	3 ft.	12 in.	8 in.	30 hp	215-260 cfm	2-3 drills	3-4 drills
11 ft.	4 ft.	10 in.	10 in.	45 hp	220 cfm	2 drills	2-3 drills
11 ft.	4 ft.	16 in.	10 in.	65 hp	560 cfm	5-6 drills	6-7 drills
12 ft.	4 ft.	12 in.	12 in.	75 hp	360 cfm	4 drills	5 drills
12 ft.	4 ft.	18 in.	12 in.	95 hp	810 cfm	8-9 drills	11-12 drills
14 ft.	5 ft.	14 in.	14 in.	115 hp	550 cfm	6-7 drills	7-8 drills
14 ft.	5 ft.	20 in.	14 in.	130 hp	1120 cfm	12 drills	14 drills

(Adapted from Ingersoll Rock Drill Co., 1888: 29 and Peele, 1918: 1065).

Table 5: Air Compressor Specifications: *Straight-line Two-stage Steam, 1890s-1920s*

Floor Space of Compressor (in feet)		Compression Cylinder Size (in inches)			Boiler Horsepower Required (hp)	Cubic Feet of Air Produced per Minute (cfm)	Number of Rockdrills Powered (At 90-110 pounds per square inch)	
Length	Width	Diameter Cyl #1	Cyl #2	Stroke			Piston Drill	Small Hammer Drill
14 ft.	4 ft.	12 in.	7 in.	14 in.	50 hp	290 cfm	3 drills	4 drills
15 ft.	5 ft.	14 in.	9 in.	14 in.	68 hp	395 cfm	4 drills	5 drills
16 ft.	5 ft.	16 in.	10 in.	16 in.	95 hp	560 cfm	5-6 drills	6-7 drills
16 ft.	5 ft.	18 in.	11 in.	16 in.	120 hp	710 cfm	6-7 drills	7-8 drills
19 ft.	6 ft.	20 in.	12 in.	20 in.	146 hp	910 cfm	9-10 drills	10-11 drills
22 ft.	6 ft.	22 in.	14 in.	20 in.	190 hp	1160 cfm	11-13 drills	12-14 drills
22 ft.	7 ft.	24 in.	14 in.	24 in.	222 hp	1380 cfm	13-15 drills	15-16 drills
25 ft.	7 ft.	26 in.	16 in.	30 in.	296 hp	1845 cfm	17-19 drills	18-20 drills
26 ft.	7 ft.	30 in.	18 in.	30 in.	395 hp	2450 cfm	23-25 drills	25-26 drills

(Adapted from Peele, 1918: 1067).

Table 6: Air Compressor Specifications: *Straight-line Single-stage Steam, 1880s-1890s*
With Outboard Flywheel

Floor Space of Compressor Not Including Flywheel (in feet)		Compression Cylinder Size (in inches)		Boiler Horsepower Required (hp)	Cubic Feet of Air Produced per Minute (cfm)	Number of Piston Rockdrills Powered (At 60 pounds per square inch)
Length	Width	Diameter	Stroke			
15 ft.	7 ft.	10 in.	16 in.	23 hp	145 cfm	2 drills
19 ft.	8 ft.	12 in.	30 in.	45 hp	295 cfm	4 drills
25 ft.	9 ft.	16 in.	30 in.	108 hp	555 cfm	8 drills
30 ft.	13 ft.	20 in.	48 ft.	140 hp	875 cfm	12 drills
35 ft.	15 ft.	28 in.	48 in.	215 hp	1370 cfm	20 drills

(Rand, 1886: 29; author's field data).

Table 7: Air Compressor Specifications: *Duplex Single-stage Steam, 1880s-1890s*

Floor Space of Compressor (in feet)		Compression Cylinder Size (in inches)		Boiler Horsepower Required (hp)	Cubic Feet of Air Produced per Minute (cfm)	Number of Piston Rockdrills Powered (At 100 pounds per square inch)
Length	Width	Diameter	Stroke			
15 ft.	7 ft.	10 in.	16 in.	45-60 hp	290 cfm	4 drills
19 ft.	8 ft.	12 in.	30 in.	90-110 hp	580-590 cfm	8-11 drills
23 ft.	11 ft.	20 in.	12 in.	175-200 hp	995-1115 cfm	16-20 drills
30 ft.	13-15 ft.	20-22 in.	30-48 ft.	280-310 hp	1745-1980 cfm	25-30 drills
35 ft.	15-17 ft.	28-30 in.	48 in.	430-520 hp	2735-2945 cfm	40-55 drills

(Adapted from Ingersoll, ca. 1888:1068).

Table 8: Air Compressor Specifications: *Duplex Single-stage Steam, 1890s-1920s*

Floor Space of Compressor (in feet)		Compression Cylinder Size Both Cylinders (in inches)		Boiler Horsepower Required (hp)	Cubic Feet of Air Produced per Minute (cfm)	Number of Rockdrills Powered (At 90-110 pounds per square inch)	
Length	Width	Diameter	Stroke			Piston Drill	Small Hammer Drill
12 ft.	7.5 ft.	11 in.	14 in.	85-95 hp	610 cfm	5-6 drills	6-7 drills
13 ft.	8.5 ft.	13 in.	16 in.	125-140 hp	875 cfm	8-9 drills	9-10 drills
14 ft.	9 ft.	15 in.	16 in.	165-185 hp	1165 cfm	11-12 drills	12-13 drills
15.5 ft.	9 ft.	14 in.	20 in.	220 hp	1070 cfm	11-13 drills	13-14 drills
17 ft.	10 ft.	15 in.	20 in.	180-250 hp	1350 cfm	13-14 drills	14-16 drills

(Adapted from Peele, 1918: 1066, Rand, 1904: 15).

Table 9: Air Compressor Specifications: *Duplex Two-stage Steam, 1890s-1920s*

Floor Space of Compressor (in feet)		Compression Cylinder Size (in inches)			Boiler Horsepower Required (hp)	Cubic Feet of Air Produced per Minute (cfm)	Number of Rockdrills Powered (At 90-110 pounds per square inch)	
Length	Width	Diameter Cyl #1	Cyl #2	Stroke			Piston Drill	Small Hammer Drill
8 ft.	5 ft.	10 in.	7 in.	8 in.	25 hp	145 cfm	1-2 drills	2 drills
8.5 ft.	5.5 ft.	12 in.	8 in.	10 in.	45 hp	250 cfm	3drills	5 drills
9.5 ft.	6 ft.	14 in.	9 in.	12 in.	65 hp.	375 cfm	4 drills	5-6 drills
11 ft.	7.5 ft.	16 in.	10 in.	14 in.	95 hp.	535 cfm	5-6 drills	6-7 drills
12 ft.	8.5 ft.	18 in.	11 in.	16 in.	125 hp	710 cfm	7 drills	8-9 drills
16 ft.	10 ft.	22 in.	13 in.	20 in.	230 hp	1320 cfm	13-14 drills	14-16 drills
19 ft.	12 ft.	28 in.	17 in.	24 in.	420 hp	2395 cfm	23-24 drills	25-26 drills
23 ft.	11 ft.	20 in.	12 in.	24 in.		1100 cfm	11-12 drills	12-13 drills
23 ft.	13 ft.	28 in.	17 in.	24 in.		2140 cfm	21-22 drills	22-24 drills
23 ft.	14 ft.	32 in.	20 in.	30 in.	420 hp	3350 cfm	33-34 drills	34-36 drills
26 ft.	11 ft.	22 in.	14 in.	30 in.		1520 cfm	15-16 drills	16-18 drills
26 ft.	13 ft.	32 in.	20 in.	30 in.		3220 cfm	33-34 drills	34-36 drills
31 ft.	15 ft.	32 in.	20 in.	36 in.		3525 cfm	35-36 drills	37-38 drills
35 ft.	16 ft.	36 in.	22 in.	42 in.		4700 cfm	48-50 drills	49-51 drills
38 ft.	17 ft.	38 in.	23 in.	48 in.		5675 cfm	57-59 drills	58-60 drills

(Adapted from Peele, 1918: 1067; Rand, 1904: 19).

Table 10: Air Compressor Specifications: *Straight-line Single-stage Belt-Driven, 1900s-1930s*

Floor Space of Compressor (in feet)		Compression Cylinder Size (in inches)		Boiler Horsepower Required (hp)	Cubic Feet of Air Produced per Minute (cfm)	Number of Rockdrills Powered (At 90-110 pounds per square inch)	
Length	Width	Diameter	Stroke			Heavy Hammer Drill	Small Hammer Drill
5 ft.	2 ft.	7 in.	6 in.	12 hp	90 cfm	None	1 drill
5.5 ft.	2 ft.	8 in.	6 in.	14 hp	122 cfm	1 drill	1 drill
6 ft.	2.5 ft.	8 in.	8 in.	22 hp	139 cfm	1 drill	1-2 drills
6.5 ft.	2.5 ft.	9 in.	8 in.	25 hp.	175 cfm	1-2 drills	2-3 drills
6.5 ft.	2.5 ft.	10 in.	8 in.	25 hp	218 cfm	1-2 drills	2-3 drills
7.5 ft.	3 ft.	10 in.	10 in.	38 hp	236 cfm	2 drills	2-3 drills
7.5 ft.	3 ft.	14 in.	10 in.	58 hp	463 cfm	4 drills	5-6 drills
9 ft.	4 ft.	12 in.	12 in.	67 hp	375 cfm	3-4 drills	4-5 drills
9 ft.	4 ft.	16 in.	12 in.	87 hp	670 cfm	6-7 drills	7-8 drills

(Adapted from Keystone, 1925: 161).

Table 11: Air Compressor Specifications: *Straight-line Two-stage Belt-Driven, 1900s-1930s*

Floor Space of Compressor (in feet)		Compression Cylinder Size (in inches)			Motor Horsepower Required (hp)	Cubic Feet of Air Produced per Minute (cfm)	Number of Rockdrills Powered (At 80-110 pounds per square inch)	
Length	Width	Diameter Cyl #1	Cyl #2	Stroke			Heavy Hammer Drill	Small Hammer Drill
8.5 ft.	2.5 ft.	12 in.	7 in.	10 in.	50 hp	240 cfm	2 drills	3 drills
10.5 ft.	2.5 ft.	14 in.	9 in.	10 in.	65 hp	330 cfm	3 drills	4 drills
11 ft.	3 ft.	16 in.	10 in.	14 in.	95 hp	535 cfm	5-6 drills	6-7 drills
12 ft.	3 ft.	18 in.	11 in.	14 in.	120 hp	680 cfm	6-7 drills	7-8 drills
12.5 ft.	3.5 ft.	20 in.	12 in.	16 in.	155 hp	870 cfm	8-9 drills	9-10 drills
13 ft.	3.5 ft.	22 in.	14 in.	16 in.	190 hp	1050 cfm	11-13 drills	12-14 drills
14 ft.	3.5 ft.	22 in.	14 in.	18 in.	210 hp	1180 cfm	12-14 drills	13-15 drills
15 ft.	3.5 ft.	24 in.	15 in.	18 in.	225 hp	1410 cfm	14-16 drills	15-17 drills
16 ft.	4 ft.	26 in.	16 in.	20 in.	295 hp	1845 cfm	19-20 drills	20-21 drills

(Adapted from Peele, 1918: 1073, and from field data).

Table 12: Air Compressor Specifications: *Duplex Single-stage Belt-Driven, 1900s-1940s*

Floor Space of Compressor (in feet)		Compression Cylinder Size (in inches)		Motor Horsepower Required (hp)	Cubic Feet of Air Produced per Minute (cfm)	Number of Rockdrills Powered (At 90-110 pounds per square inch)	
Length	Width	Diameter	Stroke			Heavy Hammer Drill	Small Hammer Drill
6 ft.	5 ft.	6 in.	8 in.	25 hp	123 cfm	1 drill	2 drill
7 ft.	5.5 ft.	7 in.	10 in.	40 hp	200 cfm	2 drills	2-3 drills
8.5 ft.	6 ft.	9 in.	12 in.	60 hp	309 cfm	3 drills	3-4 drills
10 ft.	7.5 ft.	10 in.	14 in.	85 hp	420 cfm	4 drills	4-5 drills
11.5 ft.	8 ft.	12 in.	16 in.	125 hp	625 cfm	6 drills	6-8 drills
11.5 ft.	8.5 ft.	14 in.	16 in.	175 hp	850 cfm	8-9 drills	9-10 drills
14.5 ft.	9 ft.	14 in.	20 in.	215 hp	1065 cfm	10-11 drills	12-13 drills
14.5 ft.	10 ft.	16 in.	20 in.	285 hp	1395 cfm	14-15 drills	16-18 drills

(Adapted from Rand, 1904: 26, 27).

Table 13: Air Compressor Specifications: *Duplex Two-stage Belt-Driven, 1900s-1940s*

Floor Space of Compressor (in feet)		Compression Cylinder Size (in inches)			Motor Horsepower Required (hp)	Cubic Feet of Air Produced per Minute (cfm)	Number of Rockdrills Powered (At 80-110 pounds per square inch)	
Length	Width	Diameter Cyl #1	Cyl #2	Stroke			Heavy Hammer Drill	Small Hammer Drill
6.5 ft.	5.5 ft.	10 in.	7 in.	8 in.	25 hp	145 cfm	1 drill	1-2 drills
7-8 ft.	5-6 ft.	12 in.	8 in.	10 in.	45 hp	240 cfm	1 drill	2 drills
8-9 ft.	6-7 ft.	14 in.	9 in.	12 in.	65 hp	370 cfm	3-4 drills	4-5 drills
10 ft.	7-8 ft.	16 in.	10 in.	14 in.	95 hp	535 cfm	5 drills	5-6 drills
11 ft.	8-9 ft.	18 in.	11 in.	16 in.	125 hp	700 cfm	7 drills	7-8 drills
12 ft.	8-9 ft.	22 in.	13 in.	16 in.	175 hp	900 cfm	9-11 drills	12-14 drills
13 ft.	8-9 ft.	24 in.	15 in.	18 in.	250 hp	1400 cfm	14-15 drills	16-17 drills
14 ft.	9 ft.	22 in.	14 in.	18 in.	210 hp	1180 cfm	12-14 drills	13-15 drills
14 ft.	9.5 ft.	24 in.	15 in.	18 in.	225 hp	1410 cfm	14-16 drills	15-17 drills
15 ft.	11 ft.	26 in.	16 in.	20 in.	295 hp	1845 cfm	19-20 drills	20-21 drills
19 ft.	12 ft.	28 in.	17 ft.	24 in.	420 hp	2390 cfm	23-24 drills	25-26 drills
23 ft.	14 ft.	32 in.	20 in.	30 in.	590 hp	3350 cfm	33-35 drills	35-37 drills

(Adapted from Peele, 1918: 1073; Rand, 1904: 25).

Table 14: Air Compressor Specifications: *Straight-line Single-stage Gasoline-Powered, 1910s-1930s*

Floor Space of Compressor (in feet)		Compression Cylinder Size (in inches)		Motor Horsepower Required (hp)	Cubic Feet of Air Produced per Minute (cfm)	Number of Rockdrills Powered (At 90-110 pounds per square inch)	
Length	Width	Diameter	Stroke			Heavy Hammer Drill	Small Hammer Drill
9 ft.	2.5 ft.	6 in.	8 in.	15 hp	70 cfm	1 small drill	1 drill
10 ft.	3 ft.	7 in.	10 in.	20 hp	100 cfm	1 drill	1 drill
12 ft.	3 ft.	8 in.	12 in.	30 hp	150 cfm	1 drills	1-2 drills

(Adapted from Peele, 1918: 1075).

Table 15: Air Compressor Specifications: *Angle Compound Two-stage Belt-Driven, 1920s-1940s*

Floor Space of Compressor (in feet)		Compression Cylinder Size (in inches)			Motor Horsepower Required (hp)	Cubic Feet of Air Produced per Minute (cfm)	Number of Rockdrills Powered (At 80-110 pounds per square inch)	
Length	Width	Diameter Cyl #1	Cyl #2	Stroke			Heavy Hammer Drill	Small Hammer Drill
9 ft.	4 ft.	14 in.	9 in.	10 in.	75-100 hp	445-620 cfm	4-6 drills	6-8 drills
10 ft.	6 ft.	16 in.	10 in.	12 in.	85-120 hp	545-750 cfm	5-7 drills	7-9 drills
11 ft.	7.5 ft.	18 in.	11 in.	14 in.	115-150 hp	700-940 cfm	7-9 drills	10-11 drills
11.5 ft.	8 ft.	20 in.	12 in.	14 in.	165-215 hp	1010-1300 cfm	10-13 drills	13-15 drills
11.5 ft.	8 ft.	24 in.	15 in.	18 in.	200-260 hp	1225-1575 cfm	13-15 drills	16-17 drills
12 ft.	10 ft.	22 in.	14 in.	18 in.	240-305 hp	1480-1850 cfm	16-19 drills	18-20 drills

(Adapted from Keystone, 1925: 161; Peele, 1918: 1073).

Transportation

Table 16: Dimensions & Duty of Mine Rail

Rail Type (pounds per yard)	Rail Height	Width of Base	Width of Head	Duty of Rail
Strap Rail	4 in.	1 1/2 in.	1 1/2 in.	Rail consists of iron strap nailed to the top of 2x4 boards. Such rail is temporary.
8 lb	1 1/2 in.	1 1/2 in.	1/2 in.	Temporary: for use with hand-pushed ore cars.
10 lb	1 3/4 in.	1 3/4 in.	3/4 in.	Light duty: for use with hand-pushed ore cars.
12 lb	2 in.	2 in.	1 in.	Light duty: for use with hand-pushed ore cars.
16 lb	2 3/8 in.	2 3/8 in.	1 3/16 in.	Light duty: for use with hand-pushed ore cars and short ore car trains drawn by draft animals.
20 lb	2 5/8 in.	2 5/8 in.	1 3/8 in.	Moderate duty: for use with short ore car trains drawn by draft animals or locomotives weighing 8 tons or less.
25 lb	2 3/4 in.	2 3/4 in.	1 1/2 in.	Moderate duty: for use with ore car trains drawn by draft animals or locomotives weighing at most 10 tons.
30 lb	3 1/8 in.	3 1/8 in.	1 3/4 in.	Moderate to heavy duty: for use with ore car trains drawn by draft animals or locomotives weighing at most 13 tons.
35 lb	3 1/4 in.	3 1/4 in.	1 3/4 in.	Moderate to heavy duty: for use with ore car trains drawn by locomotives weighing at most 16 tons.
40 lb	3 1/2 in.	3 1/2 in.	1 7/8 in.	Heavy duty: for use with ore car trains drawn by locomotives 15 tons or less, and for narrow-gauge railroad spurs.
45 lb	3 3/4 in.	3 3/4 in.	2 in.	Heavy duty: for use with ore car trains drawn by locomotives, and for narrow-gauge railroad spurs.
50 lb	3 7/8 in.	3 7/8 in.	2 1/8 in.	Heavy duty: for use with ore car trains drawn by locomotives, and for narrow-gauge railroad spurs.
60 lb	4 1/4 in.	4 1/4 in.	2 3/8 in.	Heavy duty: for use with ore car trains drawn by locomotives, and for narrow-gauge railroad spurs.

(Adapted from Cummins, 1973: 14-4; Keystone, 1925: 376).

Steam Boiler Specifications

Table 17: Boiler Specifications: *Type, Duty, Age Range*

Boiler Type	Boiler Design	Popularity Age Range	Sinking-Class Size Range	Production-Class Size Range
Plain Cylindrical	Water-filled tank with no flue tubes.	1800-1860s	Up to 6 ft. diam. 18 ft. L.	6 ft. diam., 20 ft. L to 8 ft. diam., diam 40 ft. L.
Flue	1-2 ft. flue tubes through shell. Smoke stack at front	1820-1870s	Up to 3 ft. diam. 14 ft. L.	4 ft. diam., 16 ft. L to 5 ft. diam., 22 ft. L.
Return-tube	Multiple 3-4 ft. flue tubes extending through shell.	1870s-1920s	Up to 3 ft. diam. 12 ft. L.	3 ft. diam, 12 ft. L to 7 ft. diam., 20 ft. L.
Scotch Marine	Firebox and flue chamber enclosed in horizontal shell. Flue tubes through shell.	1890s-1910s	All Sizes.	
Locomotive	Firebox enclosed in steel casing under shell. Flue tubes through shell, smokestack at rear.	1870s-1920s	All Sizes.	Not Manufactured.
Upright/ Vertical	Boiler shell is vertical and stands on Cast-iron base. Firebox is at base. Flue tubes through shell, smokestack on top.	1880s-1920s	All Sizes.	Not manufactured.
Water-tube	Water-tubes and header drums suspended over brick setting by steel girder frame.	1900s-1920s	Not Manufactured	All Sizes.

Table 18: Boiler Specifications: *Locomotive Boilers with no Dome*

Boiler Diameter	Boiler Length	Boiler Height	Horsepower	Boiler Weight	Foundation Construction
2 1/2-3 ft.	9 1/2 ft.	5 1/2 ft.	15 hp	4400 lbs.	Timber skids.
2 1/2-3 ft.	11 1/2 ft.	5 1/2 ft.	20 hp	5200 lbs.	Timber skids.
3 ft.	12 ft.	6 ft.	25 hp	6000 lbs.	Timber skids.
3 ft.	14 ft.	6-6 1/2 ft.	30 hp	6600 lbs.	Timber skids.
3 1/2 ft.	12 ft.	7-7 1/2 ft.	35 hp	7000 lbs.	Timber skids.
3 1/2 ft.	14 ft.	7 1/2 ft.	40 hp	8400 lbs.	Timber skids.
4 ft.	15 ft.	7 1/2 ft.	50 hp	9300 lbs.	Timber skids.
4 ft.	17 ft.	7 1/2 ft.	60 hp	11,000 lbs.	Timber skids.
4 1/2 ft.	17 ft.	8 ft.	70 hp	12,900 lbs.	Timber skids.
4 1/2 ft.	19 ft.	8 1/2 ft.	80 hp	15,000 lbs.	Timber skids.
5 ft.	18 1/2 ft.	9 1/2 ft.	90 hp	16,300 lbs.	Timber skids.
5 ft.	19 1/2 ft.	9 1/2 ft.	100 hp	17,800 lbs.	Timber skids or masonry foundation.
5 1/2 ft.	21 1/2 ft.	10 ft.	150 hp	23,000 lbs.	Timber skids or masonry foundation.

Table 19: Boiler Specifications: *Locomotive Boilers with Dome*

Boiler Diameter	Boiler Length	Boiler Height: Shell & Firebox	Horsepower (hp)	Boiler Weight	Foundation Construction
2 ft. to 2 1/2 ft.	8 1/2- 9 1/2 ft.	4-4 1/2 ft.	6 hp	2400-3100 lbs.	Timber skids.
2 1/2 ft.	10-11 ft.	5 ft.	8 hp	3300-3500 lbs.	Timber skids.
2 1/2 ft.	10-11 ft.	5 1/2 ft.	10 hp	3300-4100 lbs.	Timber skids.
2 1/2-3 ft.	10 1/2- 11 1/2 ft.	5 1/2 ft.	12 hp	3600-4500 lbs.	Timber skids.
2 1/2-3 ft.	10 1/2- 11 1/2ft.	6 ft.	15 hp	4900-5000 lbs.	Timber skids.
3 ft.	12-13 1/2 ft.	5 1/2-6 ft.	20 hp	5300-6300 lbs.	Timber skids.
3-3 1/2 ft.	13-14 ft.	6 ft.	25 hp	6000-7800 lbs.	Timber skids.
3 1/2-4 ft.	13 1/2- 15 1/2 ft.	6-6 1/2 ft.	30 hp	7200-9000 lbs.	Timber skids.
3 1/2-4 ft.	14 ft.	7 ft.	35 hp	7800 lbs.	Timber skids.
3 1/2-4 ft.	15- 16 1/2 ft.	6 1/2 ft.	40 hp	8100-9800 lbs.	Timber skids.
4 ft.	17-18 ft.	7 ft.	50 hp	10,500 lbs.	Timber skids.
4-4 1/2 ft.	18-18 1/2 ft.	7 1/2 ft.	60 hp	13,000 lbs.	Timber skids, or brick pillar at front.
4-4 1/2 ft.	19 1/2-21 ft.	8 ft.	70 hp	15,000 lbs.	Timber skids, or brick pillar at front.
5 ft.	19 1/2-21 ft.	8 1/2 ft.	80 hp	16,200 lbs.	Timber skids, or masonry foundation.
5 ft.	21-22 ft.	8 1/2 ft.	90 hp	18,000 lbs.	Timber skids, or masonry foundation.
5 ft.	22-23 ft.	9 ft.	100 hp	19,000 lbs.	Timber skids, or masonry foundation.

Table 20: Boiler Specifications: *Upright Boilers*

Boiler Diameter	Boiler Height	Horsepower (hp)	Boiler Weight	Foundation Construction
21 in.	7 ft.	3 hp	1000 lbs.	Cast-iron bedplate with no masonry.
2 ft.	6-7 1/2 ft.	4 hp	900-1500 lbs.	Cast-iron bedplate with no masonry.
2 ft. 3 in.	6 1/2-7 1/2 ft.	6 hp	1300-1700 lbs.	Cast-iron bedplate with no masonry.
2 1/2 ft.	7-8 ft.	8 hp	1600-2000 lbs.	Cast-iron bedplate with no masonry.
2 1/2 ft.	8-8 1/2 ft.	10 hp	1700-2300 lbs.	Cast-iron bedplate with no masonry.
3 ft.	8 1/2 ft.	12 hp	2300-2500 lbs.	Cast-iron bedplate with no masonry.
3 ft. - 3 1/2 ft.	8 1/2-9 ft.	15 hp	2600-2800 lbs.	Cast-iron bedplate with no masonry.
3 1/2 ft.	8 1/2-. 9 1/2 ft	20 hp	3000-3600 lbs.	Cast-iron bedplate with no masonry.
3 ft. 8 in.	9-9 1/2 ft.	25 hp	3400-4500 lbs.	Cast-iron bedplate with no masonry.
4 ft.	10 ft.	30 hp	5000 lbs.	Cast-iron bedplate with no masonry.
4 ft.	10 ft	35 hp	5300-5700 lbs.	Cast-iron bedplate with no masonry.
4 ft.	10-12 ft.	40 hp	5900 lbs.	Cast-iron bedplate with no masonry.
5 ft.	10-12 ft.	50 hp	6700 lbs.	Cast-iron bedplate on masonry.

Table 21: Boiler Specifications: *Flue Boilers*

Boiler Diameter	Boiler Length	Horsepower (hp)	Boiler Weight	Setting Construction
2 1/2-3 ft.	21 ft.	15 hp	8000 lbs.	Red brick with firebrick lined firebox.
3 ft.	26 ft.	20 hp	9100 lbs.	Red brick with firebrick lined firebox.
3-3 1/2 ft.	21-26 ft.	25 hp	9400 lbs.	Red brick with firebrick lined firebox.
3 1/2-4 ft.	25 ft.	30 hp	10,600 lbs.	Red brick with firebrick lined firebox.
3 1/2-4 ft.	27 ft.	35 hp	11,300 lbs.	Red brick with firebrick lined firebox.
4 ft.	27 ft.	40 hp	12,300 lbs.	Red brick with firebrick lined firebox.
4 ft.	33 ft.	50 hp	16,200 lbs.	Red brick with firebrick lined firebox.
4 1/2-5 ft.	35 ft.	60 hp	18,100 lbs.	Red brick with firebrick lined firebox.
5 ft.	35 ft.	70 hp	21,700 lbs.	Red brick with firebrick lined firebox.
5 ft.	19 1/2-21 ft.	80 hp	16,200 lbs.	Red brick with firebrick lined firebox.

Table 22: **Boiler Specifications:** *Return-tube Boilers with no Dome*
Dimensions of the masonry setting in the table are based on boilers with half facades. For dimensions of boiler settings with full facades, add one foot to length.

Boiler Diameter	Boiler Height	Horsepower (hp)	Boiler Weight	Setting: Length Width & Height	Setting Construction
3 ft.	7 ft.	20 hp	5000 lbs.	10 ft. L; 6 ft. W; 6 ft. H.	Red brick or stone with firebrick lined firebox.
3 1/2 ft.	8 ft.	25 hp	5800 lbs.	11 ft. L; 6 1/2 ft. W; 7 ft. H.	Red brick or stone with firebrick lined firebox.
3 1/2 ft.	10 ft.	30 hp	6600 lbs.	13 ft. L; 6 1/2 ft.W; 7 ft. H.	Red brick or stone with firebrick lined firebox.
3 1/2 ft.	12 ft	35 hp	7400 lbs.	15 ft. L; 6 1/2-. 7 ft. W; 7 ft. H	Red brick or stone with firebrick lined firebox.
3 1/2 ft.	14 ft	40 hp	8300 lbs.	16-17 ft. L; 7 ft. W; 7-7 1/2 ft. H.	Red brick or stone with firebrick lined firebox.
4 ft.	14 ft.	50 hp	10,300 lbs.	17 ft. L; 7 ft.W; 7 1/2 ft. H.	Red brick or stone with firebrick lined firebox.
4 1/2 ft.	14 ft.	60 hp	12,400 lbs.	17 ft. L; 7 1/2-8 ft. W; 71/2 ft. H.	Red brick or stone with firebrick lined firebox.
4 1/2 ft.	16 ft.	70 hp	13,700 lbs.	19 ft. L; 7 1/2-. 8 ft. W; 71/2 ft. H	Red brick or stone with firebrick lined firebox.
5 ft.	16 ft.	80 hp	16,600 lbs.	19 ft. L; 8 ft. W; 8 ft. H.	Red brick or stone with firebrick lined firebox.
5 1/2 ft.	16 ft.	100 hp	19,600 lbs.	19 ft. L; 8 1/2 ft. W; 8 ft. H.	Red brick or stone with firebrick lined firebox.
6 ft.	16 ft.	125 hp	23,100 lbs	19 ft. L; 9 ft. W; 9 ft. H.	Red brick or stone with firebrick lined firebox.
6 ft.	18 ft.	150 hp	25,000 lbs.	21 ft. L; 9-9 1/2 ft. W; 9 ft. H.	Red brick or stone with firebrick lined firebox.
6 1/2 ft.	18 ft.	180 hp	32,100 lbs.	21 ft. L; 10 ft. W; 9 ft. H.	Red brick or stone with firebrick lined firebox.
6 1/2 ft.	20 ft.	200 hp	35,000 lbs.	24 ft. L; 10 ft. W; 9 ft. H.	Red brick or stone with firebrick lined firebox.
7 ft.	20 ft.	250 hp	38,000 lbs.	23 ft. L; 10 ft. W; 10 ft. H.	Red brick or stone with firebrick lined firebox.

Table 23: Boiler Specifications: *Return-tube Boilers with Dome*

Dimensions of the masonry setting in the table are based on boilers with half facades. For dimensions of boiler settings with full facades, add one foot to given length.

Boiler Diameter	Boiler Height	Horsepower (hp)	Boiler Weight	Setting: Length Width & Height	Setting Construction
2 ft.	6 ft.	5 hp	4000 lbs	8 ft. L; 5 ft. W; 5-5 1/2 ft. H.	Red brick or stone
2 1/2 ft.	7 ft.	10 hp	4700 lbs.	9 ft. L; 5 1/2-6 ft. W; 5 1/2-6 ft. H.	Red brick or stone with firebrick lined firebox.
3 ft.	8-9 ft.	15 hp	4000-5300 lbs.	11 ft. L; 6 ft. W; 6 ft. H.	Red brick or stone with firebrick lined firebox.
3 ft.	10-11 ft.	20 hp	4500-5900 lbs.	12-13 ft. L; 6 ft. W; 6 ft. H.	Red brick or stone with firebrick lined firebox.
3 1/2 ft.	11-12 ft.	25 hp	5800-7100 lbs.	13-14 ft. L; 6 1/2 ft. W; 7 ft. H.	Red brick or stone with firebrick lined firebox.
3 1/2-4 ft.	12-13 ft.	30 hp	6200-7800 lbs.	14-15 ft. L; 6 1/2-7 ft. W; 7-7 1/2 ft. H.	Red brick or stone with firebrick lined firebox.
3 1/2-4 ft.	12-13 ft	35 hp	7000-8700 lbs.	14-15 ft. L; 6 1/2-7 ft. W; 7-7 1/2 ft. H.	Red brick or stone with firebrick lined firebox.
4 ft.	13-14 ft	40 hp	8400-9900 lbs.	16-17 ft. L; 7 ft. W; 7-7 1/2 ft. H.	Red brick or stone with firebrick lined firebox.
4 1/2-5 ft.	13 1/2-14 ft.	50 hp	10,000-11,500 lbs.	17 ft. L; 7 1/2-8 ft.W; 71/2-8 ft. H.	Red brick or stone with firebrick lined firebox.
4 1/2-5 ft.	15-16 ft.	60 hp	12,000-12,700 lbs.	18-19 ft. L; 7 1/2-8 ft. W; 71/2-8 ft. H.	Red brick or stone with firebrick lined firebox.
5 ft.	15-16 ft.	70 hp	13,500-14,500 lbs.	18-19 ft. L; 7 1/2-8 ft. W; 7 1/2-8 ft. H.	Red brick or stone with firebrick lined firebox.
5 ft.	16-17 1/2 ft.	80 hp	15,000-16,100 lbs.	19-20 ft. L; 8 ft. W; 8 ft. H.	Red brick or stone with firebrick lined firebox.
5-5 1/2 ft.	17-18 ft.	90 hp	15,700-19,100 lbs.	20-21 ft. L; 8-8 1/2 ft. W; 9 ft. H.	Red brick or stone setting with firebrick lined firebox.
5 1/2 ft.	17-18 ft.	100 hp	16,500-19,800 lbs.	20-21 ft. L; 8 1/2 ft. W; 9 ft. H.	Red brick or stone with firebrick lined firebox.
6 ft.	16-18 ft.	125 hp	21,000-24,000 lbs	19-21 ft. L; 9 ft. W; 9 ft. H.	Red brick or stone with firebrick lined firebox.
6-6 1/2 ft.	18 ft.	150 hp		21 ft. L; 9-9 1/2 ft. W; 9 ft. H.	Red brick or stone with firebrick lined firebox.
6 ft.	20 ft.	165 hp		23 ft. L; 9 ft. W; 9 ft. H.	Red brick or stone with firebrick lined firebox.
6 1/2 ft.	18 ft.	180 hp		21 ft. L; 9 1/2 ft. W; 9 ft. H.	Red brick or stone with firebrick lined firebox.
6 1/2 ft.	20 ft.	200 hp		23 ft. L; 9 1/2 ft. W; 9 ft. H.	Red brick or stone with firebrick lined firebox.
7 ft.	18 ft.	225 hp		21 ft. L; 10 ft. W; 10 ft. H.	Red brick or stone with firebrick lined firebox.
7 ft.	20 ft.	250 hp		23 ft. L; 10 ft. W; 10 ft. H.	Red brick or stone with firebrick lined firebox.

Table 24: Quantities of Red Bricks and Firebricks Required for Return-tube Boiler Settings

The quantities of bricks required for construction are based on a cast-iron half-facade setting.

Setting: Length, Width & Height	Setting Construction	Number of Red Bricks in Setting	Number of Firebricks in Setting
8 ft. L; 5 ft. W; 5-5 1/2 ft. H.	Red brick setting.	1500 Red Bricks	
9 ft. L; 5 1/2-6 ft. W; 5 1/2-6 ft. H.	Red brick setting with firebrick lined firebox.	2000 Red Bricks	300 Firebricks
11 ft. L; 6 ft. W; 6 ft. H.	Red brick setting with firebrick lined firebox.	3000 Red Bricks	500 Firebricks
12-13 ft. L; 6 ft. W; 6 ft. H.	Red brick setting with firebrick lined firebox.	4500 Red Bricks	750 Firebricks
13-14 ft. L; 6 1/2 ft. W; 7 ft. H.	Red brick setting with firebrick lined firebox.	6500 Red Bricks	1100 Firebricks
14-15 ft. L; 6 1/2-7 ft. W; 7-7 1/2 ft. H.	Red brick setting with firebrick lined firebox.	8000 Red Bricks	1200 Firebricks
14-15 ft. L; 6 1/2-7 ft. W; 7-7 1/2 ft. H.	Red brick setting with firebrick lined firebox.	8000 Red Bricks	1200 Firebricks
16-17 ft. L; 7 ft. W; 7-7 1/2 ft. H.	Red brick setting with firebrick lined firebox.	9000 Red Bricks	1400 Firebricks
17 ft. L; 71/2-8 ft.W; 7 1/2-8 ft. H.	Red brick setting with firebrick lined firebox.	10,000 Red Bricks	1600 Firebricks
18-19 ft. L; 7 1/2-8 ft. W; 7 1/2-8 ft. H.	Red brick setting with firebrick lined firebox.	11,000 Red Bricks	1700 Firebricks
19-20 ft. L; 8 ft. W; 8 ft. H.	Red brick setting with firebrick lined firebox.	14,000 Red Bricks	1800 Firebricks
20-21 ft. L; 10 ft. W; 9 ft. H.	Red brick setting with firebrick lined firebox.	15,500 Red Bricks	2100 Firebricks
23 ft. L; 10 ft. W; 9 ft. H.	Red brick setting with firebrick lined firebox.	17,000 Red Bricks	2300 Firebricks

Hoist Specifications

Table 25: General Hoist Specifications: *Type, Duty, Foundation*

Hoist Type	Hoist Class	Foundation Size	Foundation Footprint	Foundation Profile	Foundation Material
Hand Windlass	Shallow Sinking		Rectangular	Wood Frame Over Shaft	Timber
Hand Winch	Shallow Sinking	3x3 ft.	Square or Rectangular	Flat	Timber
Horse Whim: Malacate	Shallow Sinking	7 to 10 ft. Diameter	Ovoid Depression	Cable Reel Axle Located in Pit	Timber
Horse Whim: Horizontal Reel	Sinking	4x4 ft.	Rectangular	Timber Footers in Depression	Timber
Horse Whim: Geared	Sinking	4x4 ft.	Rectangular	Timber Footers in Depression	Timber
Steam Donkey	Sinking	Portable	Rectangular	None	None
Gasoline Donkey	Sinking	Portable	Rectangular	None	None
Single-drum Gasoline	Sinking	2 1/2x8 ft. to 4x141/2 ft.	Rectangular	Flat	Timber or Concrete
Single-drum Gasoline	Sinking	2 1/2x8 ft. to 4x14 1/2 ft.	T-Shaped	Flat	Timber or Concrete
Single-drum Geared to Gasoline Engine	Sinking	3x8 ft. to 8x14 1/2 ft.	L-Shaped	Flat	Timber or Concrete
Single-drum Steam	Sinking	6x6 ft. and Smaller	Rectangular	Flat	Timber or Concrete
Single-drum Steam	Light Production	6x6 ft. to 7 1/2x10 ft.	Square or Rectangular	Flat	Concrete or Masonry
Single-drum Steam	Moderate Production	7 1/2x10 ft. and Larger	Rectangular	Irregular	Concrete or Masonry
Double-drum Steam	Moderate Production	4x7 ft. to 7x12 ft.	Rectangular	Irregular	Concrete or Masonry
Double-drum Steam	Heavy Production	7x12 ft. and Larger.	Rectangular	Irregular	Concrete and Masonry
Single-drum Geared Electric	Sinking	5x6 ft. and Smaller	Square or Rectangular	Flat	Concrete
Single-drum Geared Electric	Production	6x6 ft. and Larger	Square or Rectangular	Flat	Concrete
Single-drum Direct Drive Electric	Production	5x6 ft. and Larger	Square or Rectangular	Flat	Concrete
Double-drum Geared Electric	Heavy Production	6x12 ft.	Rectangular	Irregular	Concrete
Double-drum Direct Drive Electric	Heavy Production	6x12 ft.	Rectangular	Irregular	Concrete

Table 26: Hoist Specifications: *Single-drum Geared Steam Hoist, Single Steam Cylinder, 1880s*

Bed Plate Size (ft.)	Horsepower (hp)	Hoisting Speed (feet/minute)	Hoist Load	Hoist Weight	Duty
2 1/2x31/2 ft.	2 hp	Unknown	600 lbs	1000 lbs	Light Sinking
3x4 1/2 ft.	4 hp	Unknown	1000 lbs	1500 lbs	Light Sinking
3x5 ft.	6 hp	Unknown	1500 lbs	1700 lbs	Moderate Sinking
3 1/2x5 1/2 ft.	10-15 hp	Unknown	2000-3000 lbs	2200-2900 lbs	Moderate Sinking
4x6 1/2 ft.	20-25 hp	Unknown	4500-6000 lbs	4000-5000 lbs	Moderate Sinking

(Adapted from Ingersoll, ca. 1887: 57)

Table 27: Hoist Specifications: *Steam Donkey Hoists, Single Steam Cylinder 1880s*

Bed Plate Size (ft.)	Horsepower (hp)	Boiler Size (diameter, height)	Hoist Load	Shipping Weight	Duty
3x5 ft.	4 hp	2 ft. diam. x 5 ft.	1200 lbs	3500 lbs	Light Sinking
3x5 ft.	6 hp	2 ft. diam. x 6 ft.	1500 lbs	3750 lbs	Light Sinking
3 1/2x6 ft.	8-10 hp	2.5 ft. diam x 6 ft.	2000 lbs	4250-4600 lbs	Moderate Sinking
4 x6 ft.	11-15 hp	3 ft. diam x 6 1/2 ft.	3000 lbs	5000-6500 lbs	Moderate Sinking
3 1/2x7 ft.	20 hp	3 1/2 ft. diam x 7 ft.	6000 lbs	8500 lbs	Heavy Sinking
4 1/2x7 ft.	25 hp	3 1/2 ft. diam x 7 1/2 ft.	8000 lbs	9500 lbs	Heavy Sinking

(Adapted from Ingersoll, ca. 1887: 57)

Table 28: Hoist Specifications: *Steam Donkey Hoists, Standard Duplex Cylinders 1880s-1900s*

Bed Plate Size (ft.)	Horsepower (hp)	Boiler Size (diameter, height)	Hoist Load	Shipping Weight	Duty
4x5 1/2 ft.	8-12 hp	3 ft. diam. x 6 ft.	2000-3000 lbs	5300 lbs	Moderate Sinking
4 1/2x6 ft.	16 hp	3 ft. diam. x 7 ft.	3000-4000 lbs	7500 lbs	Moderate Sinking
4 1/2x6 ft.	20 hp	3 1/2 ft. diam x 7 ft.	4000-5000 lbs	8500 lbs	Moderate Sinking
5x6 1/2 ft.	30 hp	3 1/2 ft. diam x 7 1/2 ft.	8000 lbs	9500 lbs	Moderate Sinking
6x8 ft.	40 hp	4 ft. diam x 8 ft.	10,000 lbs	15,000 lbs	Heavy Sinking
6x8 ft.	50 hp	4 ft. diam x 8 1/2 ft.	12,000 lbs	17,000 lbs	Heavy Sinking

(Adapted from Ingersoll, ca. 1887: 57; Mine & Smelter Supply, 1937: 556)

Table 29: Hoist Specifications: *Single-drum Geared Steam Hoist, 1880s-1920s*

Bed Plate Size (ft.)	Horsepower (hp)	Hoisting Speed (feet/minute)	Hoist Load	Hoist Weight	Duty
4x5 ft.	10-12 hp	225-250 feet/min.	1500 lbs	2300-3250 lbs	Sinking Duty
5x5 1/2 ft.	15 hp	265 feet/min.	2000 lbs	3800 lbs	Sinking Duty
5x6 ft.	20-30 hp	275-350 feet/min.	2000-3500 lbs	3500-5450 lbs	Sinking Duty
6x6 ft.	35 hp	350 feet/min.	3500 lbs	8800 lbs	Heavy Sinking
6x7 1/2 ft.	50 hp	400 feet/min.	4500 lbs	9400 lbs	Heavy Sinking/ Light Production
7 1/2x7 1/2 ft.	85-92 hp	480-520 feet/min.	4700-5000 lbs	8700-10,200 lbs	Light Production
7x9 ft.	75 hp	450 feet/min.	7000 lbs	18,500 lbs	Light Production
8x10 ft.	150 hp	450 feet/min.	9000-10,000 lbs	27,000-31,000 lbs	Moderate Production
9x9 ft.	130-140 hp	470-490 feet/min.	7300-7800 lbs	17,300-19,000 lbs	Light Production
10x10 ft.	160 hp	550 feet/min.	7800 lbs	26,400 lbs	Moderate Production
11x12 ft.	200-230 hp	600-750 feet/min.	7000-8500 lbs.	37,000-45,000 lbs	Moderate Production

(Adapted from Ingersoll, ca. 1887: 57; Peele, 1918: 882, 883)

Table 30: Hoist Specifications: *Double-drum Geared Steam Hoists,*
1880s-1920s

Bed Plate Size (ft.)	Horsepower (hp)	Hoisting Speed (feet/minute)	Hoist Load Unbalanced	Hoist Weight	Duty
6x7 1/2 ft.	30 hp	350 feet/min.		9000 lbs	Heavy Sinking
7 1/2x9 ft.	50 hp	400 feet/min.		15,000 lbs	Light Production
7 1/2x9 1/2 ft.	75 hp	400 feet/min.		22,500 lbs	Light Production
7 1/2x12 ft.	85-92 hp	480-520 feet/min.	4700-5000 lbs	15,000-17,500 lbs	Light Production
9x13 1/2 ft.	130-140 hp	470-490 feet/min.	7300-7800 lbs	28,600-30,000 lbs	Moderate Production
10 1/2x15 ft.	160 hp	550 feet/min.	7800 lbs	40,000 lbs	Moderate Production
11x17 ft.	200 hp	610 feet/min.	8700 lbs	51,100 lbs	Moderate Production
11x171/2 ft.	200-230 hp	730 feet/min.	7200-8300 lbs.	56,600-60,800 lbs	Moderate Production

(Adapted from Ingersoll, ca. 1887: 59; Peele, 1918: 883)

Table 31: Hoist Specifications: *Single-drum Gasoline Hoists, 1890s-*
1930s

Bed Plate Size (ft.)	Horsepower (hp)	Hoisting Speed (feet/minute)	Hoist Load	Hoist Weight	Duty
3 1/2x8 ft.	8 hp	155-225 feet/min.	930-1360 lbs	4600 lbs	Light Sinking
4 1/2x9 1/2 ft.	15 hp	180-215 feet/min.	1600-2400 lbs	8000 lbs	Sinking
5x12 ft.	25 hp	190-330 feet/min.	2000-3500 lbs	13,500 lbs	Sinking
6x13 1/2 ft.	40 hp	220-370 feet/min.	2800-4700 lbs	18,400 lbs	Sinking
6 1/2x14 ft.	50 hp	220-405 feet/min.	3200-6000 lbs	23,400 lbs	Heavy Sinking
7x14 1/2 ft.	60 hp	270 feet/min.	6000 lbs	27,800 lbs	Heavy / Light Production
9x14 1/2 ft.	60 hp	365 feet/min.	4300 lbs	27,800 lbs	Heavy Sinking

(Adapted from Mine and Smelter Supply Co., 1912: 49)

Table 32: Hoist Specifications: *Gasoline Donkey Hoists, 1920s-*
1940s

Bed Plate Size (ft.)	Horsepower (hp)	Engine Type	Hoist Load	Shipping Weight	Duty
4 1/2x6 ft.	30 hp	Four Cylinder	2500 lbs	4200 lbs	Moderate Sinking
5x6 ft.	35 hp	Four Cylinder	3000-4000 lbs	4600-4800 lbs	Moderate Sinking
5x6 1/2 ft.	50 hp	Four Cylinder	3000-5000 lbs	5000-5500 lbs	Moderate Sinking

(Adapted from Keystone, 1925: 467; Mine & Smelter Supply, 1937: 555)

Table 33: Hoist Specifications: *Double-drum Geared Electric Hoists,*
1910s-1940s

Bed Plate Size (ft.)	Horsepower (hp)	Hoisting Speed (feet/minute)	Hoist Load Unbalanced	Depth Capacity	Duty
8 1/2x13 ft.	60-80 hp	280-380 feet/min.	6000 lbs	3000 ft.	Moderate Production
8 1/2x14 1/2 ft.	80-100 hp	280-350 feet/min.	8000 lbs	3100 ft.	Moderate Production
9x16 1/2 ft.	125-150 hp	350-430 feet/min.	10,000 lbs	4100 ft.	Heavy Production
10x20 1/2 ft.	200-300 hp	375-470 feet/min.	15,000 lbs	5100 ft.	Heavy Production

(Adapted from Mine & Smelter Supply, 1937: 552)

Table 34: Hoist Specifications: *Single-drum Geared Electric Hoists, 1910s-1940s*

Bed Plate Size (ft.)	Horsepower (hp)	Hoisting Speed (feet/minute)	Hoist Load	Depth Capacity	Duty
4x5 ft.	7-15 hp	200-220 feet/min.	1000-1800 lbs	Unknown	Sinking-duty
4x6 ft.	20-30 hp	200-300 feet/min.	2750 lbs.	1850 ft.	Sinking-duty
5x6 ft.	20-30 hp	220-250 feet/min.	2400 lbs	Unknown	Sinking-duty
6x7 ft.	30-50 hp	250-300 feet/min.	2800-3500 lbs	3000 ft.	Sinking-duty
6x8 ft.	50-75 hp	300-400 feet/min.	4300-6000 lbs	3000-3400 ft.	Light Production
7x11 ft.	75 hp	400-450 feet/min.	5000-6000 lbs	2850 ft.	Light Production
7x11 1/2 ft.	75 hp	400-450 feet/min.	5500-6000 lbs	2850 ft.	Light Production
8x10 ft.	75-100 hp	350 feet/min.	5500-6000 lbs	2800 ft.	Light Production
8x8 ft.	80-100 hp	280-360 feet/min.	6000-7500 lbs	3000 ft.	Moderate Production
9x11 ft.	100 hp	300-450 feet/min.	7500-8000 lbs	2800-3100 ft.	Moderate Production
9x12 ft.	150 hp	450-500 feet/min.	9500 lbs	2800 ft.	Heavy Production
10x10 ft.	150 hp	Unknown	10,000 lbs	Unknown	Heavy Production
10x12 ft.	100 hp	550 feet/min.	6000-7500 lbs	2750 ft.	Moderate Production
10 1/2x13 ft.	125-150 hp	350-600 feet/min.	8500-10,000 lbs	2750-4000 ft.	Heavy Production
9x15 ft.	200 hp	500 feet/min.	11,000 lbs	Unknown	Heavy Production
11x12 ft.	225 hp	Unknown	15,000 lbs	Unknown	Heavy Production
12x12 ft.	300 hp	Unknown	20,000 lbs	Unknown	Heavy Production

(Adapted from Keystone, 1925: 466, 473, 476; Mine & Smelter Supply, 1937: 552-554; and field data).

Headframe Specifications

Table 35: Specifications of Headframes: *Type, Material, Class*

Headframe Type	Material	Class	Capital Investment
Tripod	Hewn Logs	Sinking	Very Low
Tripod	Light Timber	Sinking	Very Low
Two Post (Gallows Frame): Small	Timber	Sinking	Low
Two Post (Gallows Frame): Large	Timber	Production	Low to Moderate
Two Post (Gallows Frame): Large	Steel	Production	Moderate to High
Four Post: Small	Light Timber	Sinking	Low
Four Post	Timber	Production	Moderate
Six Post	Timber	Production	Moderate to High
Four and Six Post	Steel	Production	High
A-Frame	Timber	Production	Moderate to High
A-Frame	Steel	Production	High

Mine Site Artifacts

Table 36: Datable Structural Artifacts

Artifact	Description	Date Range
Cement: Natural	Pale yellow, or white to pinkish, crumbly white crust on surface.	1850s-1920
Cement: Portland	Grey to white, firm.	1910-Today
Cinderblock	Concrete cinderblock.	1930s-Today
Corrugated Steel Siding	Corrugated sheet steel.	1890s-Today
Dimension Lumber: True-Measuring	Lumber that is true to its defined size, e.g. a 2x4 is truly 2x4 in size.	1850s-1920s
Dimension Lumber: Less Than True-Measuring	Lumber that is smaller than its defined size, e.g. a 2x4 is less than 2x4 in size.	1920s-Today
Nail: Cut	Tapered square-shanked nail	1850s-1890
Nail: Wire	Common round nail	1890-Today
Tarpaper	Black or green. May be encrusted with sand.	1890s-Today
Window Glass: Amethyst	Light amethyst tint.	1860s-1920
Window Glass: Aqua	Light aqua tint.	1860s-1920
Window Glass: Selenium	Light golden-yellow tint.	WWI-1950s
Window Glass: Lime Tint	Pale lime-green tint.	1920s-1950s

Table 37: Datable Industrial Artifacts

Artifact	Description	Date Range
Air Hose: Compound	Black or Red, 1 to 2 inches diameter, rubber with built-in fibers.	1930s-Today
Air Hose: Rubberized Canvas	Black or Red, 1 to 2 inches diameter	1920s-1950s
Air Hose: Steel-Spiral-Wrapped Rubber/ Canvas	Rubberized canvas wrapped with steel wire.	1890s-1920s
Air Hose: Steel-Spiral-Wrapped Canvas	Heavy canvas wrapped with steel wire.	1870s-1900s
Automobile Parts	Chassis, body, engine parts, tires	1910s-Today
Battery Core: Dry Cell	Carbon rod 3/4 inch diameter.	1890s-1930s
Battery Core: Wet Cell	Ribbed carbon rod 3/4 inch diameter.	1880s-1910s
Blasting Powder Keg	Corrugated body with small-mouth internal-thread opening with zinc bung. Size: 10 inch diameter x 1 ft high.	1860s-1890s
Blasting Powder Keg	Corrugated body with small-mouth internal-friction opening. Size: 10 inch diameter x 1 ft high.	1890s-1910s
Boiler Water Sight Tube Glass	Heavy glass tube 1/2 inch diameter.	1870s-1920
Carbide Drum: 25 lb	Corrugated body with wide-mouth internal-thread opening. Size: 8 inch diameter x 1 ft high.	1910s-1950s
Carbide Drum: 100 lb	Corrugated body with wide-mouth internal-thread opening. Size: 1 ft diameter x 20 inches high.	1910s-1950s
Drill-Steel: Hand	Hexagonal rod less than 1 1/4 inch diam, single blade.	1850s-1910s, 1930s
Drill-Steel: Piston Drill	Hexagonal or round rod 1 1/4-1 1/2 inch diam, star bit, round butt.	1870s-1910s
Drill-Steel: Hammer Drill	Round rod 1 1/4-1 1/2 inch diam, star bit, round butt with two keys, hollow.	1910s-1950s
Drill-Steel: Hammer Drill	Square rod 1 1/4-1 1/2 inch diam, star bit, round butt, hollow.	1910s-1950s
Drill-Steel: Sinker Drill	Hexagonal 1 inch diam, star bit, hexagonal butt with collar.	1910s
Drill-Steel: Sinker Drill	Hexagonal 1 inch diam, star bit, hexagonal butt with collar, hollow.	1910s-Today
Drill-Steel: Stoper Drill	Cruciform rod less 1 1/4 inch diam, star bit, cruciform butt.	1890s-1920s
Dynamite Box	Sides joined with lock-corner tabs.	1900-1950
Dynamite Box	Sides joined with wire nails.	1890-1905
Dynamite Box	Sides joined with cut nails.	1875-1890
Electrical Insulator: Knob	Porcelain.	1890s-1940s
Electrical Insulator: Pony	Glass, aqua to green.	1890s-1960s
Electrical Wire	Textile sheath, solid copper wire.	1890s-1930s
Forge Fuel: Charcoal	Light in weight, black, and chalky.	1850s-1880s
Forge Fuel: Anthracite	Light in weight, black, lustrous, concoidal fractures.	1870s-1930s
Forge Fuel: High-Quality Bituminous Coal	Light in weight, black, and semi-lustrous, prismatic fractures	1880s-1930s
Lamp: Carbide	Brass water tank with screw-on base.	1910s-1950s
Lamp: Miner's Candlestick	Candle thimble, spike, hat hook.	1870s-1910s
Piston Rings: Compressor	Larger than 4 inch diameter	1880s-1930s
Piston Rings: Engine	Smaller than 4 inch diameter	1900s-1930s

Table 38: Datable Domestic Artifacts

Artifact	Description	Date Range
Bottle Glass: Amethyst	Light amethyst tint.	1860s-1920
Bottle Glass: Aqua	Light aqua tint.	1860s-1920
Bottle Glass: Selenium	Light golden-yellow tint.	WWI-1950s
Bottle Glass: Lime Green	Pale lime-green tint.	1920s-1950s
Bottle: Applied Lip Finish	Bottle finish applied separately.	1860s-1880s
Bottle: Hand-Finished	Side-seam terminates at neck.	1860s-1910s
Bottle: Machine-Made	Side-seam continues over top.	1900s-Today
Can: Hole-in-Cap, Lapped Side-Seam	Disk soldered on end, lapped and soldered side-seam.	1860s-1890s
Can: Hole-in-Cap, Inner-Rolled Side-Seam	Disk soldered on end, rolled and soldered side-seam.	1880s-1910s
Can: Sanitary	Crimped ends, rolled and soldered side-seam.	1900s-1930s
Can: Sanitary	Crimped ends, rolled side-seam.	1920s-Today
Can: Vent-Hole, Lapped-Side-Seam	Stamped ends, lapped and soldered side-seam, two solder dots.	1880s-early 1890s
Can: Vent-Hole, Rolled Side-Seam	Stamped ends, rolled and soldered side-seam, two solder dots.	1890s
Can: Vent-Hole, Rolled and Soldered Side-Seam	Stamped ends, rolled and soldered side-seam, one solder dot.	1900s-1930s
Can: Vent-Hole, Rolled Side-Seam	Stamped ends, rolled side-seam, one solder dot.	1910s-1950s
Can: Sardine	One-piece bottom.	1900s-Today
Can: Tobacco	Commonly called 'Prince Albert'.	1890s-Today
Ceramic, Tableware	Crazed (finely fractured) glaze.	Pre-1920
Ceramic, Tableware	Non-crazed (finely fractured) glaze.	Post-1910s

BIBLIOGRAPHY

General Mining History

Brown, Ronald *Hard Rock Miners: The Intermountain West, 1860-1920* Texas A&M Press, 1979.

Crampton, Frank A. *Deep Enough: A Working Stiff in the Western Mine Camps* University of Oklahoma Press, [1956] 1982.

Hyman, David Marks *The Romance of a Mining Venture* Larchmont Press, Cincinnati, OH, 1981.

King, Joseph E. *A Mine to Make A Mine: Financing the Colorado Mining Industry, 1859-1902* Texas A&M University Press, 1977.

Lord, Elliot *Comstock Mining and Miners* Howell-North Books, San Diego, CA, 1980 [Geological Survey, Government Printing Office, 1883].

McElvaine, Robert *The Great Depression: America, 1929-1941* Times Books, New York, NY, [1984] 1993.

Pearson, David W. *This Was Mining in the West* Schiffer Publishing Ltd., Atglen, PA, 1996.

Rice, George Graham *My Adventures with Your Money* Nevada Publications, Las Vegas, NV, 1986.

Rickard, T.A. *A History of American Mining* McGraw-Hill Book Co., Inc., New York, NY, 1932.

Shinn, Charles H. *The Story of the Mine: as Illustrated by the Great Comstock Lode of Nevada* University of Nevada Press, Reno, NV, 1984 [D. Appleton, New York, NY, 1910].

Sloane, Howard N. & Lucille L. *Pictorial History of American Mining* Crown Publishers, Inc., New York, NY, 1970.

Smith, Duane *Colorado Mining* University of New Mexico Press, Albuquerque, NM, 1977.

Smith, Duane *Rocky Mountain Mining Camps* University of Nebraska Press, Lincoln, NE, 1967.

Spence, Clark C. *Mining Engineers and the American West: the Lace Boot Brigade, 1849-1933* University of Idaho Press, Moscow, ID 1993.

Todd, Arthur C. *The Cornish Miner in America* The Arthur H. Clark Company, Spokane, WA, 1995.

Twain, Mark *Roughing It* Airmont Publishing Co., New York, NY, 1967 [1872].

Voynick, Stephen M. *Leadville: A Miner's Epic* Mountain Press Publishing Company, Missoula, MT, 1988 [1984].

Watkins, T.H. *Gold and Silver in the West: the Illustrated History of an American Dream* Bonanza Books, New York, NY, 1971.

Wyman, Mark *Hard Rock Epic: Western Miners and the Industrial Revoluion, 1860-1890* University of California Press, Berkeley, CA 1989 [1979].

Browne, J. Ross & Taylor, James W. *Reports upon the Mineral Resources of the United States* Government Printing Office, Washington DC, 1867.

Florin, Lambert *Ghost Towns of the West* Promontory Press, New York, NY, 1971.

Hills, James M. *The Mining Districts of the Western U.S.* [bibliographic source] United States Geological Survey Bulletin 507, Government Printing Office, Washington DC, 1912.

California

Aubury, Lewis E., Et Al *Register of Mines & Minerals in Inyo County, California, 1902* State Mining Bureau, San Francisco, CA, 1902.

Clark, William B. *Gold Districts of California: Bulletin 193* California Division of Mines and Geology, Sacrament, CA, 1980 [1970].

Hewett, D.F. *Geology and Mineral Resources of the Ivanpah Quadrangle, California and Nevada United States* Geological Survey Professional Paper 275, United States Government Printing Office, Washington DC., 1956.

Lingenfelter, Richard E. *Death Valley & the Amargosa: A Land of Illusion* University of California Press, Berkeley, CA, 1986.

Norman, L.A. Jr. and Stewart, Richard M. "Mines and Mineral Resources of Inyo County" *California Journal of Mines and Geology,* Division of Mines, San Francisco, CA, Jan. 1951 v.47 n.1.

Tucker, W.B., Sampson, R.J. "District Reports of Mining Engineers: Los Angeles Field Division, San Bernardino County" *Mining in California* California Minerals Bureau, July 1931, Vol.27, No.3.

Wright, Lauren A.; Stewart, Richard M.; Et Al "Mines and Mineral Resources of San Bernardino County" *California Journal of Mines and Geology,* Division of Mines, San Francisco, CA, Jan.-Feb. 1953 v.49 n.1, 2.

Colorado

Abbott, Carl; Leonard, Stephen; McComb, David *Colorado: A History of the Centennial State* University Press of Colorado, Niwot, CO, [1982] 1994.

Aldrich, John K. *Ghosts of Clear Creek County* Centennial Graphics, Lakewood, CO, 1984.

Brown, Robert L. *Colorado Ghost Towns - Past and Present* The Caxton Printers, Ltd., Caldwell, OH, 1993 [1972].

"Cripple Creek District - VI" *The Power Plants Engineering & Mining Journal* Feb. 8, 1900 p669-671.

"The Cripple Creek Gold Fields" *Engineering & Mining Journal* Oct. 21, 1893 p429.

"The Cripple Creek Gold Fields" *Engineering & Mining Journal* Dec. 9, 1893 p588.

"The Cripple Creek Mining District in Colorado" *Engineering & Mining Journal* Jan. 11, 1896 p37.

"The Cripple Creek District in 1896" *Engineering & Mining Journal* Feb. 2, 1897 p8.

"Cripple Creek Mines in 1898" *Engineering & Mining Journal* Jan. 14, 1899 p47-48.

"Cripple Creek Mines in 1899" *Engineering & Mining Journal* Jan. 30, 1900 p50.

"Cripple Creek District During 1901" *Engineering & Mining Journal* Jan. 11, 1902 p70-71.

"Cripple Creek in 1908" *Mining & Scientific Press* Jan. 2, 1909 p41.

Eberhart, Perry *Guide to the Colorado Ghost Towns and Mining Camps* Swallow Press, Athens, OH, 1987 [1959].

Emmons, W.H. & Larsen, E.S. "A Preliminary Report on the Geology and Ore Deposits of Creede, Colorado" *Contributions to Economic Geology United States Geological Survey Bulletin 530,* Government Printing Office, Washington DC., 1913.

"General Mining News: Colorado, El Paso County" *Engineering and Mining Journal* Jan. 1890-1899.

"General Mining News: Colorado, Teller County" *Engineering and Mining Journal* 1890-1921.

Georgetown Courier Among the Silver Seams of Colorado *Georgetown Courier,* Georgetown, CO, 1886.

Grimstead, Bill and Drake, Raymond *The Last Gold Rush: A Pictorial History of the Cripple Creek & Victor Gold Mining District* Pollux Press, Victor, CO, 1983.

Henderson, Charles W. *Mining in Colorado: A History of Discovery, Development, and Production* Government Printing Office, Washington, DC 1926.

Hills, Fred *The Official Manual of the Cripple Creek Mining District* Fred Hills Publishers, Colorado Springs, CO, 1900.

Hollister, Ovando J. *The Mines of Colorado* Promontory Press, New York, NY, 1974 [Samuel Bowles & Company, MA., 1867].

Koschmann, A.H. "Tenmile Mining District, Colorado" *Guidebook: Rocky Mountain Association of Geologists* Rocky Mountain Association of Geologists, 1947.

Lindgren, Waldemar and Ransome, Frederick Leslie *Geology and Gold Deposits of the Cripple Creek District, Colorado United States* Geological Survey Professional Paper 54, Government Printing Office, Washington, DC, 1906.

Lovering, T.S. *Geology and Ore Deposits of the Montezuma Quadrangle, Colorado* United States Geological Survey Professional Paper 178, United States Government Printing Office, Washington, DC, 1935.

Lovering, T.S. and Goddard, E.S. *Geology and Ore Deposits of the Front Range, Colorado* United States Geological Survey Professional Paper 223, United States Government Printing Office, Washington, DC, 1950.

Manning, J.F. *Wonderful Cripple Creek District* (text, no publisher) 1899.

Mazzulla, Fred M. *The First 100 Years: A Photographic Account of the Rip-roaring, Gold-plated Days in the World's Greatest Gold Camp* A.B. Hirschfeld Press, Denver, CO, 1956.

Division of Mines & Minerals Colorado State Archives, Denver, CO,

Mine Inspectors' Reports: Clear Creek County File # 104031 Almaden-Blazing Star Mine

Mine Inspectors' Reports: Teller County File # 104089 American Eagle Mine

Mine Inspectors' Reports: Teller County File # 104089 Anaconda Mine

Mine Inspectors' Reports: Chaffee County File # 48440 Banker Mine

Mine Inspectors' Reports: Clear Creek County File # 104032 Champion Mine

Mine Inspectors' Reports: Teller County File # 48440 Delmonico Mine

Mine Inspectors' Reports: Teller County File # 48440 Elkton Mine

Mine Inspectors' Reports: Clear Creek County File # 104032 Equator Tunnel

Mine Inspectors' Reports: Boulder County File # 104028 Fourth of July Mine

Mine Inspectors' Reports: Teller County File # 48440 Friday Shaft

Mine Inspectors' Reports: Teller County File # 48440 Glorietta Mine

Mine Inspectors' Reports: Teller County File # 48440 Gold Coin Mine

Mine Inspectors' Reports: Teller County File # 48516 Golden Cycle Mine

Mine Inspectors' Reports: Teller County File # 48441 Hoosier Mine

Mine Inspectors' Reports: Teller County File # 48441 Hull City Mine

Mine Inspectors' Reports: Teller County File # 48516 Joe Dandy Mine

Mine Inspectors' Reports: Teller County File # 48441 Longfellow Mine

Mine Inspectors' Reports: Chaffee County File # 10403 Mary Murphy Mine

Mine Inspectors' Reports: Teller County File # 48441 Mary Nevins Mine

Mine Inspectors' Reports: Clear Creek County File # 104035 Minott Shaft: Sun & Moon Mine

Mine Inspectors' Reports: Teller County File # 48441 Modoc Mine

Mine Inspectors' Reports: Teller County File # 48440Nichols Shaft

Mine Inspectors' Reports: Boulder County File # 104029 Norway Mine

Mine Inspectors' Reports: Boulder County File # 104029 Pandora Mine

Mine Inspectors' Reports: Summit County File # 104088 Paymaster Mine

Mine Inspectors' Reports: Summit County File # 104088 Pennsylvania Mine

Mine Inspectors' Reports: Teller County File # 48442 Pinnacle Mine

Mine Inspectors' Reports: Teller County File # 48440 Prince Albert Shaft

Mine Inspectors' Reports: Teller County File # 104089 Prince Albert Tunnel

Mine Inspectors' Reports: Teller County File # 104029 Revenge Mine

Mine Inspectors' Reports: Teller County File # 48442 Rigi Mine

Mine Inspectors' Reports: Teller County File # 48442 Sangre de Cristo Mine

Mine Inspectors' Reports: Teller County File # 48442 Sitting Bull Mine

Mine Inspectors' Reports: Mineral County File # 104054 Spar City Mine

Mine Inspectors' Reports: Clear Creek County File # 104035 Specie Payment Mine

Mine Inspectors' Reports: Clear Creek County File # 104035 Standard Mine

Mine Inspectors' Reports: Teller County File # 48442 Strong Mine

Mine Inspectors' Reports: Teller County File # 48442 Theresa Mine

Mine Inspectors' Reports: Teller County File # 48440 Tornado Shaft: Elkton Consolidated

Mine Inspectors' Reports: Teller County File # 48516 and 48441 Vindicator Mine

Mine Inspectors' Reports: Teller County File # 48516 Zenobia Mine

"Mining Summary: Colorado, Teller County" *Mining & Scientific Press* Jan 1899-Dec. 1921.

Moench, Robert H. And Drake, Avery Ala *Economic Geology of the Idaho Springs District, Clear Creek and Gilpin County, Colorado* United States Geological Survey Bulletin 1208, United States Government Printing Office, Washington, DC, 1966.

Moench, Robert H. And Drake, Avery Ala *Mines and Prospects, Idaho Springs District, Clear Creek and Gilpin County, Colorado* United States Geological Survey open file report 1965, 1966.

Munn, Bill *A Guide to the Mines of the Cripple Creek District* Century One Press, Colorado Springs, CO 1984.

Parker, Ben H. "Mining in Clear Creek and Gilpin Counties" *Guidebook: Rocky Mountain Association of Geologists* Rocky Mountain Association of Geologists, 1947.

Patton, Horace B. "The Montezuma Mining District of Summit County, Colorado" *Colorado Geological Survey: First Report* Smith-Brooks Printing Co., Denver, Colorado, 1909.

Pierce *History of Cripple Creek and its Great Mines* Bulletin Publishing Co. Denver, CO, 1904.

Sanborn Map Company *Victor, Teller County, Colorado, Dec. 1908* Sanborn Map Company, Brooklyn, NY, 1909.

Schader, Conrad F. *Colorado's Alluring Tin Cup: The District, Its Settlements, People, Miners Region* Alta Publications, Golden, CO 1992.

Smith, Duane *Rocky Mountain West: Colorado, Wyoming, & Montana 1859-1915* University of New Mexico Press, Albuquerque, NM, 1992.

Smith, P. David *Mountains of Silver: The Story of Colorado's Red Mountain Mining District* Pruett Publishing Co., Boulder, CO, 1994.

"Special Correspondence from Mining Centers: Cripple Creek" *Engineering & Mining Journal* May 19, 1914 p914.

"Special Correspondence: Cripple Creek" *Engineering & Mining Journal* Feb. 10, 1906 p296.

Sprague, Marshall *The King of Cripple Creek: the Life and Times of Winfield Scott Stratton* Magazine Associates, 1994.

Sprague, Marshall *Money Mountain the Story of Cripple Creek Gold* University of Nebraska Press, Lincoln, NE, [1953] 1979.

Spurr, Josiah E.; Garrey, George H.; and Ball, Sydney H. *Economic Geology of the Georgetown Quadrangle (Together with the Empire District), Colorado* United States Geological Survey Professional Paper 63, Government Printing Office, Washington, DC, 1908.

State Historic Society of Colorado *Colorado Inventory of Historic Sites* State Historic Society of Colorado, Denver, CO, 1977.

Steven, Thomas A. and Ratte, James C. *Geology and Structural Control of Ore Deposition in the Creede District, San Juan Mountains, Colorado* United States Geological Survey Professional Paper 487, United States Government Printing Office, Washington, DC, 1965.

Taylor, Robert Guilford *Cripple Creek Mining District* Filter Press, Palmer Lake, CO, 1973 [Indiana University 1966].

Tweto, Ogden *Molybdenum and Tungsten Deposits of Gold Hill, Quartz Creek, and Tin Cup Mining Districts, Colorado* United States Geological Survey Open File Report, 1944.

Watson, Jim Owner and operator of the Strong Mine in the Cripple Creek Mining District and mayor of Victor, Colorado: "The City of Mines" Personal Interview Victor, CO, September, 1998.

Montana

Becraft, G.E.; Pinkerley, D.M.; Rosenblum, S. *Geology and Mineral Deposits of the Jefferson City Quadrangle, Jefferson and Lewis and Clark Counties, Montana* United States Geological Survey Professional Paper 428, United States Government Printing Office, Washington, DC, 1963.

Klepper, M.R.; Weeks, R.A.; Ruppel, E.T. *Geology of the Southern Elkhorn Mountains, Jefferson and Broadwater Counties, Montana* United States Geological Survey Open File Report 121, 1951.

Klepper, M.R.; Weeks, R.A.; Ruppel, E.T. *Geology of the Southern Elkhorn Mountains, Jefferson and Broadwater Counties, Montana* United States Geological Survey Professional Paper 292, United States Government Printing Office, Washington, DC, 1957.

Knopf, Adolph *Ore Deposits of the Helena Mining Region, Montana* United States Geological Survey Bulletin 527, United States Government Printing Office, Washington, DC, 1913.

Koschmann, A.H. and Bergendahl, M.H. *Principle Gold Producing Districts of Montana: Geological Survey* Professional Paper Carson Enterprises, Deming, NM [No date-contemporary publication].

Miller, Donald C. *Ghost Towns of Montana* Pruett Publishing Co., Boulder, CO, 1981 [1974].

Roby, R.N.; Ackerman, W.C.; Fulkerson, F.B.; Crowley, F.A. *Mines and Mineral Deposits of Jefferson County, Montana* Montana State Bureau of Mines and Geology, Montana School of Mines, Butte, MT 1960.

Rupper, Edward T. *Geology of the Basin Quadrangle, Lewis and Clark and Powell Counties, Montana* United States Geological Survey Bulletin 1151, United States Government Printing Office, Washington, DC, 1963.

Smede, Harry W. *Geology and Igneous Petrology of the Northern Elkhorn Mountains, Jefferson and Broadwater Counties, Montana* United States Geological Survey Professional Paper 510, United States Government Printing Office, Washington, DC, 1966.

Weed, W.H. & Barrett, J. "Geology and Ore Deposits of the Elkhorn Mining District, Jefferson County, Montana" *Twenty Second Annual report of the United States Geological Survey* United States Geological Survey, Government Printing Office, Washington, DC, 1901.

Nevada

Albers, J.P. and Stewart, J.H. *Geology and Mineral Resources of Esmeralda County, Nevada* Nevada Bureau of Mines and Geology Bulletin 78, Mackay School of Mines , Reno, NV, 1972.

Albritton, Claude C., Richards, Arthur, Brokaw, Arnold L., and Reinemund, John A. *Geologic Controls of Lead and Zinc Deposits in Goodsprings (Yellow Pine) District, Nevada* United States Geological Survey Bulletin No. 1010, United States Government Printing Office, Washington DC., 1954.

Ball, Sydney H. *Notes on Ore Deposits of Southwestern Nevada and Eastern California* United States Geological Survey Bulletin 285, Government Printing Office, Washington, DC, 1905.

Ball, Sydney H. A *Geological Reconnaissance in Southwestern Nevada and Eastern California* United States Geological Survey Bulletin 308, Government Printing Office, Washington, DC, 1907.

Binyon, E.O. "Explorations of the Gold, Silver, Lead, and Zinc Properties of Eureka Corporation, Eureka County, Eureka, Nevada" *United States Bureau of Mines Report of Investigations # 3949* United States Bureau of Mines, Washington DC., 1946.

Cornwall, C.R. and Kleinhampl, Frank J. *Geology of the Bullfrog Quadrangle and Ore Deposits Related to the Bullfrog Hills Caldera, Nye County, Nevada and Inyo County, California* United States Geological Survey Professional Paper 454J, United States Government Printing Office, Washington DC., 1964.

Curtis, John S. "Abstract of the Report on the Mining Geology of the Eureka District, Nevada" *Fourth Annual Report: United States Geological Survey* United States Geological Survey, Government Printing Office, Washington, DC, 1883.

Elliot, Russell R. *Nevada's Twentieth Century Mining Boom: Tonopah, Goldfield, Ely* University of Nevada Press, Reno, NV, 1988 [1966].

Ferguson, Henry G. "The Golden Arrow, Clifford, and Ellendale Districts, Nye County, Nevada" *Contributions to Economic Geology* United States Geological Survey Bulletin 640, Government Printing Office, Washington DC., 1917.

Ferguson, Henry G. "Mining Districts of Nevada" *Economic Geology* the Economic Geology Publishing Company, March-April, 1929.

Ferguson, Henry G. *Geology of the Tybo District, Nevada* Nevada Bureau of Mines and Geology, Mackay School of Mines , Reno, NV, 1933.

Hague, Arnold "Geology of the White Pine District" *Geological Explorations Vol. III: Report of the Fortieth Parallel, Mining Industry* Clarence King, Washington DC, 1870.

Hague, Arnold *Monograph No.20: Geology of the Eureka District* United States Geological Survey, Washington DC, 1892.

Hewett, D.F. *Geology and Ore Deposits of the Goodsprings Quadrangle, Nevada* United States Geological Survey Professional Paper 162, United States Government Printing Office, Washington DC., 1931.

Hill, J.M. *Some Mining Districts in Northeastern California and Northwestern Nevada* United States Geological Survey Bulletin 594, Government Printing Office, Washington, DC, 1915.

Hill, J.M. *Mines of Battle Mountain, Reese River, Aurora, and other Western Nevada Districts* Nevada Publications, Las Vegas, NV, 1983 [U.S. Geological Survey, 1915].

The Hornsilver Herald: Pioneer Newspaper of Hornsilver District — A Born Shipper Hornsilver, Nevada, Saturday, May 9, 1908

Humphrey, Fred L. *Geology of the White Pine Mining District, White Pine County, Nevada* Nevada Bureau of Mines and Geology Bulletin 57, Mackay School of Mines , Reno, NV, 1960.

Ingalls, Walter R. "The Silver-Lead Mines of Eureka, Nevada *Engineering & Mining Journal* Dec. 7, 1907.

Kral, Victor E. *Mineral Resources of Nye County, Nevada* Nevada Bureau of Mines and Geology Bulletin 50, Mackay School of Mines, Reno, NV, 1951.

Lincoln, Francis Church *Mining Districts and Mineral Resources of Nevada* Nevada Newsletter Publishing Co., Reno, NV, 1923.

Longwell, C.R.; Pampeyan, E.H.; and Roberts, R.J. *Geology and Mineral Resources of Clark County, Nevada* Nevada Bureau of Mines and Geology Bulletin 62, Mackay School of Mines , Reno, NV, 1965.

Nevada State Bureau of Mines and Geology *Metal and Nonmetal Occurrences in Nevada* Nevada Bureau of Mines and Geology Bulletin 16, Mackay School of Mines , Reno, NV, 1932.

Nolan, Thomas B. *The Eureka Mining District, Nevada* United States Geological Survey Professional Paper 406, U.S. Government Printing Office, Washington, DC, 1962.

Paher, Stanley W. *Nevada Ghost Towns & Mining Camps* Nevada Publications, Las Vegas, NV, 1984 [1970].

Pardee, J.T. & Jones, E.L. Jr. "Deposits of Manganese Ore in Nevada" *Contributions to Economic Geology* United States Geological Survey Bulletin 710, Gov

Ransome, Frederick Leslie "Geology and Ore Deposits of the Goldfield District, Nevada" *Economic Geology* the Economic Geology Publishing Company, Vol. 5, 1910.

Ransome, Frederick Leslie and Emmons, W.H. *Geology and Ore Deposits of the Bullfrog District, Nevada* United States Geological Survey Bulletin 407, Government Printing Office, Washington DC., 1910.

Roberts, R.J., Montgomery, K.M., & Lehner, R.E. *Geology and Mineral Resources of Eureka County, Nevada* Nevada Bureau of Mines and Geology Bulletin 64, Mackay School of Mines , Reno, NV, 1967.

Ross, Clyde P. *Geology and Ore Deposits of the Reese River District, Lander County, Nevada* United States Geological Survey Bulletin 997, United States Government Printing Office, Washington DC., 1953.

Schreier, Nancy B. *Highgrade: The Story of National, Nevada* Authur H. Clark Co., Glendale, CA, 1981.

Spur, J. E. *Descriptive Geology of Nevada South of the Fortieth Parallel, and Adjacent Portions of California* United States Geological Survey Bulletin 208, Government Printing Office, Washington DC., 1903.

Stewart, John H. & McKee, Edwin H. *Geology and Mineral Deposits of Lander County, Nevada* Nevada Bureau of Mines and Geology Bulletin 88, Mackay School of Mines , Reno, NV, 1977.

Westgate, Lewis & Knopf, Adolph *Geology and Ore Deposits of the Pioche District, Nevada* United States Geological Survey Professional Paper 171, U.S. Government Printing Office, Washington, DC, 1932.

New Mexico

Clark, Ellis "The Silver Mines of Lake Valley, New Mexico" *Transactions of the American Institute of Mining Engineers* Vol. 24, 1894.

Griswold, George B. *The Mineral Deposits of Luna County, New Mexico* New Mexico Bureau of Mines and Mineral Resources, Bulletin 72, 1961.

Harley, George Townsend *The Geology and Ore Deposits of Sierra County, New Mexico* New Mexico Bureau of Mines and Mineral Resources, Bulletin 10, 1934.

Lasky, Samual G. *Geology and Ore Deposits of the Lordsburg Mining District, Hidalgo County, New Mexico* United States Geological Survey Bulletin No.885, United States Printing Office, Washington, DC, 1938.

Lasky, Samual G. And Wootton, Thomas Peltier *The Mineral Resources of New Mexico and their Economic Features* New Mexico Bureau of Mines and Mineral Resources, Bulletin 7, 1933.

Lindgren, Waldemar, Graton, Louis C., and Gordon, Charles H. *The Ore Deposits of New Mexico* United States Geological Survey Professional Paper 68, United States Printing Office, Washington, DC, 1910.

Rothrick, Howard E., Johnson, C.H., and Hahn, A.D. *Flourospar Resources of New Mexico* New Mexico Bureau of Mines and Mineral Resources, Bulletin 21, 1921.

Utah

Butler, B.S. *Geology and Ore Deposits of the San Francisco and Adjacent Districts, Utah* United States Geological Survey Professional Paper 80, Government Printing Office, Washington DC, 1913.

Earll, F. Nelson *Geology of the Central Mineral Range* Department of Geology, University of Utah [Dissertation] 1957.

Hollister, O.J. "Gold and Silver Mining in Utah" *American Institute of Mining Engineers [AIME]: Transactions,* Vol. 16, 1887-1888 New York, NY, 1888.

Lemmon, Dwight M. & Morris, Hal T. *Preliminary Geological Map of the Milford Quad, Beaver, Utah* United States Geological Survey, Open File Report 79-1471, 1979.

Lewis, Robert S. & Varney, Thomas *Bulletin 12: The Mineral Industries of Utah* University of Utah, Salt Lake City, UT, 1919.

Morris, H.T. & Lovering, T.S. *General Geology and Mines of the East Tintic Mining District, Utah and Juab Counties, Utah* United States Geological Survey Professional Paper 1024, United States Government Printing Office, Washington DC, 1979.

Nackowski, M.P. & Levy, Enrique *Bulletin 101: Mineral Resources of the Delta-Milford Area* Utah Engineering Experimental Station, University of Utah, Salt Lake City, UT, 1959.

Tower, George Warren Jr. "Geology and Mining Industry of the Tintic District, Utah" *Nineteenth Annual Report of the United States Geological Survey: Part III, Economic Geology United States* Geological Survey, Government Printing Office, Washington, DC, 1899.

MAPS

Colorado

Map of the Approved Surveys: Red Elephant, Columbian, Capital, and Ohio Mts: Clear Creek County, Colorado Published by W.C. Williams, 1879 Denver, Colorado.

Archives, University of Colorado at Boulder, Boulder, CO.

Map: Blueprint, Mine Claims Clear Creek County, Colorado [No Date, No Publisher].

Map: Blueprint, Mining Districts, Clear Creek County, Colorado [No Date, No Publisher].

Map of Clear Creek County, Colorado Drawn and Compiled by Theo. H. Lowe and F.F. Brune C.E. Idaho, Colorado Ter. 1866. Archives, University of Colorado at Boulder, Boulder, CO.

Map of Colorado Showing Location of Mining Districts, Production by County to 1934 Inclusive.

Compiled T.O. Berryovsky, Salida, Colorado 1935. Archives, University of Colorado at Boulder, Boulder, CO.

Map of the Mining Districts Surrounding the Townsite of Idaho Springs. Situated in Clear Creek County, Colorado Theo. K. Lowe C.E [1880s]. Archives, University of Colorado at Boulder, Boulder, CO.

Map of the Peru and Snake River Mining District, Summit County, and Geneva District, Clear Creek Co., Colorado Published by James Teal, C.E. George E. Kedzel, C.E. Asst. May 1882 Georgetown, Colo.

Collection of Leo Stambaugh, Georgetown, CO.

Mining Districts of Gilpin and Clear Creek Counties, Colorado Published by the Clason Map Co., Commonwealth Bldg., Denver, Colo. First Edition - 1906 Archives, University of Colorado at Boulder, Boulder, CO.

Nevada

Nolan, T. B., Marriam, C.W., & Brew, D.A. Geological Map of the Eureka Quad, Eureka and White Pine County, Nevada Map I-612, United States Geological Survey, U.S. Government Printing Office, Washington, DC, 1971.

Nolan, T. B., Marriam, C.W., & Blake, M.C. Jr. Geological Map of the Pinto Summit Quad, Eureka and White Pine County, Nevada Map I-793, United States Geological Survey, U.S. Government Printing Office, Washington, DC, 1971.

Quinlivan, W.D. & Rogers, C.L. Geological Map of the Tybo Quad, Nye County, Nevada United States Geological Survey, U.S. Government Printing Office, Washington, DC, 1974.

Emmons, S.M. Geological Atlas of the United States: Tintic Special Folio United States Geological Survey, Government Printing Office, Washington, DC, 1900.

ARCHAEOLOGY AND HISTORIC PRESERVATION

Busch, Jane "An Introduction to the Tin Can" *Historical Archaeology* Vol. 15 No.1 1981.

Conlin, Joseph R. *Bacon, Beans, and Galantines: Food and Foodways on the Western Mining Frontier* University of Nevada Press, Reno, NV, 1986.

Considerations of Historic and Cultural Resources in the Implementation of the Surface Mining Control & Reclamation Act of 1972, & General Mining Laws Oversight Hearing Before Subcommittee on Interior and Insular Affairs, U.S. House of Representatives, 102 Congress, Feb. 21, 1991, Serial #1029.

Fike, Richard E. *The Bottle Book: A Comprehensive Guide to Historic, Embossed Medicine Bottles* Peregrin Smith Books, Salt lake City, UT 1987.

Firebaugh, Gail S. "An Archaeologists Guide to the Historical Evolution of Glass Bottle Technology" *Southwestern Lore* Sept. 1989, p49.

Frankaviglia, Richard *Hard Places: Reading the Landscape of America's Historic Mining Districts* University of Iowa Press, Iowa City, IA 1991.

Hardesty, Donald L. "The Archaeological Significance of Mining Districts" Proceedings of the Workshop on Historic Mining Resources: Defining the Research Questions for Evaluation and Preservation Sponsored by the State Historic Preservation Center, South Dakota, State Historical Preservation Society, April 6-8, 1987.

Hardesty, Donald L. *The Archaeology of Mining and Miners: A View from the Silver State* The Society for Historical Archaeology, 1988.

Jones, Olive R. and Sullivan, Catherine *The Parks Canada Glass Glossary* National Historic Parks and Sites Branch, Parks Canada, Ottawa 1985.

Lubar, Steven and Kingery, David *History from Things: Essays on Material Culture* Smithsonian Institution Press, 1993.

Martin, Patrick "The Historical Archaeologist's Perspective" Proceedings of the Workshop on Historic Mining Resources: Defining the Research Questions for Evaluation and Preservation Sponsored by the State Historic Preservation Center, South Dakota, State Historical Preservation Society, April 6-8, 1987.

Miller, David B. "Preservation of Cultural Resources Related to the History of Mining in the Black Hills, The Historian's View" Proceedings of the Workshop on Historic Mining Resources: Defining the Research Questions Preservation Sponsored by the State Historic Preservation Center, South Dakota, State Historical Preservation Society, April 6-8, 1987.

Noble, Bruce Jr. and Spude, Robert *National Register Bulletin 42: Guidelines for Identifying, Evaluating, and Registering Historic Mining Properties* United States Department of the Interior, Washington DC, 1992.

Orser, Charles Jr. *Images of the Recent Past: Readings in Historical Archaeology* Alta Mira Press, Walnut Creek, CA, 1996.

Proceedings of the Workshop on Historic Mining Resources: Defining the Research Questions for Evaluation and Preservation Sponsored by the State Historic Preservation Center, South Dakota, State Historical Preservation Society, April 6-8, 1987.

Robert Robinson, Geologist, United States Geological Survey *Overseeing Abandoned Mine Land Environmental Clean up in San Juan Mountains; Personal Interview,* Lakewood, Colorado, May, 1998.

Rock, James T. "Basic Bottle Identification" Unpublished Manuscript Yreka, CA, 1990.

Rock, James T. "Cans in the Countryside" *Historical Archaeology* Vol. 18 No.2.

Rock, James T. *Tin Canisters: Their Identification* Jim Rock, Yreka, CA, 1989.

Rodman, Valerie Yvone *Modeling as a Preservation Planning Tool in Western Gold and Silver Mining Districts* Thesis Manuscript, Department of Anthropology, University of Nevada at Reno, NV, 1985.

Sagstetter, Bill and Beth *The Mining Camps Speak: A New Way to Explore the Ghost Towns of the American West* BenchMark Publishing of Colorado, Denver, CO, 1998.

Sara, Tim; Twitty, Eric; Grant, Marcus *Mining and Settlement at Timberline: Mitigative Treatment of Historic Properties Involved in a Land Exchange Between Cripple Creek & Victor Gold Mining Company and USDI-Bureau of Land Management Near Victor, Colorado* Paragon Archaeological Consultants, Inc., Denver, CO, 1998.

Schroeder, Joseph J. *Sears, Roebuck & Co.: Catalog, 1908* [reprint] Follett Publishing Co., Chicago, IL, 1969.

Spude, Robert L. "Mining Technology & Historic Preservation with Special Reference to the Black Hills" *Proceedings of the Workshop on Historic Mining Resources: Defining the Research Questions Preservation* Sponsored by the State Historic Preservation Center, South Dakota, State Historical Preservation Society, April 6-8, 1987.

Stromberg-Carlson Telephone Manufacturing Company *Telephones, Switchboards, Radio Apparatus, Supplies* Stromberg-Carlson Telephone Manufacturing Co., Rochester, NY, 1925 [trade catalog].

Swope, Karen K. *With Infinite Toil: Historical Archaeology in the Beveridge Mining District, Inyo County, California* Dissertation Manuscript, Department of Anthropology, University of California at Riverside, CA, 1993.

Torma, Carolyn "Assessing the Work to Date" Proceedings of the Workshop on Historic Mining Resources: Defining the Research Questions Sponsored by the State Historic Preservation Center, South Dakota, State Historical Preservation Society, April 6-8, 1987.

United States Department of Agriculture *Treatment of Abandoned Mine Shafts* United States Department of Agriculture, Soil Conservation Service, Washington DC, 1981.

General Mining Engineering & Technology

Baca, Florentino Berry *A Discussion of Mine Shaft Sinking* New Mexico State School of Mines, Socorro, NM, 1937 [Thesis Manuscript].

Bosqui, Frances L. "Ore Treatment at the Combination Mine, Goldfield, Nevada" *Mining & Scientific Press* Oct.6, 1906 p413.

Bramble, Charles A. *The ABC of Mining: A Handbook for Prospectors* Geology, Energy & Minerals Corporation, Santa Monica, CA, 1980 [1898].

Brunton, David W. and Davis, John A. *Modern Tunneling* John Wiley & Sons, New York, NY, 1914.

Colliery Engineer Company *Coal & Metal Miners' Pocketbook* Colliery Engineer Company, Scranton, PA, 1893.

Colliery Engineer Company *Coal & Metal Miners' Pocketbook* Colliery Engineer Company, Scranton, PA, 1905 [1893].

Colliery Engineer Company *Coal Miners' Pocketbook* McGraw-Hill Book Co., New York, NY, [1890] 1916.

Collins, Jose H. *Principles of Metal Mining* William Collins, Sons & Company, London, 1875.

Cummins, Arthur B. *Society of Mining Engineers Mining Engineering Handbook* Society of Mining Engineers, New York, NY, 1973.

Eaton, Lucien *Practical Mine Development & Equipment* McGraw-Hill Book Company, New York, NY, 1934.

Engineering & Mining Journal *Details of Practical Mining* McGraw-Hill Book Company, Inc., New York, NY, 1916.

Foster, Clement *A Textbook of Ore and Stone Mining* Charles Griffin & Co, London, 1894.

Gillette, Halbert P. *Rock Excavation: Methods and Cost* Myron C. Clark Publishing Company, New York, NY, 1907.

Goodman Manufacturing Company *Goodman Mining Handbook* Goodman Manufacturing Company, Chicago, IL, 1927.

Hoover, Herbert C. *Principles of Mining: Valuation, Organization, and Administration* McGraw-Hill Book Company, Inc., New York, NY, 1909.

Ihlseng, Magnus *A Manual of Mining* John Wiley & Sons, New York, NY, 1892.

Ihlseng, Magnus *A Manual of Mining* John Wiley & Sons, New York, NY, 1901.

Ingersoll-Rand Drill Company *Today's Most Modern Rock Drills & A Brief History of the Rock Drill Development* Ingersoll-Rand Drill Company, New York, NY, 1939.

International Textbook Company *A Textbook on Metal Mining: Preliminary Operations at Metal Mines, Metal Mining, Surface Arrangements at Metal Mines, Ore Dressing and Milling* International Textbook Company, Scranton, PA, 1899.

International Textbook Company *A Textbook on Metal Mining: Steam and Steam-Boilers, Steam Engines, Air and Air Compression, Hydromechanics and Pumping, Mine Haulage, Hoisting and Hoisting Appliances, Percussive and Rotary Boring* International Textbook Company, Scranton, PA, 1899.

International Textbook Company *Coal and Metal Miners' Pocket Book* International Textbook Company, Scranton, PA, 1905.

International Textbook Company *International Library of Technology: Hoisting, Haulage, Mine Drainage* International Textbook Company, Scranton, PA, 1906.

International Textbook Company *International Correspondence School Reference Library: Percussive and Rotary Boring, Dynamos and Motors, Ore Dressing and Milling* International Textbook Company, Scranton, PA, 1905.

International Textbook Company *International Correspondence School Reference Library: Rock Boring, Rock Drilling, Explosives and Blasting, Coal-Cutting Machinery, Timbering, Timber Trees, Trackwork* International Textbook Company, Scranton, PA, 1907.

International Textbook Company *International Library of Technology: Mine Surveying, Metal Mine Surveying, Mineral-Land Surveying, Steam and Steam Boilers, Steam Engines, Air Compression* International Textbook Company, Scranton, PA, 1924.

Jeffery Manufacturing Company *Jeffery Mine Handbook #380* Jeffery Manufacturing Co., Columbus, OH, 1924.

Lauchli, Eugene *Tunneling: Short and Long Tunnels of Small and Large Section Driven Through Hard and Soft Materials* McGraw-Hill Book Co., Inc., New York, NY, 1915.

Lewis, Robert S. *Elements of Mining* John Wiley & Sons, Inc., New York, NY, 1946 [1933].

Morrison's Mining Rights Denver, CO, 1899 [1882].

Peele, Robert *Mining Engineers' Handbook* John Wiley & Sons, New York, NY, 1918.

Peele, Robert *Mining Engineers' Handbook* John Wiley & Sons, New York, NY, 1941 [1918] 3rd Ed.

Prelini, Charles *Earth & Rock Excavation* D.Van Nostrand Co., New York, NY, 1906.

"The Ratcliff Mine" *Mining & Scientific Press* April 4, 1903 p213.

Silversmith, Julius *A Practical Handbook* Metallurgists American Mining Index, New York, NY, 1867.

Stack, Barbara *Handbook of Mining & Tunnelling Machinery* John Wiley & Sons, New York, NY, 1982.

Staley, William *Mine Plant Design* McGraw-Hill Book Co., New York, NY, 1936.

Tillson, Benjamin Franklin *Mine Plant* American Institute of Mining and Metallurgical Engineers, New York, NY, 1938.

Tinney, W.H. *Gold Mining Machinery: Its Selection, Arrangement, & Installation* D. Van Nostrand Company, New York, NY, 1906.

Tower, G.W. Jr. "What Constitutes a Mine" *Engineering & Mining Journal* Sept. 13, 1902 p343.

Whitehurst, J.W. and Cary, W.P. "Design of a Mine Plant-I" *Mining & Scientific Press* Aug. 13, 1910 p202.

Whitehurst, J.W. and Cary, W.P. "Design of a Mine Plant-II" *Mining & Scientific Press* Aug. 20, 1910 p239.

Young, George *Elements of Mining* John Wiley & Sons, New York, NY, 1923.

Young, George *Elements of Mining* John Wiley & Sons, New York, NY, 1946.

Zurn, E.N. *Coal Miners' Pocketbook* McGraw-Hill Book Co., New York, NY, [1890 Colliery Engineering Co.] 1928.

Air Compressors

"A Four-Stage Air Compressor" *Mining & Scientific Press* Dec. 19, 1903, p405.

"A Gasoline Engine for Mining Use" *Engineering & Mining Journal* March 13, 1897.

"A New Air Compressor" *Engineering & Mining Journal* Dec 6, 1902 p755.

"A New Air Compressor" *Mining & Scientific Press* Sept. 8, 1906 p300.

"A New Air Compressor" *Engineering & Mining Journal* Sept. 15, 1906 p499.

"A New Clayton Compound High Duty Air Compressor" *Engineering & Mining Journal* Dec. 9, 1893 p596.

"A New Duplex Air Compressor" *Engineering & Mining Journal* Nov. 18, 1899 p611.

"A New Type of Air Compressor" *Engineering & Mining Journal* Dec 12, 1896.

"A Reidler Duplex Air Compressor" *Engineering & Mining Journal* March 10, 1894 p221.

"A Vertical Intercooler for Air Compression" *Engineering & Mining Journal* May 24, 1902 p733.

Cleveland Rock Drill Company *Driller's Handbook* Cleveland Rock Drill Co., Cleveland, OH, 1931.

Compressed Air and Gas Institute *Compressed Air Handbook* McGraw-Hill Book Co., Inc., New York, NY, 1954.

Compressed Air Magazine *Rock Drill Data* Compressed Air Magazine, Philipsburg, NJ, 1960.

"Compressed Air Power Plant at the St. Louis Exposition" *Mining & Scientific Press* Feb. 6, 1904, p97.

Denver Rock Drill Manufacturing Company *Drill Steel* Denver Rock Drill Manufacturing Company, Denver, CO, ca. 1927 [Trade Catalog].

"Direct-Connected Oil Engine Driven Compressor" *Engineering & Mining Journal* June 27, 1914 p1297.

"Economical Air Compressor Drive" *Engineering & Mining Journal* April 13, 1918 p685.

"Enclosed Type Air Compressor" *Mining & Scientific Press* Oct. 26, 1912, p550.

"Franklin Air Compressor" *Engineering & Mining Journal* May 3, 1902.

"The Franklin Air Compressor" *Engineering & Mining Journal* June 1, 1907 p1049.

"Gasoline Engine and Air Compressor" *Engineering & Mining Journal* March 11, 1899 p295.

Hansen, Charles "Compressed Air Used in Mining" *Engineering and Mining Journal* Jan. 22, 1897 p105.

Hardsocg Wonder Drill Company *Bulletin A: Hardsocg Wonder Drill, Herron & Bury Compressors* Hardsocg Wonder Drill Company, Ottumwa, IA, 1905 [trade catalog].

Harron, Rickard, & McCone "Electric Driven Air Compressors" *Engineering & Mining Journal* Oct. 13, 1904 p593-594.

"The Hornsby-Akroyd Oil Engine and Air Compressor" *Engineering & Mining Journal* May 30, 1903 p823.

"Improved Bury Air Compressor" *Engineering & Mining Journal* April 18, 1908 p808.

Ingersoll-Rand Drill Company *Imperial Type 11 Air Compressor: Instructions for Installing and Operating* Rand Drill Co., New York, NY, 1915.

Ingersoll-Rand Drill Company *Rock Drills Data* Compressed Air Magazine, Phillipsburg, NJ., 1960.

Ingersoll Rock Drill Company *Catalog No.7: Rock Drills, Air Compressors and Air Receivers* Ingersoll Rock Drill Company, New York, NY, [1887].

J. George Leyner *Engineering Works Catalog No. 8* Carson-Harper, Denver, CO, 1906 [trade catalog].

Nelson, S.T. "Development of Reciprocating Air Compressors" *Engineering and Mining Journal* Sept. 27, 1919 p533-536.

"New Type of Compressor Intercooler" *Engineering & Mining Journal* Feb. 13, 1913 p381.

"Oil-Driven Air Compressors" *Mining & Scientific Press* May 30, 1914 p918.

"Proportions of Air Mains and Branches" *Engineering & Mining Journal* Nov. 25, 1911 p1027.

Rand Drill Company *Illustrated Catalog of the Rand Drill Company, New York, U.S.A.* Rand Drill Company, New York, NY, 1886.

Rand Drill Company *Imperial Type 10 Air Compressors* Rand Drill Co., New York, NY, 1904.

Richards, Frank "Use of Coolers in Air Compression" *Engineering and Mining Journal* June 1, 1907 p1039.

"Short Belt Drive for Compressor" *Engineering & Mining Journal* August 16, 1913 p306.

Simons, Theodore E.M., C.E. *Compressed Air: A Treatise on the Production, Transmission, and Use of Compressed* Air McGraw-Hill Book Company, Inc., New York, NY, 1921.

"Small Heavy Duty Air Compressors" *Engineering & Mining Journal* Feb. 11, 1911 p315.

Thorkelson, H.J. *Air Compression and Transmission* McGraw-Hill Book Company, Inc., New York, NY, 1912.

"Two-Stage Power-Driven Angle Compressor" *Engineering & Mining Journal* March 28, 1914 p667.

"Types of Electrically Driven Machinery in the Iron River District" *Engineering & Mining Journal* Dec. 26, 1914 p1147.

Young, George "Small Compressors Driven by Oil Engines Becoming Popular in the West" *Engineering & Mining Journal* Nov. 22, 1919 p830.

Assaying and Assay Shops

"A Double Muffle Assay Furnace" *Engineering & Mining Journal* June 15, 1905 p1138.

"A New Oil Burning Assay Furnace" *Engineering & Mining Journal* Jan. 21, 1899 p87.

"A New Ore-Sample Grinder" *Engineering & Mining Journal* May 10, 1899 p403.

Brush, George J. and Penfield, Samuel L. *Determinative Mineralogy and Blowpipe Analysis* John Wiley & Sons, New York, NY, 1904.

"The Calkins Cupel Machines" *Engineering & Mining Journal* Nov. 25, 1899 p639.

"Case Assay Furnace" *Engineering & Mining Journal* August 1, 1914 p215.

Denver Fire Clay Company *Denver Fire Clay: Manufacturers of Furnaces, Muffles, Crucibles, Scorifers, Etc.* Denver Fire Clay Company, Denver, CO, ca. 1895 [Trade Catalog].

Denver Fire Clay Company *The Denver Fire Clay Company Catalog 12* Denver Fire Clay Company, Denver, CO, 1934.

Denver Fire Clay Company *The Denver Fire Clay Company Catalog 12* Denver Fire Clay Company, Denver, CO, 1953.

"Effect of Borax in Assaying" *Engineering & Mining Journal* Oct. 3, 1908 p656.

Furman, H. Van F. *A Manual of Practical Assaying* John Wiley & Sons, New York, NY, 1893.

Keller, Edward "New Assay Furnace Tools" *Engineering & Mining Journal* April 20, 1905 p757.

"Metallics [editor's notes: assaying]" *Engineering & Mining Journal* March 28, 1908 p659.

Mine & Smelter Supply Co. Catalog No. 74: Laboratory Supplies, Scientific Apparatus, & Chemicals Mine & Smelter Supply Co., Denver, CO, 1929.

Spude, Robert L. *To Test By Fire: The Assayer in the American Mining West, 1848-1920* Dissertation: University of Illinois at Champaign-Urbana, 1989.

"The Taylor Improved Assay Furnace" *Engineering & Mining Journal* May 14, 1898 p583.

"The Thayer Portable Stamp Mill" *Engineering & Mining Journal* March 3, 1900 p263.

"Use and Care of Crucibles" *Engineering & Mining Journal* Oct. 3, 1914 p614.

Wade, W.W. "Blacksmith's Forge as an Assay Furnace" *Engineering & Mining Journal* March 9, 1912 p490.

Blacksmith and Machine Shop

"A New Drill Sharpener" *Engineering & Mining Journal* Nov. 28, 1903 p818.

"A New Drill-Sharpener [by Sullivan]" *Mining & Scientific Press* May 29, 1915 p859.

"A New Type of Drill-Sharpener [by Denver Rock Drill Manufacturing Co.]" *Mining & Scientific Press* Jan. 9, 1917 p36.

"A Rockdrill as a Blacksmith's Hammer" *Engineering & Mining Journal* Aug. 9, 1902 p181.

"Ajax Oil Forge" *Mining & Scientific Press* Jan. 14, 1905 p22.

Bacon, John Lord *Forge Practice & Heat Treatment of Steel* John Wiley & Sons, New York, NY, 1919.

Bealer, Alex W. *The Art of Blacksmithing* Castle Books, Edison, NJ., 1995 [1969].

Blandford, Percy W. *Practical Blacksmithing and Metalworking* TAB Books, Blue Ridge Summit, PA 1988.

Brown & Sharpe Mfg. Co. Catalog: Machinery and Tools Providence, RI, 1904 [Trade Catalog].

Colvin, Fred and Stanley, Frank *Running a Machine Shop* McGraw-Hill Book Co., Inc., New York, NY, 1941.

Colvin, Fred *American Machinists' Handbook & Dictionary of Terms* McGraw-Hill Book Co., Inc., New York, NY, 1935.

"Costs of Fuels for Forges" *Engineering & Mining Journal* April 29, 1911 p850.

Counsel for Small Industries in Rural Areas *The Blacksmith's Craft* Macmillan Publishing Co., New York, NY, 1952.

De Staussure, F.G. "Tool Room Care and Economy" *Engineering & Mining Journal* June 15, 1907 p1139.

"Denver Rock-Drill Sharpener" *Engineering & Mining Journal* Jan. 20, 1917 p155.

Denver Rock Drill Manufacturing Company *The Denver Drill Steel Collaring Machine* Denver Rock Drill Manufacturing Co., Denver, CO, [trade catalog] 1915.

Denver Rock Drill Manufacturing Company *Model DS-8: Denver Drill Sharpener* Denver Rock Drill Manufacturing Co., Denver, CO, [trade catalog] ca. 1920s.

Denver Rock Drill Manufacturing Company *Model 4: Sinclair Drill Sharpener* Denver Rock Drill Manufacturing Co., Denver, CO, [trade catalog] ca. 1910s.

Drew, J.M. *Farm Blacksmithing: A Manual for Farmers and Agricultural Schools* Webb Publishing Co., St. Paul, MN, 1910.

"Drill-Sharpening" *Mining & Scientific Press* May 26, 1917 p737.

"Drill-Sharpening Machine [Compressed Air Machinery Co.]" *Mining & Scientific Press* April 11, 1903.

G. S. "Drill-and Tool-Sharpening Shop at the Copper Queen Mine" *Engineering & Mining Journal* June 24, 1916 p1101.

Gillman, George H. "Ideal Shop for Sharpening Drill Steel" *Engineering & Mining Journal* Oct. 6, 1917 p585.

"Hammer Drills [and steel-sharpening]" *Mining & Scientific Press* March 23, 1907 p382.

"Handling Drill-Steel at Champion Mine" *Engineering & Mining Journal* May 4, 1912 p881.

Harron, Rickard, & McCone *The Ajax Drill Sharpener* Harron, Rickard, & McCone, San Francisco, CA, 1904.

"Hood for Portable Forge" *Engineering & Mining Journal* Oct. 6, 1917 p607.

"Imperial Drill Sharpener" *Engineering & Mining Journal* May 27, 1915 p414.

Ingersoll-Rand Co. *Leyner Drill Sharpener "I-R" Model: Instructions for Installing and Operating* Ingersoll-Rand Co., New York, NY, 1913.

"Little Giant Drill-Sharpener" *Mining & Scientific Press* Nov. 5, 1904 p311.

"Low-Pressure Oil Forge" *Engineering & Mining Journal* June 27, 1914 p1298.

McFarland, J.R. "Variable-Stroke Air Hammer" *Details of Practical Mining* McGraw-Hill Book Company, New York, NY, 1916.

"New Sullivan Drill Sharpener" *Engineering & Mining Journal* May 29, 1915 p949.

"Oil or Gas Drill-Heating Forge" *Engineering & Mining Journal* June 23, 1917 p1115.

Oke, A. Livingston "Some Smithy Appliances" *Details of Practical Mining* McGraw-Hill Book Company, New York, NY, 1916.

O'Rourke, J.E. "Operating a Steel-Sharpening Shop" *Engineering & Mining Journal* Feb. 10, 1917 p263.

Rice, Claude T. "Coke Furnace for Heating Drills" *Details of Practical Mining* McGraw-Hill Book Company, New York, NY, 1916.

Rice, Claude T. "Joplin Steel-Sharpening Kinks" *Details of Practical Mining* McGraw-Hill Book Company, New York, NY, 1916.

"Stand for the Blacksmith Shop" *Details of Practical Mining* McGraw-Hill Book Company, New York, NY, 1916.

"Stand for Machine-Drill Repairing" *Engineering & Mining Journal* May 20, 1916 p903.

Stanley, F.A. "Equipment of Mine Machine Shops" *Engineering & Mining Journal* July 14, 1914 p54.

Streeter, Donald *Professional Blacksmithing* Astral Press, Mendham, NJ, 1980.

Sullivan Machinery Company *A Handbook of Drill Steel: Its Selection, Heating, Forging and Tempering, Including Instructions for the Care and Use of Sullivan Drill Sharpeners and Drill Steel Furnaces* Sullivan Machinery Company, Chicago, IL 1930.

Van Deventer, John H. *Success in the Small Shop* McGraw-Hill Book Co., Inc., New York, NY, 1918.

"Welding Hollow Drill Steel" *Engineering & Mining Journal* Feb. 22, 1913 p421.

Weygers, Alexander *The Modern Blacksmith* Van Nostrand-Reinhold Co., New York, NY, 1974.

"The Word Drill Sharpener" *Engineering & Mining Journal* Sept. 15, 1906 p487.

Electric Power

"A New Sturtevant Generating Set" *Engineering & Mining Journal* Oct. 5, 1901 p431.

"Advantages of Electric Motors" *Mining & Scientific Press* Oct. 20, 1900 p445.

American School of Correspondence *Cyclopedia of Applied Electricity* Armour Institute of Technology, Chicago, IL 1905.

"Applications of Electricity at Cripple Creek, Colorado" *Engineering & Mining Journal* Oct. 28, 1899 p520-521.

Crocker, Francis & Schuyler, Wheeler *The Practical Management of Dynamos and Motors* D.Van Nostrand Co., New York, NY, 1896.

"Electric Installation at Silverton, Colorado" *Engineering & Mining Journal* June 23, 1895 p580.

"Electric Mining Plant at Bodie, Cal." *Engineering & Mining Journal* May 13, 1893 p439.

"Electric Power for Mining" *Engineering & Mining Journal* March 28, 1896 p304.

"Electric Power at the Golden Gate Mine, Utah" *Engineering & Mining Journal* June 25, 1898 p759.

"Electric Power Plants in the Mining Districts of Northern California" *Engineering & Mining Journal* August 31, 1901 p270.

"Electric Transmission of Power in the West and Some Hints on It" *Engineering & Mining Journal* July 23, 1892 p79.

"Electricity in Mining" *Engineering & Mining Journal* Jan. 14, 1899 p40.

"Engines for Electric Mining Plants" *Engineering & Mining Journal* May 19, 1894, p465.

Foster, Horatio *Electrical Engineers' Pocketbook* D.Van Nostrand Co., New York, NY, 1903.

Gibson, George II "Electricity in Mining" *Engineering & Mining Journal* March 1, 1902 p308.

Hazelhurst, Dr. S.R. "The Cripple Creek District, Colorado, VI-The Power Plants" *Engineering & Mining Journal* Dec. 8, 1900 p669.

Hobart, H.M. and Ellis, A.G. *High Speed Dynamo Electric Machinery* John Wiley & Sons, New York, NY, 1908.

Houston, Edwin J. and Kennelly, Arthur E. *Recent Types of Dynamo Electric Machinery* American Technical Book Co., New York, NY, 1898.

Kloeffler, Royce G.; Brenneman, Jesse L.; and Kerchner, Russell M. *Direct Current Machinery* Macmillan Co., New York, NY, 1934.

Mine & Smelter Supply Co. Catalog No. 76: Electrical Supplies & Equipment Mine & Smelter Supply Co., Denver, CO, 1930.

Puchstein, A.F. and Lloyd, T.C. *Alternating Current Machines* John Wiley & Sons, New York, NY, 1942.

Raber, W.F. *The Arkansas Valley Railway, Light, & Power Company* State Bureau of Mines, Denver, CO, 1911.

Renz, Robert E. "The Electric Equipment of the Yak Tunnel" *Engineering & Mining Journal* May 25, 1907 p985-990.

Rickard, T.A. "Electricity or Steam For Hoisting" *Engineering & Mining Journal* May 2, 1904 p881.

Saunders, William [untitled discussion of electric power] *Engineering & Mining Journal* Jan. 2, 1892.

Seager, James "Electrical Mine Machinery Troubles" *Engineering & Mining Journal* Feb.4, 1911 p274-276.

Twitty, Eric "From Steam Engines to Electric Motors: Electrification in the Cripple Creek Mining District" *The Mining History Journal* Fifth Annual Journal, 1998.

"Types of Electrically Driven Machinery in the Iron River District" *Engineering & Mining Journal* Dec. 26, 1914 p1147.

"Westinghouse Electric Mine Plant" *Engineering & Mining Journal* Jan. 13, 1894 p33-34.

"Westinghouse Type 'C' Induction Motors" *Mining & Scientific Press* April 7, 1900 p372.

Worcester, S.A. "Electric Hoisting and Pumping in the Cripple Creek District" *Engineering & Mining Journal* May 22, 1909 p1057.

Gasoline Engines

Audel, Theo. & Co. *Audel's Gas Engine Manual: A Practical Treatise* Theo. Audel & Co. Publishers, New York, NY, 1908.

Clerk, Dugald *The Gas and Oil Engine* John Wiley & Sons, New York, NY, 1902.

Curtis, A.R. "The Small Gas Engine" *Engineering & Mining Journal* Sept. 8, 1900, p285.

"Gas Engines in Mining Plants" *Engineering & Mining Journal* July 3, 1897 p2.

"Gas Engines for Mining Work" *Engineering & Mining Journal* Nov. 2, 1895 p414.

"Gas Engine Hoists for Small Mines" *Engineering & Mining Journal* Oct. 24, 1896 p386.

"Horsby-Akroyd Oil Engine" *Mining & Scientific Press* Dec. 31, 1904 p442.

Hutton, Frederick R. *The Gas Engine: A Treatise on the Internal Combustion Engine* John Wiley & Sons, New York, NY, 1904.

"The McGeorge Gas Engine" *Engineering & Mining Journal* Jan. 5, 1895 p10.

"The Priestman Oil Engine from an American Standpoint" *Engineering & Mining Journal* July 23, 1892 p81.

"Questions and Answers [pertaining to gas engines and mining]" *Engineering & Mining Journal* June. 11, 1898 p706.

"Questions and Answers [pertaining to gas engines and mining]" *Engineering & Mining Journal* Oct. 21, 1899 p492.

"Steam and Gas Engines" *Mining & Scientific Press* Nov. 24, 1900 p546.

Walker, Edward "The History of the Petroleum Engine" *Engineering & Mining Journal* Aug. 13, 1892 p154.

"The Weber Gas and Gasolene Engine" *Engineering & Mining Journal* March 31, 1894 p299.

Headframes

"A Small Head-Frame" *Mining & Scientific Press* June 6, 1914 p928.

Altenbach, J. Scott Personal Interview: Albuquerque, New Mexico Sept. 27, 1998. Altenbach is a mining historian who erected a two-post gallows headframe from 8 X 8 timbers in 1998.

"Automatic Bucket Dumping Devices" *Mining & Scientific Press* June 15, 1905 p40.

Barbour, Percy E. "Details of a Wooden Headframe" *Engineering & Mining Journal* Aug. 19, 1911 p344.

Binckley, George S. "The Rational Design of Head Frames for Mines" *Mining & Scientific Press* June 10, 1905 p374.

Botsford, H.L. "A Timber Headframe" *Engineering & Mining Journal* June 10, 1911 p1148.

Botsford, H.L. "Small Timber Headframe" *Engineering & Mining Journal* Dec. 28, 1912 p1215.

Botsford, H.L. "Types of Headframes" *Engineering & Mining Journal* Oct. 13, 1913 p690.

Botsford, H.L. "Small Four Post Headframe" *Details of Practical Mining* McGraw-Hill Book Co., New York, NY, 1916.

Burr, Floyd "Design of Headframes" *Engineering & Mining Journal* April 7, 1917 p611.

"Concentrates [headframe notes]" *Mining & Scientific Press* Oct.1, 1904 p220.

Denny, G.A. "A-Type Timber Headframe" *Engineering & Mining Journal* March 30, 1914 p1100.

"Fairbanks-Morse Automatic Ore Dump" *Engineering & Mining Journal* April 26, 1902.

Fitch, Charles H. "Head Works Framing - Primary Notions" *Mining & Scientific Press* June 6, 1903 p335.

Fitch, Charles H. "Head Works Framing" *Mining & Scientific Press* June 13, 1903 p380.

"Florence-Goldfield Headframe" *Engineering & Mining Journal* Oct. 14, 1911 p739.

Forsyth, Alexander "The Headframes of Shafts at Cripple Creek" *Engineering & Mining Journal* March 7, 1903 p366.

Halloran, A.H. "The Leonard Headframe" *Mining & Scientific Press* June 30, 1906 p432.

"Headframe for a Prospect Shaft" *Engineering & Mining Journal* Nov. 11, 1911 p980.

Hodge, W.R. "A Simple Timber Headframe" *Engineering & Mining Journal* Dec. 2, 1911 p1074.

Le Veque, G.E. "Tripod Headframe and Novel Bucket" *Engineering & Mining Journal* April 23, 1913 p755.

McFarland, J.R. "Devices for Bucket Dumping" *Engineering & Mining Journal* March 29, 1913 p659.

Mentzel, Charles "Prospecting Headframe" *Engineering & Mining Journal* Oct. 5, 1912.

"Photographs from the Field: Headframe at Oliver Shaft of Calumet & Arizona Mining Co." *Engineering & Mining Journal* March 10, 1917 p420.

"Prospecting Equipment" *Mining & Scientific Press* Jan. 23, 1904.

Rogers, A.P. "Bucket and Chute for Shaft Sinking" *Engineering & Mining Journal* June 11, 1910 p1214.

"Searchlight District, Nevada" *Mining & Scientific Press* May 16, 1903 p5.

"Simple Head Frame Construction" *Mining & Scientific Press* Feb. 6, 1904 p94.

"Small Timber Headframe" *Engineering & Mining Journal* April 8, 1916 p. 645.

"Some Mine Headframes and Hoists" *Mining & Scientific Press* Aug. 16, 1904 p264.

"Something More of Headframes" *Mining & Scientific Press* Dec. 30, 1905.

"Some Types of Headframes" *Mining & Scientific Press* Dec. 16, 1905 p410.

"Steel Headframe at the Gwinn Mine [California]" *Mining & Scientific Press* Jan. 2, 1904 p5.

"Steel Headframe of the Parrot Mine, Butte, Montana" *Engineering & Mining Journal* June 21, 1902 p862.

"Steel Headgear at Cripple Creek, Colo." *Mining & Scientific Press* Dec. 8, 1900 p573.

"Steel vs. Timber Headframes" *Mining & Scientific Press* July 11, 1903 p18.

"The St. Eugene Headframe at Moyie, B.C." *Engineering & Mining Journal* Aug. 26, 1911 p392.

"Types of Head Frames" *Mining & Scientific Press* Feb. 21, 1903.

"Variations in [headframe] Construction" *Mining & Scientific Press* Aug. 22, 1903.

Worcester, S.A. "Using the Ore Bucket" *Engineering & Mining Journal* March 12, 1910 p552.

Worcester, S.A. "A Cripple Creek Ore and Waste Handling Plant" *Engineering & Mining Journal* June 5, 1912 p1173.

Hoists and Hoisting

"A Coal Mine Hoisting Engine" *Engineering & Mining Journal* Oct. 22, 1898 p491.

"A Compact Hoisting Engine" *Engineering & Mining Journal* Jan. 9, 1897.

"A Double Gasoline Hoist" *Engineering & Mining Journal* Jan. 26, 1901.

"A Gasoline Hoisting Engine" *Engineering & Mining Journal* May 1, 1897.

"A Gas Engine [Double Drum] Hoisting Plant" *Engineering & Mining Journal* Sept. 12, 1896 p247.

"A Gasoline [Double Drum] Hoisting Engine" *Engineering & Mining Journal* Dec. 31, 1898.

"A Gasoline [Double Drum] Hoisting Engine" *Engineering & Mining Journal* Aug. 5, 1899 p161.

"A [Witte] Gasoline Mine Hoist" *Mining & Scientific Press* Oct. 27, 1906 p518.

"A Hoisting Engine Built in Arizona" *Engineering & Mining Journal* Jan. 17, 1903 p124.

"A Motor-Driven Mine Hoist" *Mining & Scientific Press* August 8, 1903 p81.

"A Motor-Driven Mine Hoist" *Engineering & Mining Journal* August 29, 1903.

"A New Electric Mining Hoist" *Engineering & Mining Journal* Feb. 4, 1893 p101.

"A New Electric Mine Hoist" *Engineering & Mining Journal* March 2, 1895 p201.

"A New Gasoline Hoist" *Engineering & Mining Journal* March 26, 1900.

"A New Prospecting [Gasoline] Hoist" *Engineering & Mining Journal* Aug. 3, 1901 p142.

"A New Type of Hoisting Engine" *Engineering & Mining Journal* Oct. 163, 1909 p790.

"A Small Direct-Acting Hoist" *Mining & Scientific Press* June 25, 1904 p427.

"A 130 Horsepower [Gasoline] 'Union' Mining Hoist" *Mining & Scientific Press* Nov. 4, 1899.

"Allis-Chalmers Electric Hoist" *Engineering & Mining Journal* March 2, 1907 p435.

"Alternating Current Electric Motor Hoist" *Mining & Scientific Press* July 9, 1904 p25.

"An American Hoisting Engine for Japan" *Engineering & Mining Journal* Dec. 3, 1898 p671.

"An Electric Mining Hoist" *Engineering & Mining Journal* March 3, 1904 p305.

"Automatic Band Friction Hoist" *Engineering & Mining Journal* Aug. 20, 1892 p177.

"Boom Swinging Gear for [Donkey] Hoisting Engines" *Mining & Scientific Press* Sept. 24, 1904 p208.

Carter, E.E. "Chain Driven Convertible Hoist" *Details of Practical Mining* McGraw-Hill Book Company, New York, NY, 1916.

"Character of Hoisting Plants" *Mining & Scientific Press* April 25, 1903 p258.

Cole, T.F. "A Single Engine Hoisting Plant" *Engineering & Mining Journal* Sept. 5, 1896 p220.

"Converting Steam Hoists to Electric" *Engineering & Mining Journal* Aug. 29, 1914 p390.

"Corliss Hoisting Engine, Anaconda Mines, Montana" *Engineering & Mining Journal* Aug. 10, 1896 p345.

Del Mar, Algernon "Disposal of Waste at a Prospect Shaft" *Engineering & Mining Journal* June 4, 1910 p1151.

"Device for Handling Prospect Bucket" *Engineering & Mining Journal* March 15, 1913 p565.

"Direct-Acting Hoists" *Mining & Scientific Press* Dec. 15, 1900 p585.

"Direct Current Electric Mining Hoist" *Mining & Scientific Press* March 5, 1904 p165.

"Double Drum Hoist and Electric Motor" *Engineering & Mining Journal* Jan. 14, 1899 p44.

"Electric Hoist at Calumet & Arizona" *Engineering & Mining Journal* Jan. 20, 1912 p162.

"Electric Hoist at Dawson, Penn." *Engineering & Mining Journal* Nov. 13, 1909 p975.

"Electric Mine Hoists" *Engineering & Mining Journal* Feb. 1, 1902 p183.

"Electric Mine Hoists" *Mining & Scientific Press* Oct. 13, 1900 p433.

"Electric Reel Hoists" *Engineering & Mining Journal* Sept. 21, 1907 p540.

"Electrically Operated Hoist with Reels" *Mining & Scientific Press* Aug. 3, 1907 p156.

"The Fairbanks-Morse Gasoline Hoister" *Engineering & Mining Journal* May 14, 1898 p585.

"Gas Engine Hoist" *Mining & Scientific Press* March 4, 1905 p133.

"The Gasoline Engine for Hoisting and Pumping" *Engineering & Mining Journal* Dec. 26, 1896.

Hall, Leon M. "Electric Hoists on the Comstock" *Engineering & Mining Journal* August 3, 1901 p140.

Hamilton, J.W.H. "Electric Mining Hoists" *Engineering & Mining Journal* Sept. 22, 1906 p537-540.

"Hoisting Engine" *Mining & Scientific Press* Nov. 26, 1904 p357.

"Hoisting Engine for the Wharton Iron Mines, New Jersey" *Engineering & Mining Journal* Feb. 13, 1897 p163.

"Hoisting Plant for the Newton Slate Quarry, New Jersey" *Engineering & Mining Journal* Jan. 12, 1901 p53.

"Hoisting 10,000 Tons in 14 Hours" *Mining & Scientific Press* April 17, 1915 p635.

"The Hunt Steam and Electric Hoists" *Engineering & Mining Journal* March 18, 1899 p323.

"The Improved Witte Gasoline Hoist" *Engineering & Mining Journal* Nov. 17, 1900 p583.

"Joplin Type of Horse Whim" *Engineering & Mining Journal* August 3, 1912 p205.

"The Latest Weber Gasoline Hoist" *Engineering & Mining Journal* Nov. 30, 1901.

"Metallics [Editorial Notes: on hoisting]" *Engineering & Mining Journal* July 31, 1909 p226.

Muir, Douglass "Unwatering and Equipping Untimbered shafts" *Details of Practical Mining* McGraw-Hill Book Co., New York, NY, 1916.

Nelson, Sven T. "Steam Hoists for Shallow Mines" *Engineering & Mining Journal* Nov. 25, 1911 p1027.

"Nevada-Douglas Electric Hoist" *Mining & Scientific Press* Jan. 27, 1912 p191.

"New Single and Double Hoisting Engines" *Engineering & Mining Journal* Feb. 27, 1897.

O'Neil, Frederick, W. "Recent Developments in the Design of Hoisting Engines" *Engineering & Mining Journal* April 17, 1917 p602-607.

"Photographs from the Field: the Way Shafts are Started at Butte, Montana" *Engineering & Mining Journal* Nov. 14, 1914 p882.

"The Reynolds Horse Whim" *Engineering & Mining Journal* March 4, 1893 p203.

Rosenblatt, G.B. "Automatic Mine Hoisting" *Mining & Scientific Press* June 14, 1913 p897.

Rushmore, D.B. "Electric Hoists for Mines" *Engineering & Mining Journal* Dec. 16, 1916 p1177.

Russell, H.A. "Electric Hoists for Mine Service" *Mining & Scientific Press* Feb. 8, 1913 p236-240.

Stovall, Dennis H. "Where the Horse Whim Fails" *Mining & Scientific Press* Nov. 4, 1911 p583.

"Temporary Whim for Hoisting" *Engineering & Mining Journal* June 15, 1907 p1133.

"Test Pit Windlass" *Engineering & Mining Journal* Dec. 14, 1912 p1121.

Thompson, James R. "The Hoisting Problem" *Engineering & Mining Journal* Jan. 26, 1905 p173.

"Variations in Hoisting Machinery" *Mining & Scientific Press* Feb. 14, 1903.

Walker, S.F. "Electric Hoisting in Mining Operations" *Engineering & Mining Journal* Nov. 19, 1910 p1014-1016.

Warren, H.M. "Use of Electricity in Anthracite Mining" *Engineering & Mining Journal* March 2, 1907 p433.

"The Weber Gasoline Hoisting Engine" *Engineering & Mining Journal* May, 1900.

"The Witte Gasoline Hoisting Engine" *Engineering & Mining Journal* July 30, 1898 p131.

"The Witte Gasoline Mine Hoist" *Engineering & Mining Journal* Nov. 23, 1901 p672.

Hoisting Vehicles

"Joplin Ore Buckets" *Details of Practical Mining* McGraw-Hill Book CompaNY,, New York, NY, 1916.

"Miscellaneous Notes (editorial material)" *Engineering & Mining Journal* Oct. 23, 1897 p481.

"Concentrates (editorial material)" *Mining & Scientific Press* April 4, 1903 p211.

"Crossheads for Bucket Hoisting" *Engineering & Mining Journal* June 19, 1915 p1078.

"Drill-steel Bucket" *Engineering & Mining Journal* Aug. 7, 1909 p272.

"Joplin Ore Buckets" *Engineering & Mining Journal* Vol 94 No 6 p254.

McFarlane, George C. "Automatic Dumping Skip for Vertical Shafts" *Engineering & Mining Journal* June 26, 1909 p1281.

Proctor, Olin S. "Direct Automatic Bucket Tipple" *Engineering & Mining Journal* Vol 94 No 24 p1120.

Mining Machinery, General

"A Gasoline Engine Mining Plant" *Engineering & Mining Journal* July 10, 1897 p41.

"A Gasoline Hoisting Plant" *Mining & Scientific Press* May 11, 1901 p220.

Bailey, Lynn *Supplying the Mining World: the Mining Equipment Manufacturers of San Francisco 1850-1900* Western Lore Press, Tucson, AZ, 1996.

Cleveland Pneumatic Tool Co. *Air Hose, Air Hose Couplings, Air Hose Clamps, Air Hose Nipples, Air Hose Menders* Cleveland Pneumatic Tool Co., Cleveland, OH, ca. 1905 [Trade Catalog].

Cleveland Pneumatic Tool Co. *Cleveland Air Hammer Drills* Cleveland Pneumatic Tool Co., Cleveland, OH, ca. 1905 [Trade Catalog].

Colorado Iron Works *Catalog of the Colorado Iron Works: Manufacturers of Mill and Mining Machinery* Colorado Iron Works, Denver, CO, 1882 [Trade Catalog].

Colorado Iron Works *Catalog No. 11: Accessories for Mines* Colorado Iron Works, Denver, CO, 1882 [Trade Catalog].

Denver Equipment Company *Bulletin No. 3800: Complete Mill Equipment from Crusher to Filter* Denver Equipment Company, Denver, CO ca 1935 [Trade Catalog].

Denver Equipment Company *Catalog of Reconditioned and New Equipment* Denver Equipment Company, Denver, CO 1941 [Trade Catalog].

Denver Rock Drill Manufacturing Company *Drill Steel* Denver Rock Drill Manufacturing Company, Denver, CO ca. 1927 [Trade Catalog].

Dooley Brothers *Miners' Drilling Machines, Tools, Mine Supplies, Mine Tools, Etc.* Dooley Brothers, Peoria, IL 1919 [Trade Catalog].

E.I. DuPont de Nemours & Company *DuPont Blasting Powder* E.I. DuPont de Nemours & Co., Wilmington, DE 1917.

E.I. DuPont de Nemours & Company *DuPont Blasting Powder* E.I. DuPont de Nemours & Co., Wilmington, DE 1934.

"Gasoline Engine Mining Plant" *Engineering & Mining Journal* May 7, 1898 p555.

Hall, Leon M. "Modernizing the Comstock Lode" *Mining & Scientific Press* March 17, 1906 p183.

Hendrie & Bolthoff Manufacturing Company *Compliments of the Hendrie & Bolthoff Manufacturing Co.* Smith-Brooks Printing Co., Denver, CO 1897.

Howard, L.G. "Hoisting Works in the Park City District, Utah-I" *Mining & Scientific Press* Oct. 9, 1915 p545.

Ingersoll-Rand *Drill More at Less Cost with Jackbits* Ingersoll-Rand, New York, NY, ca. late 1930s [Trade Catalog].

Institute of Makers of Explosives *Safety in the Handling & Use of Explosives* Institute of Makers of Explosives, Washington DC, 1934.

J. George Leyner Engineering Works Co. *Leyner Rock Terrier: One-Man Compressed Air Drill: Catalog No.8* J. George Leyner Engineering Works Co., Denver, CO, 1906 [Trade Catalog].

Jeffrey Manufacturing Company *Jeffrey Power Drills for Rock and Coal* Jeffrey Manufacturing Co., Columbus, OH, ca. 1905 [Trade Catalog].

Jeffrey Manufacturing Company *Jeffrey Material Handling and Mining Machinery: General Catalog Number 85* Jeffrey Manufacturing Co., Columbus, OH, 1922.

Keystone Consolidated Publishing Company Inc. *The Mining Catalog: 1925 Metal-Quarry Edition* Keystone Consolidated Publishing Company Inc., (no location given), 1925.

The Martin Hardsocg Company *Catalog H: Spring 1912* The Martin Hardsocg Company, Pittsburgh, PA, 1912 [Trade Catalog].

Metal Quarry Catalogs: 1935/1936 McGraw-Hill Catalog Service, New York, NY, 1936.

Mine & Smelter Supply Co. *Catalog No. 22: Machinery and General Supplies* J.D. Abraham Publishing Co., 1912.

Mine & Smelter Supply Co. *Catalog No. 72 Machinery & Supplies* R.R. Donnelly & Sons Co., 1926.

Mine & Smelter Supply Co. *Catalog No. 92 Supplies & Equipment* R.R. Donnelly & Sons Co., 1937.

Rehfuss, Louis A. and Rehfuss, W. Clifford "Portable Mining Equipment for Prospectors" *Engineering &Mining Journal* June 10, 1916 p1025.

Pacific Mill & Mine Supply Company *Pacific Mill & Mine Supply Company: Catalog No. 30* Pacific Mill & Mine Supply Co., San Francisco, CA, 1920.

"Power Equipment" *Mining & Scientific Press* Jan. 23, 1904 p55.

Rice, Claude T. "Tonopah-Belmont Surface Plant" *Engineering & Mining Journal* April 29, 1911 p853.

Ross, W.B. "An Air-Cooled Gasolene Hoist and Compressor" *Engineering & Mining Journal* July 1, 1911 p16.

Sprague Electric Co. *Sprague Flexible Armored Hose for Compressed Air or Steam* Sprague Electric Co., New York, NY, 1910 [Trade Catalog].

Stromberg-Carlson Telephone Manufacturing Co. *Stromberg-Carlson: Telephones, Switchboards, Radio Apparatus, Supplies* Stromberg-Carlson Telephone Mfg Co., Rochester, NY, 1925 [Trade Catalog].

Shafts and Tunnels

Bernard, Clinton P. "Hinges for Shaft Doors" Details of Practical Mining McGraw-Hill Book Co., New York, NY, 1916.

"Concentrates" *Mining & Scientific Press* March 10, 1900 p259.

Edyvean, E.H. "Safety Bonnet for Shaft Opening" *Engineering & Mining Journal* May 16, 1914 p1001.

Linke, Harold A. "Light Shaft timbering" *Details of Practical Mining* McGraw-Hill Book Co., New York, NY, 1916.

"Mesabi Shaft Collars" *Engineering & Mining Journal* Sept. 20, 1913 p546.

"Rectangular Concrete Shaft Lining" *Engineering & Mining Journal* Nov. 23, 1914 p959.

"Sinking or Tunneling" *Engineering & Mining Journal* Dec. 9, 1899 p700.

"The Best Shape for a Shaft.—I" *Mining & Scientific Press* Aug. 11, 1906 p167.

"The Best Shape for a Shaft.—II" *Mining & Scientific Press* Sept. 1, 1906 p256.

Steam Boilers & Steam Power

Barton, Denys Bradford *The Cornish Beam Engine: A Survey of Its History and Development in the Mines of Cornwall and Devon from Before 1800 to Present Day, with Something of its Use Elsewhere in Britain and Abroad* Wordens of Cornwall, Ltd. 1966.

Blake & Knowles Steam Pump Works *Knowles Boiler Feed & Tank Pumps: Single & Duplex* Blake & Knowles Steam Pump Works, New York, NY, [trade catalog] ca. 1910.

"Choice and Care of Grate-Bars" *Engineering & Mining Journal* Vol 97, No8.

Cole, T.F. "A Single Engine Hoisting Plant" *Engineering & Mining Journal* Sept. 5, 1896.

Croft, Terrell *Steam Boilers* McGraw-Hill Book Co., New York, NY, 1921.

Gebhardt, G.F. *Steam Power Plant Engineering* John Wiley & Sons, New York, NY, 1925.

Greeley, Horace; Case, Leon; Howland, Edward; Gough, John B.; Ripley, Philip; Perkins, E.B.; Lyman, J.B.; Brisbane, Albert; Hall, E.E. "Babcock and Wilcox Boiler" *The Great Industries of the United States* J.B. Burr, Hartford, CT, 1872.

Greeley, Horace; Case, Leon; Howland, Edward; Gough, John B.; Ripley, Philip; Perkins, E.B.; Lyman, J.B.; Brisbane, Albert; Hall, E.E. "Steam and the Steam Engine" *The Great Industries of the United States* J.B. Burr, Hartford, CT, 1872.

Halen, George B. and Swett, George W. *The Design of Steam Boilers and Pressure Vessels* John Wiley & Sons, New York, NY, 1923.

Hawkins, N. *New Catechism of the Steam Engine: With Chapters on Gas, Oil, and Hot Air Engines* Theo. Audel & Co. Publishers, New York, NY, 1900.

Holbrook, E.A. "Wood as a Fuel for Mine Boiler Plants" *Engineering & Mining Journal* April 10, 1915, p645-647.

"Kingsley Water-Tube Boiler" *Engineering & Mining Journal* Oct. 28, 1893, p447.

Kleinhans, Frank B. *Locomotive Boiler Construction* Norman W. Henley Publishing Co., New York, NY, 1915.

Linstrom, C.B. & Clemens, A.B. *Steam Boilers and Equipment* International Textbook Co., Scranton, PA, 1928.

Lukens Steel Company *Lukens Steel Company* Lukens Steel Co., Coatesville, PA, 1924.

The M.J. O'Fallon Supply Company *Catalog No.8 Steam, Mill, & Water Supplies* R.R. Donnelly & Sons Co., 1923.

MacNaughton, Edgar. *Elementary Steam Power Engineering* John Wiley & Sons, New York, NY, 1933.

"New Standard Reynolds Corliss Engine" *Mining & Scientific Press* Dec. 5, 1893, p373.

"The Nordberg Corliss Engine" *Engineering & Mining Journal* May 3, 1890, p497.

Thurston, R.H. *A Manual of Steam Boilers: Their Design, Construction, and Operation* John Wiley & Sons, New York, NY, 1901.

"The Philadelphia Water Tube Safety Boiler" *Engineering & Mining Journal* August 14, 1897, p191.

Potter, Andrey A. and Calderwood, James P. *Elements of Steam and Gas Engineering* McGraw-Hill Book Co., Inc., New York, NY, 1938.

Rigg, Arthur *A Practical Treatise on the Steam Engine* E.&F.N. Spon, London, 1878.

Rogers, Warren U. "The Care of Small Steam Boilers" *Engineering & Mining Journal* Dec. 8, 1909, p1217-1219.

"The Sederholm [Water Tube] Boiler" *Engineering & Mining Journal* Oct. 23, 1897, p491.

Stein, Ernest "A Home-Made Boiler Feed Water Heater" *Engineering & Mining Journal* June 23, 1904, p1009.

Tilden, E.G. "Burning Wood Under Boilers" *Engineering & Mining Journal* March 6, 1909, p499.

Topeka Steam Boiler Works *Topeka Steam Boiler Works Catalog B Steam Boilers, Power Plant Equipment, Plumbing and Heating Supplies* R.R. Donnelly Sons Co. 1923.

"The Ward [Water Tube] Boiler" *Engineering & Mining Journal* Oct. 28, 1893, p447.

Structures, Foundations, and Materials

Bauer, Edward E. *Plain Concrete* McGraw-Hill Book Co., New York, NY, 1949.

"Clinched Steel Pipes" *Mining & Scientific Press* Jan. 6, 1900 p6.

"Concentrates [foundation notes]" *Mining & Scientific Press* Sept. 10, 1904.

"Concentrates [ore bin notes]" *Mining & Scientific Press* Sept. 10, 1904.

Croft, Terrell *Machinery Foundations and Erection* McGraw-Hill Book Co., New York, NY, 1923.

"Durability of Wood Pipe" *Engineering & Mining Journal* April 23, 1913 p755.

Archaeological Excavation Records, Site 5TL0351, Feature 86, Hoist Foundation, Cripple Creek, Teller County, Colorado, Sept. 1994 Cripple Creek & Victor Gold Mining Company, Paragon Archaeological Consultants, Inc., Denver, CO.

"Editorial [surface plant notes]" *Mining & Scientific Press* Feb. 18, 1905 p98.

"Foundation Bolts for Steam Engines" *Mining & Scientific Press* Jan. 16, 1904 p41.

Glover, John George *The Development of American Industries: Their Economic Significance* Prentice-Hall, New York, McGraw-Hill Book Co., New York, NY, 1912.

Glover, Margaret L. *Log Structures: Criteria for Their Description, Evaluation, and Management as Cultural Resources* Thesis Manuscript, Department of Anthropology, Portland State University, Portland, OR, 1982.

"Iron and Steel Buildings for Mining Plants" *Engineering & Mining Journal* March 13, 1897 p256.

Lunt, Horace "Foundations for Prospect Machinery" *Details of Practical Mining* McGraw-Hill Book Co., New York, NY, 1916 p9.

Ketchum, Milo S. C.E. *The Design of Mine Structures* McGraw-Hill Book Co., New York, N.Y., 1912.

McLaughlin, R.P. "Manufacture of Portland Cement in California" *Mining & Scientific Press* March 21, 1903 p180.

Meade, Richard *Portland Cement, Its Composition, Raw Materials, Manufacture, Testing and Analysis* Chemical Publishing Co., New York, NY, 3rd ed. (no date given, but probably 1924).

Mills, Adelbert [1915], Rader, Lloyd *Materials of Construction, Their Manufacture and Properties* John Wiley & Sons, New York, NY, 5th ed. 1939.

"New Type of Copper Queen Change House" *Engineering & Mining Journal* July 25, 1914 p163.

Oke, A. Livingston "Timber Foundations for Engines" *Details of Practical Mining* McGraw-Hill Book Co., New York, NY, 1916 p8.

"Oxyacetylene Welding" *Engineering & Mining Journal* August 1, 1914 p213.

Rice, Claude T. "Magazine for Storing and Thawing Dynamite" *Engineering & Mining Journal* Aug. 31, 1912 p397.

Rice, Claude T. "Powder House with Concrete Roof" *Engineering & Mining Journal* March 9, 1912 p490.

Roth, Leland M. *A Concise History of American Architecture* Harper & Row Publishers, New York, NY, 1979.

Roys, Francis *Materials of Engineering Construction* The Ronald Press Co., New York, NY, 1930.

Sanders, Wilbur E. "Improved Type of Ore Bin" *Details of Practical Mining* McGraw-Hill Book Company, New York, NY, 1916.

Searle, Alfred *Cement, Concrete, & Bricks* D. Van Nostrand Co., New York, NY, 2nd ed.1926.

"Thawing Dynamite [thawhouses]" *Engineering & Mining Journal* March 2, 1907 p428.

"Tonopah Orehouses" *Engineering & Mining Journal* May 27, 1911 p1048.

Wallace, William "Typical Coal Trestles" *Engineering & Mining Journal* Feb. 21, 1914 p419.

Werner, Orville II Registered Professional Engineer, President CTL-Thompson Inc. Denver, CO Personal Interview Denver, CO April 3, 1997.

Young, Joseph *A Brief Outline of the History of Cement* Lehigh Portland Cement Co., Lehigh, PA, 1955.

Transportation In and Around the Mine

"A Compressed Air Locomotive" *Mining & Scientific Press* Dec. 30, 1899.

"A New Electric Mine Locomotive" *Engineering & Mining Journal* Dec 17, 1892 p582.

De Zomboria, C.E. "Aerial Tramways" *Mining & Scientific Press* Sept 26, 1903.

"Electric Haulage in Coal Mines" *Engineering & Mining Journal* April 29, 1893 p395.

"Electric Haulage in Metal Mines" *Engineering & Mining Journal* Feb. 25, 1904 p324.

"Electric Mine Locomotives" *Mining & Scientific Press* Dec. 8, 1900.

General Electric Company *Electric Mine Locomotives: Catalog No. 1045* General Electric Power and Mining Department, Chicago, IL, 1904 [Trade Catalog].

Hansen, Charles "Compressed Air as Used in Mining" *Engineering & Mining Journal* March 9, 1895 p220.

Hirsch, Richard "Compressed Air Haulage Systems" *Engineering & Mining Journal* March 15, 1902 p376.

International Textbook Company *Mine Haulage: Rope Haulage in Coal Mines, Locomotive Haulage in Coal Mines, Mine Haulage Systems, Calculations, and Cars* International Textbook Company, Scranton, PA, 1926.

"Mining Summary: Colorado, Teller County" *Mining & Scientific Press* Sept. 8, 1900 p289.

"Post and Cap Joint for Dumping Trestle" *Details of Practical Mining* McGraw-Hill Book Company, New York, NY, 1916.

"Review of Mining: Cripple Creek, Colorado" *Mining & Scientific Press* July 7, 1915 p181.

Stoltz, Guy C. "Tram Car for the Prospector" *Engineering & Mining Journal* April 2, 1910 p696.

Trennert, Robert A. "From Gold Ore to Bat Guano: Aerial Tramways in the West" *The Mining History Journal* 1997.

"Typical Coal Trestles" *Details of Practical Mining* McGraw-Hill Book Company, New York, NY, 1916.

"Vitiation of Mine Air by Use of Gasoline Locomotives" *Engineering & Mining Journal* April 8, 1916 p645.

"Wood Trestle for Motor Tramming" *Details of Practical Mining* McGraw-Hill Book Company, New York, NY, 1916.

Zern, E.N. *Mine Cars and Mine Tracks* West Virginia University, Morgantown, WV, 1917.

Ventilation

"A Handy Windlass [and ventilation]" *Mining & Scientific Press* Oct. 5, 1907 p429.

"A Mine Ventilating Plant" *Engineering & Mining Journal* July 5, 1902 p19.

"A New Hand-Blower" *Engineering & Mining Journal* June 2, 1904 p893.

Beckett, F.S. "A Handy Windlass" *Mining & Scientific Press* Oct. 5, 1907 p429.

"The Buffalo Steel Pressure Blower" *Engineering & Mining Journal* Nov. 21, 1896 p489.

"The Buffalo Double Crank Hand Blower" *Engineering & Mining Journal* June 10, 1899 p683.

"Cheap Mine Ventilation" *Mining & Scientific Press* Dec. 12, 1916 p886.

Davenport, L.D. "Canvas Ventilation Pipes" *Engineering & Mining Journal* Vol 94 No 22 p1024.

"Improvised Ventilating Machinery" *Mining & Scientific Press* April 11, 1903 p229.

Johnson, Thomas "The Ventilation of Deep Levels-II" *Mining & Scientific Press* Oct. 17, 1903 p250.

Montgomery, William *Theory & Practice of Mine Ventilation* Jeffrey Manufacturing Co., Columbus, OH, 1926.

Myrick, C.M. "Improvised Ventilation Machinery" *Mining & Scientific Press* April 11, 1903 p229.

"The Squirrel Cage Fan for Mine Ventilation" *Engineering & Mining Journal* March 26, 1910 p674.

Swift, T. "Adobe Stove for Tunnel Ventilation" *Details of Practical Mining* McGraw-Hill Book Company, New York, NY, 1916.

"Tunnel and Level Ventilation" *Details of Practical Mining* McGraw-Hill Book Company, New York, NY, 1916.

Weeks, Walter S. *Ventilation of Mines* McGraw-Hill Book Company, New York, NY, 1926.

Worster, S.A. "Ventilation at Cripple Creek" *Engineering & Mining Journal* Jan. 17, 1914 p183.

Worster, S.A. "Pressure Ventilating System Used in Cripple Creek Mines" *Engineering & Mining Journal* June 5, 1915 p981.

Worster, S.A. "Pressure Ventilation in Cripple Creek" *Details of Practical Mining* McGraw-Hill Book Company, New York, NY, 1916.

ENDNOTES

Chapter 2: Building the Surface Plant

1. King, Joseph E. *A Mine to Make A Mine: Financing the Colorado Mining Industry, 1859-1902* Texas A&M University Press, 1977, p35.

2. Ibid, p56.

3. Ibid.

4. Hyman, David Marks *The Romance of a Mining Venture* Larchmont Press, Cincinnati, OH. 1981 p1.

King, Joseph E. *A Mine to Make A Mine: Financing the Colorado Mining Industry, 1859-1902* Texas A&M University Press, 1977 p63.

5. Hyman, David Marks *The Romance of a Mining Venture* Larchmont Press, Cincinnati, OH. 1981 p9.

6. Ibid, p10.

King, Joseph E. *A Mine to Make A Mine: Financing the Colorado Mining Industry, 1859-1902* Texas A&M University Press, 1977 p65.

7. Rice, George Graham *My Adventures with Your Money* Nevada Publications, Las Vegas, NV. 1986 p9.

8. Ibid, p10.

9. King, Joseph E. *A Mine to Make A Mine: Financing the Colorado Mining Industry, 1859-1902* Texas A&M University Press, 1977 p118.

10. Lingenfelter, Richard E. *Death Valley & the Amargosa: A Land of Illusion* University of California Press, Berkeley, CA., 1986 p146.

11. Ibid.

12. Ibid, p147.

13. King, Joseph E. *A Mine to Make A Mine: Financing the Colorado Mining Industry, 1859-1902* Texas A&M University Press, 1977 p29.

14. Ibid, p30.

15. Ibid, p33.

16. Ibid, p33, 79.

17. Spence, Clark C. *Mining Engineers and the American West: the Lace Boot Brigade, 1849-1933* University of Idaho Press, Moscow, ID, 1993 p139.

18. Ibid, p18.

19. Ibid, p18.

20. Ibid, p18.

21. Ibid, p18.

22. Ibid, p246.

23. Sprague, Marshall *Money Mountain the Story of Cripple Creek Gold* University of Nebraska Press, Lincoln, NE, 1979 p111.

Sprague, Marshall *The King of Cripple Creek: The Life and Times of Winfield Scott Stratton* Magazine Associates, 1994 p10.

24. Sprague, Marshall *Money Mountain the Story of Cripple Creek Gold* University of Nebraska Press, Lincoln, NE, 1979 p112.

25. Crampton, Frank A. *Deep Enough: A Working Stiff in the Western Mine Camps* University of Oklahoma Press, [1956] 1982 p4.

26. Ibid, p18.

27. Spence, Clark C. *Mining Engineers and the American West: The Lace Boot Brigade, 1849-1933* University of Idaho Press, Moscow, ID, 1993 p43, 59, 285, 286, 298.

Kennedy, David M. *Over Here: the First World War and American Society* Oxford University Press, New York, NY, 1980 p116.

28. Spence, Clark C. *Mining Engineers and the American West: The Lace Boot Brigade, 1849-1933* University of Idaho Press, Moscow, ID, 1993 p12, 33.

Todd, Arthur C. *The Cornish Miner in America* The Arthur H. Clark Company, Spokane, WA, 1995 p58.

29. Todd, Arthur C. *The Cornish Miner in America* The Arthur H. Clark Company, Spokane, WA, 1995 p182.

30. Spence, Clark C. *Mining Engineers and the American West: The Lace Boot Brigade, 1849-1933* University of Idaho Press, Moscow, ID, 1993 p12, 84.

31. Young, Otis E. *Black Powder and Hand Steel: Miners and Machines on the Old Western Frontier* University of Oklahoma Press, 1975 p3.

32. Todd, Arthur C. *The Cornish Miner in America* The Arthur H. Clark Company, Spokane, WA, 1995 p119.

Young, Otis E. *Black Powder and Hand Steel: Miners and Machines on the Old Western Frontier* University of Oklahoma Press, 1975 p92.

33. Barton, Denys Bradford *The Cornish Beam Engine* Wordens of Cornwall, Ltd. 1966 p185, 219.

Todd, Arthur C. *The Cornish Miner in America* The Arthur H. Clark Company, Spokane, WA, 1995 p119.

Young, Otis E. *Black Powder and Hand Steel: Miners and Machines on the Old Western Frontier* University of Oklahoma Press, 1975 p137.

34. Todd, Arthur C. *The Cornish Miner in America* The Arthur H. Clark Company, Spokane, WA, 1995 p18-22.

Young, Otis E. *Black Powder and Hand Steel: Miners and Machines on the Old Western Frontier* University of Oklahoma Press, 1975 p4.

35. International Textbook Company *A Textbook on Metal Mining: Preliminary Operations at Metal Mines, Metal Mining, Surface Arrangements at Metal Mines, Ore Dressing and Milling* International Textbook Company, Scranton, PA, 1899 A40 p7-10.

Peele, Robert *Mining Engineers' Handbook* John Wiley & Sons, New York, NY, 1918 p449.

"Sinking or Tunneling" *Engineering & Mining Journal* Dec. 9, 1899 p700.

36. Ihlseng, Magnus *A Manual of Mining* John Wiley & Sons, New York, NY, 1892 p22.

International Textbook Company *A Textbook on Metal Mining: Preliminary Operations at Metal Mines, Metal Mining, Surface Arrangements at Metal Mines, Ore Dressing and Milling* International Textbook Company, Scranton, PA, 1899 A40 p7.

Lewis, Robert S. *Elements of Mining* John Wiley & Sons, Inc., New York, NY, 1946 p46.

Silversmith, Julius *A Practical Handbook Metallurgists* American Mining Index, New York,NY, 1867 p46.

37. Baca, Florentino Berry *A Discussion of Mine Shaft Sinking* New Mexico State School of Mines, Socorro, NM. 1937.

Eaton, Lucien *Practical Mine Development & Equipment* McGraw-Hill Book Company, New York, NY, 1934 p4.

38. Colliery Engineer Company *Coal & Metal Miners' Pocketbook* Colliery Engineer Company, Scranton, PA, 1893 p257.

International Textbook Company *A Textbook on Metal Mining: Preliminary Operations at Metal Mines, Metal Mining, Surface Arrangements at Metal Mines, Ore Dressing and Milling* International Textbook Company, Scranton, PA, 1899 A40 p8.

39. "Concentrates" *Mining & Scientific Press* March 10, 1900 p259.

International Textbook Company *A Textbook on Metal Mining: Preliminary Operations at Metal Mines, Metal Mining, Surface Arrangements at Metal Mines, Ore Dressing and Milling* International Textbook Company, Scranton, PA, 1899 A40 p10.

International Textbook Company *Coal and Metal Miners' Pocket Book* International Textbook Company, Scranton, PA, 1905 p257.

Peele, Robert *Mining Engineers' Handbook* John Wiley & Sons, New York, NY, 1918 p450.

"Sinking or Tunneling" *Engineering & Mining Journal* Dec. 9, 1899 p700.

40. Morrison's Mining Rights Denver, CO, 1899 p17, 20.

Peele, Robert *Mining Engineers' Handbook* John Wiley & Sons, New York, NY, 1918 p1474.

41. Morrison's Mining Rights Denver, CO, 1899 p24.

42. International Textbook Company *A Textbook on Metal Mining: Preliminary Operations at Metal Mines, Metal Mining, Surface Arrangements at Metal Mines, Ore Dressing and Milling* International Textbook Company, Scranton, PA, 1899 A42 p1.

43. Colliery Engineer Company *Coal & Metal Miners' Pocketbook* Colliery Engineer Company, Scranton, PA, 1905 p262.

Croft, Terrell *Machinery Foundations and Erection* McGraw-Hill Book Co., New York, NY, 1923 p382.

Hawkins, N. *New Catechism of the Steam Engine: With Chapters on Gas, Oil, and Hot Air Engines* Theo. Audel & Co. Publishers, New York, NY, 1900 p35.

Ihlseng, Magnus *A Manual of Mining* John Wiley & Sons, New York, NY, 1892 p62.

International Textbook Company *International Library of Technology: Hoisting, Haulage, Mine Drainage* International Textbook Company, Scranton, PA, 1906 p46.

Lunt, Horace "Foundations for Prospect Machinery" *Details of Practical Mining* McGraw-Hill Book Co., New York, NY, 1916 p9.

Oke, A. Livingston "Timber Foundations for Engines" *Details of Practical Mining* McGraw-Hill Book Co., New York, NY, 1916 p8.

44. Croft, Terrell *Machinery Foundations and Erection* McGraw-Hill Book Co., New York, NY, 1923 p382.

Lunt, Horace "Foundations for Prospect Machinery" *Details of Practical Mining* McGraw-Hill Book Co., New York, NY, 1916 p9.

Oke, A. Livingston "Timber Foundations for Engines" *Details of Practical Mining* McGraw-Hill Book Co., New York, NY, 1916 p8.

Physical evidence encountered in mining districts indicates that miners in the mountain states favored timber cribbing for temporary machinery foundations. Evidence includes exposed foundations, and an archaeological excavation of a timber hoist foundation during September, 1994 by Western Cultural Resource Management (File: Site 5TL0351, Feature 86) in the Cripple Creek Mining District.

Chapter 3: The Surface Plants for Mine Tunnels

1. "Concentrates" *Mining & Scientific Press* March 10, 1900 p259.

International Textbook Company *A Textbook on Metal Mining: Preliminary Operations at Metal Mines, Metal Mining, Surface Arrangements at Metal Mines, Ore Dressing and Milling* International Textbook Company, Scranton, PA, 1899 A40 p10.

2. Peele, Robert *Mining Engineers' Handbook* John Wiley & Sons, New York, NY, 1918 p459.

Young, George *Elements of Mining* John Wiley & Sons, New York, NY, 1923 p463.

3. Dooley Brothers *Miners' Drilling Machines, Tools, Mine Supplies, Mine Tools, Etc.* Dooley Brothers, Peoria, IL, 1919 p73.

4. Collins, Jose H. *Principles of Metal Mining* William Collins, Sons & Company, London, 1875 p71.

International Textbook Company *International Library of Technology: Hoisting, Haulage, Mine Drainage* International Textbook Company, Scranton, PA, 1906 A54 p44.

Lewis, Robert S. *Elements of Mining* John Wiley & Sons, Inc., New York, NY, 1946 p336.

Peele, Robert *Mining Engineers' Handbook* John Wiley & Sons, New York, NY, 1918 p838.

Young, George *Elements of Mining* John Wiley & Sons, New York, NY, 1923 p485.

5. International Textbook Company *A Textbook on Metal Mining: Preliminary Operations at Metal Mines, Metal Mining, Surface Arrangements at Metal Mines, Ore Dressing and Milling* International Textbook Company, Scranton, PA, 1899 A40 p53.

Young, George *Elements of Mining* John Wiley & Sons, New York, NY, 1923 p192.

Zern, E.N. *Mine Cars and Mine Tracks* West Virginia University, Morgantown, WV, 1917 p31.

In addition to the above references, the conclusion in the text is supported by determinations from the analysis of numerous historic mine sites in the Great Basin and mountain states.

6. International Textbook Company *A Textbook on Metal Mining: Preliminary Operations at Metal Mines, Metal Mining, Surface Arrangements at Metal Mines, Ore Dressing and Milling* International Textbook Company, Scranton, PA, 1899 A40 p53.

International Textbook Company *International Correspondence School Reference Library: Rock Boring, Rock Drilling, Explosives and Blasting, Coal-Cutting Machinery, Timbering, Timber Trees, Trackwork* International Textbook Company, Scranton, PA, 1907 A48 p4.

Peele, Robert *Mining Engineers' Handbook* John Wiley & Sons, New York, NY, 1918 p846.

Zern, E.N. *Mine Cars and Mine Tracks* West Virginia University, Morgantown, WV, 1917 p33.

The author has documented evidence of strap rail at small and remote mines in the Rocky Mountains and Great Basin. All of the operations had been abandoned prior to 1890, and they exhibited evidence of having been poorly funded.

7. Hoover, Herbert C. *Principles of Mining* McGraw-Hill Book Co., New York, NY, 1909 p150.

International Textbook Company *International Correspondence School Reference Library: Rock Boring, Rock Drilling, Explosives and Blasting, Coal-Cutting Machinery, Timbering, Timber Trees, Trackwork* International Textbook Company, Scranton, PA, 1907 A48 p13.

Peele, Robert *Mining Engineer's Handbook* John Wiley & Sons, New York, NY, 1918 p184.

Young, George *Elements of Mining* John Wiley & Sons, New York, NY, 1946 p87.

8. Bealer, Alex W. *The Art of Blacksmithing* Castle Books, Edison, NJ., 1995 p47.

Ihlseng, Magnus *A Manual of Mining* John Wiley & Sons, New York, NY, 1901 p415.

Young, George *Elements of Mining* John Wiley & Sons, New York, NY, 1923.

Conclusions regarding vernacular forges have been determined from the analysis of historic mine sites in California, Colorado, Montana, Nevada, New Mexico, and Utah.

9. Bacon, John Lord *Forge Practice & Heat Treatment of Steel* John Wiley & Sons, New York, NY, 1919 p2.

Bealer, Alex W. *The Art of Blacksmithing* Castle Books, Edison, NJ., 1995 p131.

Drew, J.M. *Farm Blacksmithing: A Manual for Farmers and Agricultural Schools* Webb Publishing Co., St. Paul, MN, 1910 p9.

International Textbook Company *Coal and Metal Miners' Pocket Book* International Textbook Company, Scranton, PA 1905 p172.

Thurston, R.H. *A Manual of Steam Boilers: Their Design, Construction, and Operation* John Wiley & Sons, New York, NY, 1901 p155.

10. Artifact assemblages associated with blacksmith work areas have been characterized from the analysis of historic prospect adit and shaft sites in California, Colorado, Montana, Nevada, New Mexico, and Utah.

11. "Cheap Mine Ventilation" *Mining & Scientific Press* Dec. 12, 1916 p886.

Collins, Jose H. *Principles of Metal Mining* William Collins, Sons & Company, London 1875 p116.

International Textbook Company *A Textbook on Metal Mining: Preliminary Operations at Metal Mines, Metal Mining, Surface Arrangements at Metal Mines, Ore Dressing and Milling* International Textbook Company, Scranton, PA, 1899 A41 p137.

International Textbook Company *Coal and Metal Miners' Pocket Book* International Textbook Company, Scranton, PA 1905 p383, 385.

Myrick, C.M. "Improvised Ventilation Machinery" *Mining & Scientific Press* April 11, 1903 p229.

Peele, Robert *Mining Engineers' Handbook* John Wiley & Sons, New York, NY, 1918 p1040.

Swift, T. "Adobe Stove for Tunnel Ventilation" *Details of Practical Mining* McGraw-Hill Book Company, New York, NY, 1916.

The author documented a furnace ventilation system at the Imperial Mine, Frisco Mining District, Utah.

12. "The Buffalo Steel Pressure Blower" *Engineering & Mining Journal* Nov. 21, 1896 p489.

"The Buffalo Double Crank Hand Blower" *Engineering & Mining Journal* June 10, 1899 p683.

The author observed forge bellows ventilation systems at prospect tunnels in the Delamar, Revielle, and White Pine districts in Nevada; and in the Tin Cup district in Colorado.

13. Peele, Robert *Mining Engineers' Handbook* John Wiley & Sons, New York, NY, 1918 p252.

14. International Textbook Company *A Textbook on Metal Mining: Preliminary Operations at Metal Mines, Metal Mining, Surface Arrangements at Metal Mines, Ore Dressing and Milling* International Textbook Company, Scranton, PA, 1899 A40 p42.

15. Colvin, Fred and Stanley, Frank *Running a Machine Shop* McGraw-Hill Book Co., Inc., New York, NY, 1941.

Eaton, Lucien *Practical Mine Development & Equipment* McGraw-Hill Book Company, New York, NY, 1934.

G. S. "Drill-and Tool-Sharpening Shop at the Copper Queen Mine" *Engineering & Mining Journal* June 24, 1916 p1101.

O'Rourke, J.E. "Operating a Steel-Sharpening Shop" *Engineering & Mining Journal* Feb. 10, 1917 p263.

16. Characteristics for shops at medium-sized mines that operated prior to approximately 1890 came from field analyses of mine sites in California, Colorado, Montana, Nevada, New Mexico, and Utah. Some generalizations were also made from historic mining texts and periodical articles.

17. The basic characteristics for shops erected by well-financed mining companies prior to approximately 1890 came predominantly from field analyses of mine sites in California, Colorado, Montana, Nevada, New Mexico, and Utah. Some generalizations were also made from historic mining texts and periodical articles.

18. Eaton, Lucien *Practical Mine Development & Equipment* McGraw-Hill Book Company, New York, NY, 1934 p112.

Tinney, W.H. *Gold Mining Machinery: Its Selection, Arrangement, & Installation* D. Van Nostrand Company, New York, NY, 1906 p266.

Young, George *Elements of Mining* John Wiley & Sons, New York, NY, 1923 p485.

19. Bealer, Alex W. *The Art of Blacksmithing* Castle Books, Edison, NJ., 1995 p63.

Drew, J.M. *Farm Blacksmithing: A Manual for Farmers and Agricultural Schools* Webb Publishing Co., St. Paul, MN, 1910 p53.

Eaton, Lucien *Practical Mine Development & Equipment* McGraw-Hill Book Company, New York, NY, 1934 p131.

20. International Textbook Company *International Library of Technology: Mine Surveying, Metal Mine Surveying, Mineral-Land Surveying, Steam and Steam Boilers, Steam Engines, Air Compression* International Textbook Company, Scranton, PA, 1924 A24 p1.

21. Rice, Claude T. "Joplin Steel-Sharpening Kinks" *Details of Practical Mining* McGraw-Hill Book Company, New York, NY, 1916.

22. Ingersoll-Rand Drill Company *Today's Most Modern Rock Drills & A Brief History of the Rock Drill Development* Ingersoll-Rand Drill Company, New York, NY, 1939 p3.

Stack, Barbara *Handbook of Mining & Tunneling Machinery* John Wiley & Sons, New York, NY, 1982 p26.

Young, George, *1946 Elements of Mining* John Wiley & Sons, New York, NY, p111.

23. Cleveland Rock Drill Co. *Driller's Handbook* Cleveland Rock Drill Co., Cleveland, OH 1931 p5.

Ingersoll-Rand Drill Company *Today's Most Modern Rock Drills & A Brief History of the Rock Drill Development* Ingersoll-Rand Drill Company, New York, NY, 1939 p3.

Stack, Barbara *Handbook of Mining & Tunneling Machinery* John Wiley & Sons, New York, NY,1982 p30.

24. Rice, Claude T. "Joplin Steel-Sharpening Kinks" *Details of Practical Mining* McGraw-Hill Book Company, New York, NY, 1916.

"Stand for the Blacksmith Shop" *Details of Practical Mining* McGraw-Hill Book Company, New York, NY, 1916.

25. Oke, A. Livingston "Some Smithy Appliances" *Details of Practical Mining* McGraw-Hill Book Company, New York, NY, 1916.

The author has documented backing blocks in mine shops in California, Colorado, and Montana.

26. "Drill-Sharpening Machine [Compressed Air Machinery Co.]" *Mining & Scientific Press* April 11, 1903.

"Little Giant Drill-Sharpener" *Mining & Scientific Press* Nov. 5, 1904 p311.

27. "Drill-Sharpening Machine [Compressed Air Machinery Co.]" *Mining & Scientific Press* April 11, 1903.

"Hammer Drills [and steel-sharpening]" *Mining & Scientific Press* March 23, 1907 p382.

"The Word Drill Sharpener" *Engineering & Mining Journal* Sept. 15, 1906 p487.

28. "Drill-Sharpening Machine [Compressed Air Machinery Co.]" *Mining & Scientific Press* April 11, 1903.

"Hammer Drills [and steel-sharpening]" *Mining & Scientific Press* March 23, 1907 p382.

"Little Giant Drill-Sharpener" *Mining & Scientific Press* Nov. 5, 1904 p311.

29. "A New Drill-Sharpener [by Sullivan]" *Mining & Scientific Press* May 29, 1915 p859.

"A New Type of Drill-Sharpener [by Denver Rock Drill Manufacturing Co.]" *Mining & Scientific Press* Jan. 9, 1917 p36.

"Denver Rock-Drill Sharpener" *Engineering & Mining Journal* Jan. 20, 1917 p155.

"Imperial Drill Sharpener" *Engineering & Mining Journal* May 27, 1915 p414.

"New Sullivan Drill Sharpener" *Engineering & Mining Journal* May 29, 1915 p949.

30. "Imperial Drill Sharpener" *Engineering & Mining Journal* May 27, 1915 p414.

"New Sullivan Drill Sharpener" *Engineering & Mining Journal* May 29, 1915 p949.

31. "Imperial Drill Sharpener" *Engineering & Mining Journal* May 27, 1915 p414.

"A New Drill-Sharpener [by Sullivan]" *Mining & Scientific Press* May 29, 1915 p859.

Keystone Consolidated Publishing Company Inc. *The Mining Catalog: 1925 Metal-Quarry Edition* Keystone Consolidated Publishing Company Inc., 1925 p307.

32. "Drill-Sharpening Machine [Compressed Air Machinery Co.]" *Mining & Scientific Press* April 11, 1903.

"Hammer Drills [and steel-sharpening]" *Mining & Scientific Press* March 23, 1907 p382.

"Little Giant Drill-Sharpener" *Mining & Scientific Press* Nov. 5, 1904 p311.

"The Word Drill Sharpener" *Engineering & Mining Journal* Sept. 15, 1906 p487.

33. "Ajax Oil Forge" *Mining & Scientific Press* Jan. 14, 1905 p22.

Gillman, George H. "Ideal Shop for Sharpening Drill Steel" *Engineering & Mining Journal* Oct. 6, 1917 p585.

G. S. "Drill-and Tool-Sharpening Shop at the Copper Queen Mine" *Engineering & Mining Journal* June 24, 1916 p1101.

"Low-Pressure Oil Forge" *Engineering & Mining Journal* June 27, 1914 p1298.

"Oil or Gas Drill-Heating Forge" *Engineering & Mining Journal* June 23, 1917 p1115.

Rice, Claude T. "Coke Furnace for Heating Drills" *Details of Practical Mining* McGraw-Hill Book Company, New York, NY, 1916.

34. G. S. "Drill-and Tool-Sharpening Shop at the Copper Queen Mine" *Engineering & Mining Journal* June 24, 1916 p1101.

Harron, Rickard, & McCone *The Ajax Drill Sharpener* Harron, Rickard, & McCone, San Francisco, CA 1904.

McFarland, J.R. "Variable-Stroke Air Hammer" *Details of Practical Mining* McGraw-Hill Book Company, New York, NY, 1916.

Rice, Claude T. "Joplin Steel-Sharpening Kinks" *Details of Practical Mining* McGraw-Hill Book Company, New York, NY, 1916.

The Author has examined the remains of power hammers at numerous medium-sized and large historic mines in Colorado, Montana, Nevada, and Utah.

35. The remains of blacksmith shops have been characterized from field analyses of mine sites in California, Colorado, Montana, Nevada, New Mexico, and Utah.

36. International Textbook Company *A Textbook on Metal Mining: Preliminary Operations at Metal Mines, Metal Mining, Surface Arrangements at Metal Mines, Ore Dressing and Milling* International Textbook Company, Scranton, PA, 1899 A41 p133.

International Textbook Company *Coal and Metal Miners' Pocket Book* International Textbook Company, Scranton, PA 1905 p381.

Lewis, Robert S. *Elements of Mining* John Wiley & Sons, Inc., New York, NY, 1946 p454.

Peele, Robert *Mining Engineers' Handbook* John Wiley & Sons, New York, NY, 1918 p1038.

Young, George *Elements of Mining* John Wiley & Sons, New York, NY, 1923 p255.

37. International Textbook Company *A Textbook on Metal Mining: Preliminary Operations at Metal Mines, Metal Mining, Surface Arrangements at Metal Mines, Ore Dressing and Milling* International Textbook Company, Scranton, PA, 1899 A41 p133.

Worster, S.A. "Pressure Ventilation in Cripple Creek" *Details of Practical Mining* McGraw-Hill Book Company, New York, NY, 1916.

38. International Textbook Company *A Textbook on Metal Mining: Preliminary Operations at Metal Mines, Metal Mining, Surface Arrangements at Metal Mines, Ore Dressing and Milling* International Textbook Company, Scranton, PA, 1899 A41 p141.

39. Collins, Jose H. *Principles of Metal Mining* William Collins, Sons & Company, London 1875 p115.

Ihlseng, Magnus *A Manual of Mining* John Wiley & Sons, New York, NY, 1892 p187.

International Textbook Company *A Textbook on Metal Mining: Preliminary Operations at Metal Mines, Metal Mining, Surface Arrangements at Metal Mines, Ore Dressing and Milling* International Textbook Company, Scranton, PA, 1899 A41 p133.

Silversmith, Julius *A Practical Handbook* Metallurgists American Mining Index, New York, NY, 1867 p46, 49.

Determinations from field analyses of productive adit mines in California, Colorado, Utah, Nevada, and Montana are concurrent with the propensity for mining engineers to develop an ore body through multiple openings to provide ventilation and greater access to the ore.

40. International Textbook Company *A Textbook on Metal Mining: Preliminary Operations at Metal Mines, Metal Mining, Surface Arrangements at Metal Mines, Ore Dressing and Milling* International Textbook Company, Scranton, PA, 1899 A41 p138.

Peele, Robert *Mining Engineers' Handbook* John Wiley & Sons, New York, NY, 1918 p1028.

41. International Textbook Company *A Textbook on Metal Mining: Preliminary Operations at Metal Mines, Metal Mining, Surface Arrangements at Metal Mines, Ore Dressing and Milling* International Textbook Company, Scranton, PA, 1899 A41 p138.

Johnson, Thomas "The Ventilation of Deep Levels-II" *Mining & Scientific Press* Oct. 17, 1903 p250.

Peele, Robert *Mining Engineers' Handbook* John Wiley & Sons, New York, NY, 1918 p1028.

Young, George *Elements of Mining* John Wiley & Sons, New York, NY, 1923 p261.

42. "Tunnel and Level Ventilation" *Details of Practical Mining* McGraw-Hill Book Company, New York, NY, 1916.

43. Lord, Elliot *Comstock Mining and Miners* Howell-North Books, San Diego, CA, 1980 p391.

44. Thorkelson, H.J. *Air Compression and Transmission* McGraw-Hill Book Company, Inc., New York, NY, 1912 p63.

45. Keystone Consolidated Publishing Company Inc. *The Mining Catalog: 1925 Metal-Quarry Edition* Keystone Consolidated Publishing Company Inc., 1925 p559.

46. Compressed Air Magazine *Rock Drill Data* Compressed Air Magazine, Philipsburg, NJ 1960 p107.

47. Worster, S.A. "Ventilation at Cripple Creek" *Engineering & Mining Journal* Jan. 17, 1914 p183.

48. Ibid.

49. Ibid.

50. In his memoirs entitled *Deep Enough* (University of Oklahoma Press, 1982) mining engineer, Frank Crampton, recounts his work at numerous mines in the Great Basin and the Southwest where he had miners drill by hand as late as the 1910s. W.S. Stratton designed the first production-class plant at his wealthy Independence Mine in the Cripple Mining District, near Victor, Colorado in the mid 1890s, and there he did much hand-drilling himself. In addition, the analysis of mine sites that operated between the 1890s and 1910s in California, Colorado, Montana, Nevada, New Mexico, and Utah reflects that a many operations did not rely on rockdrills for boring blast-holes.

51. Gillette, Halbert P. *Rock Excavation: Methods and Cost* Myron C. Clark Publishing Company, New York, NY, 1907 p15.

Hoover, Herbert C. *Principles of Mining* McGraw-Hill Book Co., New York, NY, 1909 p150.

International Correspondence Schools *Rock Boring, Blasting, Coal Cutting, Trackwork* International Textbook Co., Scranton, Pennsylvania, 1907 p13.

Peele, Robert *Mining Engineer's Handbook* John Wiley & Sons, New York, NY, 1918 p184, 213.

Young, George *Elements of Mining* John Wiley & Sons, New York, NY, 1946 p87.

52. Lord, Elliot *Comstock Mining and Miners* Howell-North Books, San Diego, CA, 1980 p335.

Wyman, Mark *Hard Rock Epic: Western Mining and the Industrial Revolution, 1860-1910* University of California Press, Berkeley, CA 1989 p84.

53. International Textbook Company *Coal and Metal Miners' Pocket* Book International Textbook Company, Scranton, PA 1905 p195.

Keystone Consolidated Publishing Company Inc. *The Mining Catalog: 1925 Metal-Quarry Edition* Keystone Consolidated Publishing Company Inc., (no location given), 1925 p129.

Thorkelson, H.J. *Air Compression and Transmission* McGraw-Hill Book Company, Inc., New York, NY, 1912 p181.

Whitehurst, J.W. and Cary, W.P. "Design of a Mine Plant-I" *Mining & Scientific Press* Aug. 13, 1910 p202.

Whitehurst, J.W. and Cary, W.P. "Design of a Mine Plant-II" *Mining & Scientific Press* Aug. 20, 1910 p239.

54. International Textbook Company *Coal and Metal Miners' Pocket Book* International Textbook Company, Scranton, PA 1905 195.

Whitehurst, J.W. and Cary, W.P. "Design of a Mine Plant-I" *Mining & Scientific Press* Aug. 13, 1910 p202.

Whitehurst, J.W. and Cary, W.P. "Design of a Mine Plant-II" *Mining & Scientific Press* Aug. 20, 1910 p239.

55. International Textbook Company *Coal and Metal Miners' Pocket Book* International Textbook Company, Scranton, PA, 1905 p195.

Nelson, S.T. "Development of Reciprocating Air Compressors" *Engineering and Mining Journal* Sept. 27, 1919 p533-536.

Rand Drill Company *Illustrated Catalogue of the Rand Drill Company, New York, U.S.A.* Rand Drill Company, New York, NY, 1886.

Thorkelson, H.J. *Air Compression and Transmission* McGraw-Hill Book Company, Inc., New York, NY, 1912 p181.

Whitehurst, J.W. and Cary, W.P. "Design of a Mine Plant-I" *Mining & Scientific Press* Aug. 13, 1910 p202.

Whitehurst, J.W. and Cary, W.P. "Design of a Mine Plant-II" *Mining & Scientific Press* Aug. 20, 1910 p239.

56. Eaton, Lucien *Practical Mine Development & Equipment* McGraw-Hill Book Company, New York, NY, 1934 p186.

Whitehurst, J.W. and Cary, W.P. "Design of a Mine Plant-I" *Mining & Scientific Press* Aug. 13, 1910 p202.

Whitehurst, J.W. and Cary, W.P. "Design of a Mine Plant-II" *Mining & Scientific Press* Aug. 20, 1910 p239.

57. "A New Clayton Compound High Duty Air Compressor" *Engineering & Mining Journal* Dec. 9, 1893 p596.

58. International Textbook Company *A Textbook on Metal Mining: Steam and Steam-Boilers, Steam Engines, Air and Air Compression, Hydromechanics and Pumping, Mine Haulage, Hoisting and Hoisting Appliances, Percussive and Rotary Boring* International Textbook Company, Scranton, PA, 1899 A20 p20.

International Textbook Company *International Library of Technology: Mine Surveying, Metal Mine Surveying, Mineral-Land Surveying, Steam and Steam Boilers, Steam Engines, Air Compression* International Textbook Company, Scranton, PA, 1924 A25 p24.

59. International Textbook Company *Coal and Metal Miners' Pocket Book* International Textbook Company, Scranton, PA, 1905 p196.

International Textbook Company *A Textbook on Metal Mining: Steam and Steam-Boilers, Steam Engines, Air and Air Compression, Hydromechanics and Pumping, Mine Haulage, Hoisting and Hoisting Appliances, Percussive and Rotary Boring* International Textbook Company, Scranton, PA, 1899 A20 p18, 25.

International Textbook Company *International Library of Technology: Mine Surveying, Metal Mine Surveying, Mineral-Land Surveying, Steam and Steam Boilers, Steam Engines, Air Compression* International Textbook Company, Scranton, PA, 1924 A25 p17.

Lewis, Robert S. *Elements of Mining* John Wiley & Sons, Inc., New York, NY, 1946 p442.

Peele, Robert *Mining Engineers' Handbook* John Wiley & Sons, New York, NY, 1918 p1061.

Rand Drill Company *Illustrated Catalog of the Rand Drill Company, New York, U.S.A.* Rand Drill Company, New York, NY, 1886 p19.

Thorkelson, H.J. *Air Compression and Transmission* McGraw-Hill Book Company, Inc., New York, NY, 1912 p75.

60. International Textbook Company *A Textbook on Metal Mining: Steam and Steam-Boilers, Steam Engines, Air and Air Compression, Hydromechanics and Pumping, Mine Haulage, Hoisting and Hoisting Appliances, Percussive and Rotary Boring* International Textbook Company, Scranton, PA, 1899 A20 p23, 25.

International Textbook Company *International Library of Technology: Mine Surveying, Metal Mine Surveying, Mineral-Land Surveying, Steam and Steam Boilers, Steam Engines, Air Compression* International Textbook Company, Scranton, PA, 1924 A25 p27, 31.

Thorkelson, H.J. *Air Compression and Transmission* McGraw-Hill Book Company, Inc., New York, NY, 1912 p82.

61. Peele, Robert *Mining Engineers' Handbook* John Wiley & Sons, New York, NY, 1918 p1062.

62. Eaton, Lucien *Practical Mine Development & Equipment* McGraw-Hill Book Company, New York, NY, 1934 p187.

Ingersoll Rock Drill Company *Catalog No.7: Rock Drills, Air Compressors and Air Receivers* Ingersoll Rock Drill Company, New York, NY, [1887] p36.

International Textbook Company *Coal and Metal Miners' Pocket Book* International Textbook Company, Scranton, PA, 1905 p194.

International Textbook Company International *Library of Technology: Mine Surveying, Metal Mine Surveying, Mineral-Land Surveying, Steam and Steam Boilers, Steam Engines, Air Compression* International Textbook Company, Scranton, PA, 1924 A25 p30.

Peele, Robert *Mining Engineers' Handbook* John Wiley & Sons, New York, NY, 1918 p1066.

Rand Drill Company *Illustrated Catalog of the Rand Drill Company, New York, U.S.A.* Rand Drill Company, New York, NY, 1886 p21.

Simons, Theodore E.M., C.E. *Compressed Air: A Treatise on the Production, Transmission, and Use of Compressed Air* McGraw-Hill Book Company, Inc., New York, NY, 1921 p138.

Thorkelson, H.J. *Air Compression and Transmission* McGraw-Hill Book Company, Inc., New York, NY, 1912 p86.

63. International Textbook Company *A Textbook on Metal Mining* International Textbook Company, Scranton, PA, 1899 A20 p32.

International Textbook Company *Coal and Metal Miners' Pocket Book* International Textbook Company, Scranton, PA, 1905 p194.

International Textbook Company *International Library of Technology: Mine Surveying, Metal Mine Surveying, Mineral-Land Surveying, Steam and Steam Boilers, Steam Engines, Air Compression* International Textbook Company, Scranton, PA, 1924 A25 p24.

Peele, Robert *Mining Engineers' Handbook* John Wiley & Sons, New York, NY, 1918 p1066.

Richards, Frank "Use of Coolers in Air Compression" *Engineering and Mining Journal* June 1, 1907 p1039.

Simons, Theodore E.M., C.E. *Compressed Air: A Treatise on the Production, Transmission, and Use of Compressed Air* McGraw-Hill Book Company, Inc., New York, NY, 1921 p43.

Thorkelson, H.J. *Air Compression and Transmission* McGraw-Hill Book Company, Inc., New York, NY, 1912 p89.

64. Keystone Consolidated Publishing Company Inc. *The Mining Catalog: 1925 Metal-Quarry Edition* Keystone Consolidated Publishing Company Inc., (no location given), 1925 p125.

Peele, Robert *Mining Engineers' Handbook* John Wiley & Sons, New York, NY, 1918 p1062.

Simons, Theodore E.M., C.E. *Compressed Air: A Treatise on the Production, Transmission, and Use of Compressed Air* McGraw-Hill Book Company, Inc., New York, NY, 1921 p59.

Thorkelson, H.J. *Air Compression and Transmission* McGraw-Hill Book Company, Inc., New York, NY, 1912 p90.

65. Gillette, Halbert P. *Rock Excavation: Methods and Cost* Myron C. Clark Publishing Company, New York, NY, 1907 p49.

Peele, Robert *Mining Engineers' Handbook* John Wiley & Sons, New York, NY, 1918 p1066.

Simons, Theodore E.M., C.E. *Compressed Air: A Treatise on the Production, Transmission, and Use of Compressed Air* McGraw-Hill Book Company, Inc., New York, NY, 1921 p58.

Thorkelson, H.J. *Air Compression and Transmission* McGraw-Hill Book Company, Inc., New York, NY, 1912 p91.

The Author has examined historic mine sites featuring evidence of single-stage and compound straight-line compressors in Colorado, Montana, Nevada, New Mexico, and Utah.

66. Rand Drill Company *Illustrated Catalog of the Rand Drill Company, New York, U.S.A.* Rand Drill Company, New York, NY, 1886 p23.

67. International Textbook Company *A Textbook on Metal Mining: Preliminary Operations at Metal Mines, Metal Mining, Surface Arrangements at Metal Mines, Ore Dressing and Milling* International Textbook Company, Scranton, PA, 1899 A40 p11.

Ihlseng, Magnus *A Manual of Mining* John Wiley & Sons, New York, NY, 1892 p60.

68. Croft, Terrell *Machinery Foundations and Erection* McGraw-Hill Book Co., New York, NY, 1923 p282.

Tinney, W.H. *Gold Mining Machinery: Its Selection, Arrangement, & Installation* D. Van Nostrand Company, New York, NY, 1906 p33.

69. Croft, Terrell *Machinery Foundations and Erection* McGraw-Hill Book Co., New York, NY, 1923 p372, 398.

70. Croft, Terrell *Machinery Foundations and Erection* McGraw-Hill Book Co., New York, NY, 1923 p247.

International Textbook Company *A Textbook on Metal Mining: Preliminary Operations at Metal Mines, Metal Mining, Surface Arrangements at Metal Mines, Ore Dressing and Milling* International Textbook Company, Scranton, PA, 1899 A51 p49.

"Foundation Bolts for Steam Engines" *Mining & Scientific Press* Jan. 16, 1904 p41.

71. "Foundation Bolts for Steam Engines" *Mining & Scientific Press* Jan. 16, 1904 p41.

Artifacts and foundations examined at mine sites in Colorado, Montana, Nevada, and Utah reflect the use of the four types of anchor bolts discussed in the text for machine foundations.

72. The Author has documented timber compressor foundations at a greater number of small, poorly financed operations than at mines with moderate financing.

73. The characteristics of steam-driven compressor foundations have been determined from the analyses of numerous mine sites pre-dating 1900 in California, Colorado, Montana, Nevada, New Mexico, and Utah.

74. Bauer, Edward E. *Plain Concrete* McGraw-Hill Book Co., New York, NY, 1949 p1.

McLaughlin, R.P. "Manufacture of Portland Cement in California" *Mining & Scientific Press* March 21, 1903 p180.

Meade, Richard *Portland Cement, Its Composition, Raw Materials, Manufacture, Testing and Analysis* Chemical Publishing Co., New York, NY, (circa 1924) p1, 14.

Mills, Adelbert [1915], Rader, Lloyd *Materials of Construction, Their Manufacture and Properties* John Wiley & Sons, New York, NY, 5th ed. 1939 p292.

Roys, Francis *Materials of Engineering Construction* The Ronald Press Co., New York, NY, 1930 p77.

Young, Joseph *A Brief Outline of the History of Cement* Lehigh Portland Cement Co., Lehigh, PA, 1955 p19.

75. Meade, Richard *Portland Cement, Its Composition, Raw Materials, Manufacture, Testing and Analysis* Chemical Publishing Co., New York, NY, (circa 1924) p14.

76. McLaughlin, R.P. "Manufacture of Portland Cement in California" *Mining & Scientific Press* March 21, 1903 p180.

Young, Joseph *A Brief Outline of the History of Cement* Lehigh Portland Cement Co., Lehigh, PA, 1955 p19.

77. Roys, Francis *Materials of Engineering Construction* The Ronald Press Co., New York, NY, 1930 p77.

Werner, Orville II Registered Professional Engineer, President CTL-Thompson Inc. Denver, CO, *Personal Interview* Denver, CO, April 3, 1997.

78. Peele, Robert *Mining Engineers' Handbook* John Wiley & Sons, New York, NY, 1918 p1072.

79. Simons, Theodore E.M., C.E. *Compressed Air: A Treatise on the Production, Transmission, and Use of Compressed Air* McGraw-Hill Book Company, Inc., New York, NY, 1921 p144.

Analyses of historic mine sites in Colorado, Montana, Nevada, and Utah determined that electric compressors were usually belt-driven, except for the units at a few large operations.

80. "Direct-Connected Oil Engine Driven Compressor" *Engineering & Mining Journal* June 27, 1914 p1297.

"Gasoline Engine and Air Compressor" *Engineering & Mining Journal* March 11, 1899 p295.

Eaton, Lucien *Practical Mine Development & Equipment* McGraw-Hill Book Company, New York, NY, 1934 p188.

"Oil-Driven Air Compressors" *Mining & Scientific Press* May 30, 1914 p918.

Peele, Robert *Mining Engineers' Handbook* John Wiley & Sons, New York, NY, 1918 p1074.

Young, George "Small Compressors Driven by Oil Engines Becoming Popular in the West" *Engineering & Mining Journal* Nov. 22, 1919 p830.

81. "Direct-Connected Oil Engine Driven Compressor" *Engineering & Mining Journal* June 27, 1914 p1297.

"Gasoline Engine and Air Compressor" *Engineering & Mining Journal* March 11, 1899 p295.

"Oil-Driven Air Compressors" *Mining & Scientific Press* May 30, 1914 p918.

Young, George "Small Compressors Driven by Oil Engines Becoming Popular in the West" *Engineering & Mining Journal* Nov. 22, 1919 p830.

Material written by the above mining engineers regarding petroleum engine compressor foundations is concurrent with field data from the analyses of mine sites in California, Nevada, and Utah.

82. Electric and petroleum engine compressor foundation characteristics were determined from the analyses of mine sites ranging in age from the late 1890s to the 1930s.

83. Eaton, Lucien *Practical Mine Development & Equipment* McGraw-Hill Book Company, New York, NY, 1934 p192.

International Textbook Company *A Textbook on Metal Mining: Preliminary Operations at Metal Mines, Metal Mining, Surface Arrangements at Metal Mines, Ore Dressing and Milling* International Textbook Company, Scranton, PA, 1899 A20 p36.

International Textbook Company *International Library of Technology: Mine Surveying, Metal Mine Surveying, Mineral-Land Surveying, Steam and Steam Boilers, Steam Engines, Air Compression* International Textbook Company, Scranton, PA, 1924 A25 p65.

Peele, Robert *Mining Engineers' Handbook* John Wiley & Sons, New York, NY, 1918 p1081.

Rand Drill Company *Illustrated Catalog of the Rand Drill Company, New York, U.S.A.* Rand Drill Company, New York, NY, 1886 p31.

Simons, Theodore E.M., C.E. *Compressed Air: A Treatise on the Production, Transmission, and Use of Compressed Air* McGraw-Hill Book Company, Inc., New York, NY, 1921 p158.

Thorkelson, H.J. *Air Compression and Transmission* McGraw-Hill Book Company, Inc., New York, NY, 1912 p159.

Whitehurst, J.W. and Cary, W.P. "Design of a Mine Plant-I" *Mining & Scientific Press* Aug. 13, 1910 p202.

Whitehurst, J.W. and Cary, W.P. "Design of a Mine Plant-II" *Mining & Scientific Press* Aug. 20, 1910 p239.

84. Tinney, W.H. *Gold Mining Machinery: Its Selection, Arrangement, & Installation* D. Van Nostrand Company, New York, NY, 1906 p152.

85. Division of Mines & Minerals Colorado State Archives, Denver, CO.

Mine Inspectors' Reports: Teller County File # 48516 and 48441 Vindicator Mine

86. Colliery Engineer Company *Coal & Metal Miners' Pocketbook* Colliery Engineer Company, Scranton, PA, 1893 p267.

Tinney, W.H. *Gold Mining Machinery: Its Selection, Arrangement, & Installation* D. Van Nostrand Company, New York, NY, 1906 p153.

Whitehurst, J.W. and Cary, W.P. "Design of a Mine Plant-I" *Mining & Scientific Press* Aug. 13, 1910 p202.

Whitehurst, J.W. and Cary, W.P. "Design of a Mine Plant-II" *Mining & Scientific Press* Aug. 20, 1910 p239.

The author has examined well-engineered mines featuring large-diameter air piping above and below ground, and poorly engineered air systems at small, poorly financed operations.

87. The author encountered wood rails and the remains of wooden ore cars in several medium-sized and large Great Basin mines that operated during the 1880s and 1890s.

88. "Electric Haulage in Metal Mines" *Engineering & Mining Journal* Feb. 25, 1904 p324.

International Textbook Company *International Correspondence School Reference Library: Rock Boring, Rock Drilling, Explosives and Blasting, Coal-Cutting Machinery, Timbering, Timber Trees, Trackwork* International Textbook Company, Scranton, PA, 1907 A48 p4.

Lewis, Robert S. *Elements of Mining* John Wiley & Sons, Inc., New York, NY, 1946 p336.

Tinney, W.H. *Gold Mining Machinery: Its Selection, Arrangement, & Installation* D. Van Nostrand Company, New York, NY, 1906 p228.

The characterizations of mine rail lines mentioned in the text have been confirmed by field data collected at historic mine sites in California, Colorado, Montana, Nevada, and New Mexico.

89. International Textbook Company *International Library of Technology: Hoisting, Haulage, Mine Drainage* International Textbook Company, Scranton, PA, 1906 A54 p47.

International Textbook Company *Mine Haulage: Rope Haulage in Coal Mines, Locomotive Haulage in Coal Mines, Mine Haulage Systems, Calculations, and Cars* International Textbook Company, Scranton, PA, 1926 p2.

Peele, Robert *Mining Engineers' Handbook* John Wiley & Sons, New York, NY, 1918 p864.

Young, George *Elements of Mining* John Wiley & Sons, New York, NY, 1923 p151.

Young, Otis E. *Western Mining* University of Oklahoma Press, 1987 p162.

90. "Electric Haulage in Coal Mines" *Engineering & Mining Journal* April 29, 1893 p395.

"Electric Haulage in Metal Mines" *Engineering & Mining Journal* Feb. 25, 1904 p324.

"Electric Mine Locomotives" *Mining & Scientific Press* Dec. 8, 1900.

General Electric Company *Electric Mine Locomotives: Catalog No. 1045* General Electric Power and Mining Department, Chicago, IL, 1904 p5.

91. General Electric Company *Electric Mine Locomotives: Catalog No. 1045* General Electric Power and Mining Department, Chicago, IL, 1904 p23.

Peele, Robert *Mining Engineers' Handbook* John Wiley & Sons, New York, NY, 1918 p862, 871.

92. Colliery Engineer Company *Coal Miners' Pocketbook* McGraw-Hill Book Co., New York, NY, 1916 p767.

International Textbook Company *International Library of Technology: Hoisting, Haulage, Mine Drainage* International Textbook Company, Scranton, PA, 1906 A55 p6.

International Textbook Company *International Correspondence School Reference Library: Rock Boring, Rock Drilling, Explosives and Blasting, Coal-Cutting Machinery, Timbering, Timber Trees, Trackwork* International Textbook Company, Scranton, PA, 1907 A48 p2.

International Textbook Company *Mine Haulage: Rope Haulage in Coal Mines, Locomotive Haulage in Coal Mines, Mine Haulage Systems, Calculations, and Cars* International Textbook Company, Scranton, PA, 1926 p1.

Young, George *Elements of Mining* John Wiley & Sons, New York, NY, 1923 p192.

Zern, E.N. *Mine Cars and Mine Tracks* West Virginia University, Morgantown, WV, 1917 p32.

93. Hansen, Charles "Compressed Air as Used in Mining" *Engineering & Mining Journal* March 9, 1895 p220.

Hoover, Herbert C. *Principles of Mining: Valuation, Organization, and Administration* McGraw-Hill Book Company, Inc., New York, NY, 1909 p135.

International Textbook Company *International Library of Technology: Hoisting, Haulage, Mine Drainage* International Textbook Company, Scranton, PA, 1906 A55 p22, 35.

Peele, Robert *Mining Engineers' Handbook* John Wiley & Sons, New York, NY, 1918 p868.

Young, George *Elements of Mining* John Wiley & Sons, New York, NY, 1923 p151.

94. "Concentrates [ore bin notes]" *Mining & Scientific Press* Sept. 10, 1904.

Peele, Robert *Mining Engineers' Handbook* John Wiley & Sons, New York, NY, 1918 p978.

Sanders, Wilbur E. "Improved Type of Ore Bin" *Details of Practical Mining* McGraw-Hill Book Company, New York, NY, 1916.

95. "Concentrates [ore bin notes]" *Mining & Scientific Press* Sept. 10, 1904.

Peele, Robert *Mining Engineers' Handbook* John Wiley & Sons, New York, NY, 1918 p978.

Sanders, Wilbur E. "Improved Type of Ore Bin" *Details of Practical Mining* McGraw-Hill Book Company, New York, NY, 1916.

"Tonopah Orehouses" *Engineering & Mining Journal* May 27, 1911 p1048.

The Author characterized flat-bottomed and sloped-floor ore bins at productive mine sites in Colorado, Montana, Nevada, New Mexico, and Utah.

96. Lord, Elliot *Comstock Mining and Miners* Howell-North Books, San Diego, CA, 1980 p221.

Evidence of ore having been hand-sorted and sacked has been recorded at several mines worked during the 1870s and early 1880s at Tybo, Nevada, and at early mines in the Lookout District in eastern California.

97. International Textbook Company *A Textbook on Metal Mining: Preliminary Operations at Metal Mines, Metal Mining, Surface Arrangements at Metal Mines, Ore Dressing and Milling* International Textbook Company, Scranton, PA, 1899 A42 p13.

"Tonopah Orehouses" *Engineering & Mining Journal* May 27, 1911 p1048.

A study was made of ore sorting houses at the Mary Murphy Mine in the Chalk Creek Mining District, Colorado; the Cresson, Damon Main, Glorietta, Hull City, Joe Dandy, Longfellow, Mabel M., Mary Nevins, Nellie V., Mollie Kathleen, Nichols, Pinnacle, Prince Albert, Sangre de Cristo, Sitting Bull, Strong, Victor, and Vindicator Mines in Cripple Creek, Colorado; and in Colorado's Tin Cup, Peru Creek and Silverton mining districts.

98. The remains of ore bins and sorting houses have been characterized from the analyses of productive mines in California, Colorado, Montana, Nevada, New Mexico, and Utah.

99. Ihlseng, Magnus *A Manual of Mining* John Wiley & Sons, New York, NY, 1892 p137.

100. Bailey, Lynn *Supplying the Mining World: the Mining Equipment Manufacturers of San Francisco 1850-1900* Western Lore Press, Tucson, AZ, 1996 p116.

101. Bailey, Lynn *Supplying the Mining World: the Mining Equipment Manufacturers of San Francisco 1850-1900* Western Lore Press, Tucson, AZ, 1996 p116.

Trennert, Robert A. "From Gold Ore to Bat Guano: Aerial Tramways in the West" *The Mining History Journal* 1997 p4.

102. Trennert, Robert A. "From Gold Ore to Bat Guano: Aerial Tramways in the West" *The Mining History Journal* 1997 p5.

103. Ihlseng, Magnus *A Manual of Mining* John Wiley & Sons, New York, NY, 1892 p137.

Trennert, Robert A. "From Gold Ore to Bat Guano: Aerial Tramways in the West" *The Mining History Journal* 1997 p5.

104. Ihlseng, Magnus *A Manual of Mining* John Wiley & Sons, New York, NY, 1892 p138.

International Textbook Company *Coal and Metal Miners' Pocket Book* International Textbook Company, Scranton, PA, 1905 p122.

Lewis, Robert S. *Elements of Mining* John Wiley & Sons, Inc., New York, NY, 1946 p372.

Peele, Robert *Mining Engineers' Handbook* John Wiley & Sons, New York, NY, 1918 p1563.

Trennert, Robert A. "From Gold Ore to Bat Guano: Aerial Tramways in the West" *The Mining History Journal* 1997 p6.

105. Ihlseng, Magnus *A Manual of Mining* John Wiley & Sons, New York, NY, 1892 p138.

International Textbook Company *Coal and Metal Miners' Pocket Book* International Textbook Company, Scranton, PA, 1905 p122.

Keystone Consolidated Publishing Company Inc. *The Mining Catalog: 1925 Metal-Quarry Edition* Keystone Consolidated Publishing Company Inc., 1925 p502.

Peele, Robert *Mining Engineers' Handbook* John Wiley & Sons, New York, NY, 1918 p1556.

Trennert, Robert A. "From Gold Ore to Bat Guano: Aerial Tramways in the West" *The Mining History Journal* 1997 p6.

106. Bailey, Lynn *Supplying the Mining World: the Mining Equipment Manufacturers of San Francisco 1850-1900* Western Lore Press, Tucson, AZ, 1996 p124.

Keystone Consolidated Publishing Company Inc. *The Mining Catalog: 1925 Metal-Quarry Edition* Keystone Consolidated Publishing Company Inc., 1925 p500.

Trennert, Robert A. "From Gold Ore to Bat Guano: Aerial Tramways in the West" *The Mining History Journal* 1997 p6.

107. Bailey, Lynn *Supplying the Mining World: the Mining Equipment Manufacturers of San Francisco 1850-1900* Western Lore Press, Tucson, AZ, 1996 p124.

Keystone Consolidated Publishing Company Inc. *The Mining Catalog: 1925 Metal-Quarry Edition* Keystone Consolidated Publishing Company Inc., 1925 p500.

Peele, Robert *Mining Engineers' Handbook* John Wiley & Sons, New York, NY, 1918 p1581.

Trennert, Robert A. "From Gold Ore to Bat Guano: Aerial Tramways in the West" *The Mining History Journal* 1997 p5.

The general characteristics for tramway terminals were derived by comparing the above sources with the examination of historic mines in Colorado, Montana, and Nevada.

108. The author observed several aerial tram terminals at which beveled gears were added to the sheave axles to power generators. The mines include the Mary Murphy, the Pennsylvania in the Peru Creek District, and the Black Prince and Iowa in the Silverton District, Colorado.

109. Keystone Consolidated Publishing Company Inc. *The Mining Catalog: 1925 Metal-Quarry Edition* Keystone Consolidated Publishing Company Inc., 1925 p500.

Lewis, Robert S. *Elements of Mining* John Wiley & Sons, Inc., New York, NY, 1946 p372.

Peele, Robert *Mining Engineers' Handbook* John Wiley & Sons, New York, NY, 1918 p1589.

Single rope tramways were examined in California, Colorado, New Mexico, and Nevada.

110. The tramway at the Pennsylvania Mine's upper haulage tunnels relied on an antiquated single cylinder steam hoist for power, and the tramway at the Revenge Mine in Boulder County, Colorado used an old double cylinder steam hoist. The tramway at the Radcliff Mine in the Panamint Mountains, California relied on a gas hoist, and the tramway at the Big Four Mine incorporated a windlass.

Chapter 4: Gear Oil & Steam Power: The Surface Plants for Shafts

1. Croft, Terrell *Machinery Foundations and Erection* McGraw-Hill Book Co., New York, NY, 1923 p282.

International Textbook Company *A Textbook on Metal Mining: Preliminary Operations at Metal Mines, Metal Mining, Surface Arrangements at Metal Mines, Ore Dressing and Milling* International Textbook Company, Scranton, PA, 1899 A40 p11, A42 p2.

Peele, Robert *Mining Engineers' Handbook* John Wiley & Sons, New York, NY, 1918 p2095.

Tinney, W.H. *Gold Mining Machinery: Its Selection, Arrangement, & Installation* D. Van Nostrand Company, New York, NY, 1906 p33.

2. Twain, Mark *Roughing It* Airmont Publishing Co., New York, NY, 1967 p119.

3. International Textbook Company *A Textbook on Metal Mining: Preliminary Operations at Metal Mines, Metal Mining, Surface Arrangements at Metal Mines, Ore Dressing and Milling* International Textbook Company, Scranton, PA, 1899 A40 p19.

Peele, Robert *Mining Engineers' Handbook* John Wiley & Sons, New York, NY, 1918 p263.

In the mountain states and Great Basin, the author encountered numerous shafts featuring both open and closed cribbing. The shafts ranged in age from the 1870s to the 1910s.

4. Colliery Engineer Company *Coal & Metal Miners' Pocketbook* Colliery Engineer Company, Scranton, PA, 1893 p109.

Eaton, Lucien *Practical Mine Development & Equipment* McGraw-Hill Book Company, New York, NY, 1934 p9.

International Textbook Company *Coal and Metal Miners' Pocket Book* International Textbook Company, Scranton, PA, 1905 p259.

International Textbook Company *A Textbook on Metal Mining: Preliminary Operations at Metal Mines, Metal Mining, Surface Arrangements at Metal Mines, Ore Dressing and Milling* International Textbook Company, Scranton, PA, 1899 A40 p15.

Peele, Robert Mining *Engineers' Handbook* John Wiley & Sons, New York, NY, 1918 p263 p249.

Silversmith, Julius *A Practical Handbook* Metallurgists American Mining Index, New York, NY, 1867 p47.

Young, George *Elements of Mining* John Wiley & Sons, New York, NY, 1923 p461.

5. Mining companies used cages at least as early as the late 1860s. The Author documented a mine in the Tybo / Keystone Mining District in Nevada that used a cage during the late 1860s. In addition, Eliot Lord's Comstock Mining & Miners depicts cages in use during the 1860s.

6. Wyman, Mark *Hard Rock Epic: Western Mining and the Industrial Revolution, 1860-1910* University of California Press, Berkeley, CA, 1989 p99.

7. Silversmith, Julius *A Practical Handbook* Metallurgists American Mining Index, New York, NY, 1867 p57.

The main shaft at the Rye Patch Mine in the Rye Patch District, and the Bunker Hill Mine in Tybo, Nevada featured two compartment shafts like the type described in the text. Miners sank the Rye Patch Shaft during the late 1860s or early 1870s, and miners sank the Bunker Hill shaft around 1880.

8. International Textbook Company *Coal and Metal Miners' Pocket Book* International Textbook Company, Scranton, PA 1905 p261.

Peele, Robert *Mining Engineers' Handbook* John Wiley & Sons, New York, NY., 1918 p263 p251.

Young, George *Elements of Mining* John Wiley & Sons, New York, NY., 1923 p171 p462.

9. International Textbook Company *Coal and Metal Miners' Pocket Book* International Textbook Company, Scranton, PA, 1905 p262.

"Joplin Ore Buckets" *Details of Practical Mining* McGraw-Hill Book Company, New York, NY, 1916.

"Joplin Ore Buckets" *Engineering & Mining Journal* Vol 94 No 6 p254.

Rogers, A.P. "Bucket and Chute for Shaft Sinking" *Engineering & Mining Journal* June 11, 1910 p1214.

Worcester, S.A. "Using the Ore Bucket" *Engineering & Mining Journal* March 12, 1910 p552.

Nearly all small shaft mines and prospects analyzed in association with this work exhibited evidence of having relied on ore buckets as hoisting vehicles.

10. Division of Mines & Minerals Colorado State Archives, Denver, CO

Mine Inspectors' Reports: Teller County File # 48440 Prince Albert Shaft.

11. Barton, Denys Bradford *The Cornish Beam Engine* Wordens of Cornwall, Ltd. 1966 p191.

Hoover, Herbert C. *Principles of Mining: Valuation, Organization, and Administration* McGraw-Hill Book Company, Inc., New York, NY., 1909 p132.

International Textbook Company *A Textbook on Metal Mining: Steam and Steam-Boilers, Steam Engines, Air and Air Compression, Hydromechanics and Pumping, Mine Haulage, Hoisting and Hoisting Appliances, Percussive and Rotary Boring* International Textbook Company, Scranton, PA, 1899 A23 p85.

Ketchum, Milo S. C.E. *The Design of Mine Structures* McGraw-Hill Book Co., New York, NY., 1912 p33.

McFarlane, George C. "Automatic Dumping Skip for Vertical Shafts" *Engineering & Mining Journal* June 26, 1909 p1281.

Proctor, Olin S. "Direct Automatic Bucket Tipple" *Engineering & Mining Journal* Vol. 94 No 24 p1120.

Todd, Arthur C. *The Cornish Miner in America* The Arthur H. Clark Company, Spokane, WA, 1995 p91.

Young, George *Elements of Mining* John Wiley & Sons, New York, NY., 1923 p171.

Nearly all of the shaft mines active during the 1930s examined by the author for this work exhibited evidence of relying on either ore buckets or on skips as hoisting vehicles.

12. Colliery Engineer Company *Coal & Metal Miners' Pocketbook* Colliery Engineer Company, Scranton, PA, 1893 p109.

Ihlseng, Magnus *A Manual of Mining* John Wiley & Sons, New York, NY, 1892 p98.

International Textbook Company *A Textbook on Metal Mining: Steam and Steam-Boilers, Steam Engines, Air and Air Compression, Hydromechanics and Pumping, Mine Haulage, Hoisting and Hoisting Appliances, Percussive and Rotary Boring* International Textbook Company, Scranton, PA, 1899 A23 p75.

International Textbook Company *International Library of Technology: Hoisting, Haulage, Mine Drainage* International Textbook Company, Scranton, PA, 1906 A53 p20.

Lord, Elliot *Comstock Mining and Miners* Howell-North Books, San Diego, CA, 1980 p224.

Peele, Robert *Mining Engineers' Handbook* John Wiley & Sons, New York, NY, 1918 p263 p251.

Rogers, A.P. "Bucket and Chute for Shaft Sinking" *Engineering & Mining Journal* June 11, 1910 p1214.

Worcester, S.A. "Using the Ore Bucket" *Engineering & Mining Journal* March 12, 1910 p552.

Young, George *Elements of Mining* John Wiley & Sons, New York, NY, 1923 p171 p201.

13. Barton, Denys Bradford *The Cornish Beam Engine* Wordens of Cornwall, Ltd., 1966 p195.

Eaton, Lucien *Practical Mine Development & Equipment* McGraw-Hill Book Company, New York, NY, 1934 p13.

Peele, Robert *Mining Engineers' Handbook* John Wiley & Sons, New York, NY., 1918 p263 p251.

Young, George *Elements of Mining* John Wiley & Sons, New York, NY, 1923 p461.

All shaft compartments that accommodated cages analyzed in conjunction with this study were at a minimum four-by-four to four-by-five feet in area, and they featured evidence of having possessed guide rails.

14. The author derived the description of the Locan Shaft's Cornish pump from analysis of archaeological remains at the mine site.

15. Eaton, Lucien *Practical Mine Development & Equipment* McGraw-Hill Book Company, New York, NY, 1934 p19.

International Textbook Company *A Textbook on Metal Mining: Preliminary Operations at Metal Mines, Metal Mining, Surface Arrangements at Metal Mines, Ore Dressing and Milling* International Textbook Company, Scranton, PA, 1899 A40 p19.

International Textbook Company *Coal and Metal Miners' Pocket Book* International Textbook Company, Scranton, PA, 1905 p271.

Linke, Harold A. "Light Shaft timbering" *Details of Practical Mining* McGraw-Hill Book Co., New York, NY, 1916.

"Mesabi Shaft Collars" *Engineering & Mining Journal* Sept. 20, 1913 p546.

Most the open shafts examined in association with this work featured shaft-sets constructed primarily with six-by-six to ten-by-ten inch timbers. The Locan Shaft in Eureka, Nevada featured four compartments supported with twelve-by-twelve timber sets.

16. "Headframe for a Prospect Shaft" *Engineering & Mining Journal* Nov. 11, 1911 p980.

Ihlseng, Magnus *A Manual of Mining* John Wiley & Sons, New York, NY, 1892 p57.

International Textbook Company *A Textbook on Metal Mining: Preliminary Operations at Metal Mines, Metal Mining, Surface Arrangements at Metal Mines, Ore Dressing and Milling* International Textbook Company, Scranton, PA, 1899 A40 p16.

International Textbook Company *International Library of Technology: Hoisting, Haulage, Mine Drainage* International Textbook Company, Scranton, PA, 1906 A50 p1.

Lewis, Robert S. *Elements of Mining* John Wiley & Sons, Inc., New York, NY, 1946 p185.

"Metallics [Editorial Notes: on hoisting]" *Engineering & Mining Journal* July 31, 1909 p226.

Peele, Robert *Mining Engineers' Handbook* John Wiley & Sons, New York, NY, 1918 p263 p252, p880.

"Prospecting Equipment" *Mining & Scientific Press* Jan. 23, 1904.

Stovall, Dennis H. "Where the Horse Whim Fails" *Mining & Scientific Press* Nov. 4, 1911 p583.

17. The author's field investigations of historic mines in the Rocky Mountains, the Great Basin, and the Southwest reflect the trend of a near total absence of steam hoisting systems pre-dating the mid 1870s, and an increase in their popularity during the early 1880s.

18. Bailey, Lynn *Supplying the Mining World: the Mining Equipment Manufacturers of San Francisco 1850-1900* Western Lore Press, Tucson, AZ, 1996 p92, 113.

Bramble, Charles A. *The ABC of Mining* Geology, Energy & Minerals Corporation, Santa Monica, CA 1980 p103.

"Headframe for a Prospect Shaft" *Engineering & Mining Journal* Nov. 11, 1911 p980.

Ihlseng, Magnus *A Manual of Mining* John Wiley & Sons, New York, NY, 1892 p56.

International Textbook Company *A Textbook on Metal Mining: Preliminary Operations at Metal Mines, Metal Mining, Surface Arrangements at Metal Mines, Ore Dressing and Milling* International Textbook Company, Scranton, PA, 1899 A40 p16.

International Textbook Company *International Library of Technology: Hoisting, Haulage, Mine Drainage* International Textbook Company, Scranton, PA, 1906 A50 p3.

"Metallics [Editorial Notes: on hoisting]" *Engineering & Mining Journal* July 31, 1909 p226.

Peele, Robert *Mining Engineers' Handbook* John Wiley & Sons, New York, NY, 1918 p252, 880.

Stovall, Dennis H. "Where the Horse Whim Fails" *Mining & Scientific Press* Nov. 4, 1911 p583.

19. Peele, Robert *Mining Engineers' Handbook* John Wiley & Sons, New York, NY, 1918 p880.

Stoltz, Guy C. "Tram Car for the Prospector" *Engineering & Mining Journal* April 2, 1910 p696.

The Author characterized malacate horse whims from an examination of mines in Colorado and Nevada.

20. Bailey, Lynn *Supplying the Mining World: the Mining Equipment Manufacturers of San Francisco 1850-1900* Western Lore Press, Tucson, AZ, 1996 p77.

International Textbook Company *International Library of Technology: Hoisting, Haulage, Mine Drainage* International Textbook Company, Scranton, PA, 1906 A50 p3.

The Author characterized horizontal reel horse whims from an examination of mines in California and Nevada.

21. Bailey, Lynn *Supplying the Mining World: the Mining Equipment Manufacturers of San Francisco 1850-1900* Western Lore Press, Tucson, AZ, 1996 p92, 113.

International Textbook Company *International Library of Technology: Hoisting, Haulage, Mine Drainage* International Textbook Company, Scranton, PA, 1906 A50 p3.

Peele, Robert *Mining Engineers' Handbook* John Wiley & Sons, New York, NY, 1918 p880.

The author characterized geared horse whims from a review of historic photos and trade advertisements, and from analyses of remains at prospect shafts in Colorado and Nevada.

22. Peele, Robert *Mining Engineers' Handbook* John Wiley & Sons, New York, NY, 1918 p252.

"Prospecting Equipment" *Mining & Scientific Press* Jan. 23, 1904.

23. International Textbook Company *A Textbook on Metal Mining: Steam and Steam-Boilers, Steam Engines, Air and Air Compression, Hydromechanics and Pumping, Mine Haulage, Hoisting and Hoisting Appliances, Percussive and Rotary Boring* International Textbook Company, Scranton, PA, 1899 A23 p9.

International Textbook Company *International Library of Technology: Hoisting, Haulage, Mine Drainage* International Textbook Company, Scranton, PA, 1906 A50 p6.

24. International Textbook Company *International Library of Technology: Hoisting, Haulage, Mine Drainage* International Textbook Company, Scranton, PA, 1906 A50 p17, 32.

25. "Headframe for a Prospect Shaft" *Engineering & Mining Journal* Nov. 11, 1911 p980.

International Textbook Company *International Library of Technology: Hoisting, Haulage, Mine Drainage* International Textbook Company, Scranton, PA, 1906 A50 p7.

Peele, Robert *Mining Engineers' Handbook* John Wiley & Sons, New York, NY, 1918 p252, 882.

Analyses of historic shaft sites active between the 1880s and 1910s in Colorado, Montana, Nevada, and New Mexico support conclusions that mining companies typically used steam hoists less than six-by-six feet in area for deep prospecting.

26. Croft, Terrell *Steam Boilers* McGraw-Hill Book Co., New York, NY, 1921 p44.

Ingersoll Rock Drill Company *Catalog No.7: Rock Drills, Air Compressors and Air Receivers* Ingersoll Rock Drill Company, New York, NY, [1887] p51.

International Textbook Company *A Textbook on Metal Mining: Steam and Steam-Boilers, Steam Engines, Air and Air Compression, Hydromechanics and Pumping, Mine Haulage, Hoisting and Hoisting Appliances, Percussive and Rotary Boring* International Textbook Company, Scranton, PA, 1899 A18 p30.

International Textbook Company *International Library of Technology: Mine Surveying, Metal Mine Surveying, Mineral-Land Surveying, Steam and Steam Boilers, Steam Engines, Air Compression* International Textbook Company, Scranton, PA, 1924 A23 p16.

Linstrom, C.B. & Clemens, A.B. *Steam Boilers and Equipment* International Textbook Co., Scranton, PA, 1928 p23.

The M.J. O'Fallon Supply Company *Catalog No.8 Steam, Mill, & Water Supplies* R.R. Donnelly & Sons Co., 1923 p395.

Peele, Robert *Mining Engineers' Handbook* John Wiley & Sons, New York, NY, 1918 p2083.

Potter, Andrey A. and Calderwood, James P. *Elements of Steam and Gas Engineering* McGraw-Hill Book Co., Inc., New York, NY, 1938 p67.

Rand Drill Company *Illustrated Catalog of the Rand Drill Company, New York, U.S.A.* Rand Drill Company, New York, NY, 1886 p43.

Rogers, Warren U. "The Care of Small Steam Boilers" *Engineering & Mining Journal* Dec. 8, 1909, p1217-1219.

Thurston, R.H. *A Manual of Steam Boilers: Their Design, Construction, and Operation* John Wiley & Sons, New York, NY, 1901 p25.

Most of the steam-powered prospect operations examined by the author used locomotive boilers. Several probably used upright units, and a few operations may have used Pennsylvania boilers.

27. Croft, Terrell *Steam Boilers* McGraw-Hill Book Co., New York, NY, 1921 p48.

International Textbook Company *A Textbook on Metal Mining: Steam and Steam-Boilers, Steam Engines, Air and Air Compression, Hydromechanics and Pumping, Mine Haulage, Hoisting and Hoisting Appliances, Percussive and Rotary Boring* International Textbook Company, Scranton, PA, 1899 A18 p34.

Kleinhans, Frank B. *Locomotive Boiler Construction* Norman W. Henley Publishing Co., New York, NY, 1915 p12.

Rand Drill Company Illustrated *Catalog of the Rand Drill Company, New York, U.S.A.* Rand Drill Company, New York, NY, 1886 p47.

Rogers, Warren U. "The Care of Small Steam Boilers" *Engineering & Mining Journal* Dec. 8, 1909, p1217-1219.

Tinney, W.H. *Gold Mining Machinery: Its Selection, Arrangement, & Installation* D. Van Nostrand Company, New York, NY, 1906 p50.

28. Linstrom, C.B. & Clemens, A.B. *Steam Boilers and Equipment* International Textbook Co., Scranton, PA, 1928 p28.

The Ouray Mine in Breckenridge, Colorado used a Pennsylvania boiler to power a hoist, and the Imperial Mine in Frisco, Utah possessed a Pennsylvania boiler to run a compressor and ventilating fan.

29. Colliery Engineer Company *Coal & Metal Miners' Pocketbook* Colliery Engineer Company, Scranton, PA, 1893 p262.

International Textbook Company *A Textbook on Metal Mining: Steam and Steam-Boilers, Steam Engines, Air and Air Compression, Hydromechanics and Pumping, Mine Haulage, Hoisting and Hoisting Appliances, Percussive and Rotary Boring* International Textbook Company, Scranton, PA, 1899 A18 p28.

The M.J. O'Fallon Supply Company *Catalog No.8 Steam, Mill, & Water Supplies* R.R. Donnelly & Sons Co., 1923 p402.

Peele, Robert *Mining Engineers' Handbook* John Wiley & Sons, New York, NY, 1918 p2083.

Potter, Andrey A. and Calderwood, James P. *Elements of Steam and Gas Engineering* McGraw-Hill Book Co., Inc., New York, NY, 1938 p66.

Thurston, R.H. *A Manual of Steam Boilers: Their Design, Construction, and Operation* John Wiley & Sons, New York, NY, 1901 p31.

30. Bailey, Lynn *Supplying the Mining World: the Mining Equipment Manufacturers of San Francisco 1850-1900* Western Lore Press, Tucson, AZ, 1996 p48, 104.

Ihlseng, Magnus *A Manual of Mining* John Wiley & Sons, New York, NY, 1892 p62.

Ingersoll Rock Drill Company *Catalog No.7: Rock Drills, Air Compressors and Air Receivers* Ingersoll Rock Drill Company, New York, NY, [1887] p55.

Sara, Tim; Twitty, Eric; Grant, Marcus *Mining and Settlement at Timberline* Paragon Archaeological Consultants, Inc., Denver, CO, 1998 p4-35.

31. "A Gas Engine Hoisting Plant" *Engineering & Mining Journal* Sept. 12, 1896 p247.

"A Gasoline Hoisting Engine" *Engineering & Mining Journal* May 1, 1897.

"A New Gasoline Hoist" *Engineering & Mining Journal* March 26, 1900.

"Gas Engine Hoist" *Mining & Scientific Press* March 4, 1905 p133.

International Textbook Company *International Library of Technology: Hoisting, Haulage, Mine Drainage* International Textbook Company, Scranton, PA, 1906 A50 p30.

Peele, Robert *Mining Engineers' Handbook* John Wiley & Sons, New York, NY, 1918 p905.

"Questions and Answers [pertaining to gas engines and mining]" *Engineering & Mining Journal* June. 11, 1898 p706.

"Questions and Answers [pertaining to gas engines and mining]" *Engineering & Mining Journal* Oct. 21, 1899 p492.

Rehfuss, Louis A. and Rehfuss, W. Clifford "Portable Mining Equipment for Prospectors" *Engineering & Mining Journal* June 10, 1916 p1025.

"Steam and Gas Engines" *Mining & Scientific Press* Nov. 24, 1900 p546.

Stovall, Dennis H. "Where the Horse Whim Fails" *Mining & Scientific Press* Nov. 4, 1911 p583.

"The Witte Gasoline Hoisting Engine" *Engineering & Mining Journal* July 30, 1898 p131.

"The Witte Gasoline Mine Hoist" *Engineering & Mining Journal* Nov. 23, 1901 p672.

Young, George *Elements of Mining* John Wiley & Sons, New York, NY, 1923 p173.

Young, George *Elements of Mining* John Wiley & Sons, New York, NY, 1946 p203.

32. Hoover, Herbert C. *Principles of Mining: Valuation, Organization, and Administration* McGraw-Hill Book Company, Inc., New York, NY, 1909 p130.

33. "A Gas Engine Hoisting Plant" *Engineering & Mining Journal* Sept. 12, 1896 p247.

"A New Gasoline Hoist" *Engineering & Mining Journal* March 26, 1900.

Lewis, Robert S. *Elements of Mining* John Wiley & Sons, Inc., New York, NY, 1946 [1933].

During field research for this work, the author encountered several mines in Nevada and Utah that used their gas hoists to run additional mine plant appliances.

34. Hoover, Herbert C. *Principles of Mining: Valuation, Organization, and Administration* McGraw-Hill Book Company, Inc., New York, NY, 1909 p130.

Nearly all of the mines featuring gas hoists analyzed for this study operated after 1900, and were predominantly located in the Great Basin and Southwest. Only a few small mines and prospects in the Rocky Mountains relied on gas hoists.

35. Botsford, H.L. "Types of Headframes" *Engineering & Mining Journal* Oct. 13, 1913 p690.

Forsyth, Alexander "The Headframes of Shafts at Cripple Creek" *Engineering & Mining Journal* March 7, 1903 p366.

Mentzel, Charles "Prospecting Headframe" *Engineering & Mining Journal* Oct. 5, 1912.

Peele, Robert *Mining Engineers' Handbook* John Wiley & Sons, New York, NY, 1918 p252, 926.

36. Botsford, H.L. "A Timber Headframe" *Engineering & Mining Journal* June 10, 1911 p1148.

Botsford, H.L. "Small Four Post Headframe" *Details of Practical Mining* McGraw-Hill Book Co., New York, NY, 1916.

Colliery Engineer Company *Coal & Metal Miners' Pocketbook* Colliery Engineer Company, Scranton, PA, 1893 p109.

Denny, G.A. "A-Type Timber Headframe" *Engineering & Mining Journal* March 30, 1914 p1100.

"Headframe for a Prospect Shaft" *Engineering & Mining Journal* Nov. 11, 1911 p980.

International Textbook Company *Coal and Metal Miners' Pocket Book* International Textbook Company, Scranton, PA, 1905 p262, 275, 397.

Mentzel, Charles "Prospecting Headframe" *Engineering & Mining Journal* Oct. 5, 1912.

Muir, Douglass "Unwatering and Equipping Untimbered shafts" *Details of Practical Mining* McGraw-Hill Book Co., New York, NY, 1916.

Peele, Robert *Mining Engineers' Handbook* John Wiley & Sons, New York, NY, 1918 p936.

"Simple Head Frame Construction" *Mining & Scientific Press* Feb. 6, 1904 p94.

"Small Timber Headframe" *Engineering & Mining Journal* April 8, 1916 p645.

"Types of Head Frames" *Mining & Scientific Press* Feb. 21, 1903.

"Variations in [headframe] Construction" *Mining & Scientific Press* Aug. 22, 1903.

The characteristics and applications of sinking-class headframes were also determined by analyzing the remains at prospect shafts in the Great Basin and mountain states.

37. Burr, Floyd "Design of Headframes" *Engineering & Mining Journal* April 7, 1917 p611.

Colliery Engineer Company *Coal Miners' Pocketbook* McGraw-Hill Book Co., New York, NY, 1916 p581.

Fitch, Charles H. "Head Works Framing - Primary Notions" *Mining & Scientific Press* June 6, 1903 p335.

Fitch, Charles H. "Head Works Framing" *Mining & Scientific Press* June 13, 1903 p380.

Forsyth, Alexander "The Headframes of Shafts at Cripple Creek" *Engineering & Mining Journal* March 7, 1903 p366.

Ihlseng, Magnus *A Manual of Mining* John Wiley & Sons, New York, NY, 1892 p91.

International Textbook Company *A Textbook on Metal Mining: Steam and Steam-Boilers, Steam Engines, Air and Air Compression, Hydromechanics and Pumping, Mine Haulage, Hoisting and Hoisting Appliances, Percussive and Rotary Boring* International Textbook Company, Scranton, PA, 1899 A23 p103.

International Textbook Company *Coal and Metal Miners' Pocket Book* International Textbook Company, Scranton, PA, 1905 p262, 275, 397.

International Textbook Company *International Library of Technology: Hoisting, Haulage, Mine Drainage* International Textbook Company, Scranton, PA, 1906 A53 p31.

Ketchum, Milo S. C.E. *The Design of Mine Structures* McGraw-Hill Book Co., New York, NY, 1912 p41.

Lewis, Robert S. *Elements of Mining* John Wiley & Sons, Inc., New York, NY, 1946 p216.

Muir, Douglass "Unwatering and Equipping Untimbered shafts" *Details of Practical Mining* McGraw-Hill Book Co., New York, NY, 1916.

Peele, Robert *Mining Engineers' Handbook* John Wiley & Sons, New York, NY, 1918 p930.

"Prospecting Equipment" *Mining & Scientific Press* Jan. 23, 1904.

Staley, William *Mine Plant Design* McGraw-Hill Book Co., New York, NY, 1936.

"Types of Head Frames" *Mining & Scientific Press* Feb. 21, 1903.

Nearly all of the headframes analyzed in conjunction with this work featured backbraces in the locations and at the angles recommended by mining engineers.

38. Altenbach, J. Scott Personal Interview: Albuquerque, New Mexico Sept. 27, 1998. The description of building a headframe was adapted from an interview with Altenbach, who built a two-post gallows headframe in 1998.

"Simple Head Frame Construction" *Mining & Scientific Press* Feb. 6, 1904 p94, discusses sinking-class headframe foundations.

Statements regarding the placement of headframe foundations are supported by the remains examined at small mines and prospect operations in the Great Basin and mountain states.

39. The description of typical hoist houses was assembled from data collected from historic photographs and analyses of historic mine sites in the Great Basin and mountain states.

40. Bernard, Clinton P. "Hinges for Shaft Doors" *Details of Practical Mining* McGraw-Hill Book Co., New York, NY, 1916.

Division of Mines & Minerals Colorado State Archives, Denver, CO

Mine Inspectors' Reports: Teller County File # 48440 Elkton Mine.

Edyvean, E.H. "Safety Bonnet for Shaft Opening" *Engineering & Mining Journal* May 16, 1914 p1001.

International Textbook Company *A Textbook on Metal Mining: Preliminary Operations at Metal Mines, Metal Mining, Surface Arrangements at Metal Mines, Ore Dressing and Milling* International Textbook Company, Scranton, PA, 1899 A40 p41.

Analysis of prospect shaft collars in the Great Basin and mountain states revealed that many operations used angled trap doors, a few operations used flat trap doors, and a paucity of outfits, many of them operating prior to 1890, had unprotected shaft collars.

41. Altenbach, J. Scott Personal Interview: Albuquerque, New Mexico Sept. 27, 1998. The description of emptying an ore bucket was adapted from a demonstration conducted by Altenbach and other individuals during the National Association of Abandoned Mine Lands 20th Annual Conference in Albuquerque, New Mexico in 1998.

Le Veque, G.E. "Tripod Headframe and Novel Bucket" *Engineering & Mining Journal* April 23, 1913 p755.

Muir, Douglass "Unwatering and Equipping Untimbered shafts" *Details of Practical Mining* McGraw-Hill Book Co., New York, NY, 1916.

The author analyzed sinking-class headframes with dumping aprons at prospect shafts in the Cripple Creek Mining District, Colorado.

42. The remains of horse whims have been characterized from the analysis of at least seven prospect shafts featuring this type of hoisting system, and from analyses of historic photographs and advertisements.

43. Ihlseng, Magnus *A Manual of Mining* John Wiley & Sons, New York, NY, 1892 p60.

International Textbook Company *A Textbook on Metal Mining: Preliminary Operations at Metal Mines, Metal Mining, Surface Arrangements at Metal Mines, Ore Dressing and Milling* International Textbook Company, Scranton, PA, 1899 A42 p2.

44. The remains of gas hoists have been characterized from mines examined in the Rocky Mountains and Great Basin.

45. Eberhart, Perry *Guide to the Colorado Ghost Towns and Mining Camps* Swallow Press, Athens, OH., 1987 p453.

Munn, Bill *A Guide to the Mines of the Cripple Creek District* Century One Press, Colorado Springs, CO, 1984 p47.

The author determined the characteristics of the Gold Coin plant from analysis of the site.

46. Crampton, Frank A. *Deep Enough: A Working Stiff in the Western Mine Camps* University of Oklahoma Press, 1982.

47. Division of Mines & Minerals Colorado State Archives, Denver, CO

Mine Inspectors' Reports: Teller County File # 104089 American Eagle Mine.

Mine Inspectors' Reports: Teller County File # 48440 Delmonico Mine.

Mine Inspectors' Reports: Teller County File # 48441 Hull City Mine.

Mine Inspectors' Reports: Teller County File # 48516 and 48441 Vindicator Mine

Sanborn Map Company Victor, Teller County, Colorado, Dec. 1908 Sanborn Map Company, Brooklyn, NY, 1909.

Field investigations conduced by the author confirm that the mines listed above featured three-compartment shafts and double-drum hoists, except for the Delmonico, which was served by a single-drum hoist.

48. Bailey, Lynn *Supplying the Mining World: the Mining Equipment Manufacturers of San Francisco 1850-1900* Western Lore Press, Tucson, AZ, 1996 p14.

Barton, Denys Bradford *The Cornish Beam Engine* Wordens of Cornwall, Ltd. 1966 p185-188.

Lord, Elliot *Comstock Mining and Miners* Howell-North Books, San Diego, CA, 1980 p221.

Todd, Arthur C. *The Cornish Miner in America* The Arthur H. Clark Company, Spokane, WA, 1995 p119.

49. Barton, Denys Bradford *The Cornish Beam Engine* Wordens of Cornwall, Ltd. 1966 p201.

Collins, Jose H. *Principles of Metal Mining* William Collins, Sons & Company, London 1875 p85.

International Textbook Company *A Textbook on Metal Mining: Steam and Steam-Boilers, Steam Engines, Air and Air Compression, Hydromechanics and Pumping, Mine Haulage, Hoisting and Hoisting Appliances, Percussive and Rotary Boring* International Textbook Company, Scranton, PA, 1899 A23 p9.

Shinn, Charles H. *The Story of the Mine: as Illustrated by the Great Comstock Lode of Nevada* University of Nevada Press, Reno, NV, 1984 p217.

Silversmith, Julius *A Practical Handbook* Metallurgists American Mining Index, New York, NY, 1867 p48.

Single cylinder steam hoists have been characterized by analyses of historic photos depicting mines between the 1860s and 1880s, and from the author's field investigations at historic mines closed prior to 1885 in several Nevada districts.

50. Young, Otis E. *Western Mining* University of Oklahoma Press, 1987 p107.

Landing chairs have been examined at many shaft collars and stations underground by the author, including a mine in the Tybo/ Keystone District, Nevada abandoned by the early 1870s. Eliot Lord's Comstock Mining & Miners contains photos of landing chairs at mines during the 1870s.

51. Bailey, Lynn *Supplying the Mining World: the Mining Equipment Manufacturers of San Francisco 1850-1900* Western Lore Press, Tucson, AZ, 1996 p45.

52. "Headframe for a Prospect Shaft" *Engineering & Mining Journal* Nov. 11, 1911 p980.

International Textbook Company *A Textbook on Metal Mining: Steam and Steam-Boilers, Steam Engines, Air and Air Compression, Hydromechanics and Pumping, Mine Haulage, Hoisting and Hoisting Appliances, Percussive and Rotary Boring* International Textbook Company, Scranton, PA, 1899 A23 p9.

International Textbook Company *International Library of Technology: Hoisting, Haulage, Mine Drainage* International Textbook Company, Scranton, PA, 1906 A50 p6.

Peele, Robert *Mining Engineers' Handbook* John Wiley & Sons, New York, NY, 1918 p882.

Nelson, Sven T. "Steam Hoists for Shallow Mines" *Engineering & Mining Journal* Nov. 25, 1911 p1027.

53. Ingersoll Rock Drill Company *Catalog No.7: Rock Drills, Air Compressors and Air Receivers* Ingersoll Rock Drill Company, New York, NY, [1887] p57.

Peele, Robert *Mining Engineers' Handbook* John Wiley & Sons, New York, NY, 1918 p882, 900.

Nearly all of the shafts post-dating 1880 examined in association with this study were equipped with geared hoists.

54. "Direct-Acting Hoists" *Mining & Scientific Press* Dec. 15, 1900 p585.

Mine & Smelter Supply Co. *Catalog No. 22: Machinery and General Supplies* J.D. Abraham Publishing Co., 1912.

Shinn, Charles H. *The Story of the Mine: as Illustrated by the Great Comstock Lode of Nevada* University of Nevada Press, Reno, NV, 1984 p215.

The author analyzed mines featuring double-drum steam hoists in Colorado's Cripple Creek and Creede districts.

55. Bailey, Lynn *Supplying the Mining World: the Mining Equipment Manufacturers of San Francisco 1850-1900* Western Lore Press, Tucson, AZ, 1996 p122.

Lord, Elliot *Comstock Mining and Miners* Howell-North Books, San Diego, CA, 1980 [Geological Survey, Government Printing Office, 1883] p224.

The number of mines examined by the author for this work that featured large first-motion double-drum hoists were few in number, and all were located in well-developed mining districts.

56. Colliery Engineer Company *Coal & Metal Miners' Pocketbook* Colliery Engineer Company, Scranton, PA, 1905 p394.

Eaton, Lucien *Practical Mine Development & Equipment* McGraw-Hill Book Company, New York, NY, 1934 p195.

International Textbook Company *International Library of Technology: Hoisting, Haulage, Mine Drainage* International Textbook Company, Scranton, PA, 1906 A52 p17.

Keystone Consolidated Publishing Company Inc. *The Mining Catalog: 1925 Metal-Quarry Edition* Keystone Consolidated Publishing Company Inc., (no location given), 1925 p451.

Lewis, Robert S. *Elements of Mining* John Wiley & Sons, Inc., New York, NY, 1946 p193.

Young, George *Elements of Mining* John Wiley & Sons, New York, NY, 1923 p175.

Young, George *Elements of Mining* John Wiley & Sons, New York, NY, 1946 p209.

57. Eaton, Lucien *Practical Mine Development & Equipment* McGraw-Hill Book Company, New York, NY, 1934 p193.

International Textbook Company *A Textbook on Metal Mining: Steam and Steam-Boilers, Steam Engines, Air and Air Compression, Hydromechanics and Pumping, Mine Haulage, Hoisting and Hoisting Appliances, Percussive and Rotary Boring* International Textbook Company, Scranton, PA, 1899 A23 p30.

International Textbook Company *International Library of Technology: Hoisting, Haulage, Mine Drainage* International Textbook Company, Scranton, PA, 1906 A50 p11.

Keystone Consolidated Publishing Company Inc. *The Mining Catalog: 1925 Metal-Quarry Edition* Keystone Consolidated Publishing Company Inc., (no location given), 1925 p445.

Lewis, Robert S. *Elements of Mining* John Wiley & Sons, Inc., New York, NY, 1946 p191.

Young, George *Elements of Mining* John Wiley & Sons, New York, NY, 1923 p174.

Young, George *Elements of Mining* John Wiley & Sons, New York, NY, 1946 p203.

58. Cole, T.F. "A Single Engine Hoisting Plant" *Engineering & Mining Journal* Sept. 5, 1896.

International Textbook Company *International Library of Technology: Mine Surveying, Metal Mine Surveying, Mineral-Land Surveying, Steam and Steam Boilers, Steam Engines, Air Compression* International Textbook Company, Scranton, PA, 1924 A23 p39.

Thompson, James R. "The Hoisting Problem" *Engineering & Mining Journal* Jan. 26, 1905 p173.

Very few mines that operated prior to 1890s, recorded in association with this study, featured more than one boiler. However, several medium-sized mines and many large mines that had been developed after around 1890 had extra boilers.

59. Barton, Denys Bradford *The Cornish Beam Engine* Wordens of Cornwall, Ltd. 1966 p115.

Collins, Jose H. *Principles of Metal Mining* William Collins, Sons & Company, London, 1875 p104.

Croft, Terrell *Steam Boilers* McGraw-Hill Book Co., New York, NY, 1921 p15.

60. International Textbook Company *A Textbook on Metal Mining: Steam and Steam-Boilers, Steam Engines, Air and Air Compression, Hydromechanics and Pumping, Mine Haulage, Hoisting and Hoisting Appliances, Percussive and Rotary Boring* International Textbook Company, Scranton, PA, 1899 A18 p11, 45.

International Textbook Company *International Library of Technology: Mine Surveying, Metal Mine Surveying, Mineral-Land Surveying, Steam and Steam Boilers, Steam Engines, Air Compression* International Textbook Company, Scranton, PA, 1924 A24 p2.

The M.J. O'Fallon Supply Company *Catalog No.8 Steam, Mill, & Water Supplies* R.R. Donnelly & Sons Co., 1923 p389.

61. Blake & Knowles Steam Pump Works *Knowles Boiler Feed & Tank Pumps: Single & Duplex* Blake & Knowles Steam Pump Works, New York, NY, [trade catalog] ca. 1910.

62. Rand Drill Company *Illustrated Catalog of the Rand Drill Company, New York, U.S.A.* Rand Drill Company, New York, NY, 1886 p43.

63. Hall, Leon M. "Modernizing the Comstock Lode" *Mining & Scientific Press* March 17, 1906 p183.

Holbrook, E.A. "Wood as a Fuel for Mine Boiler Plants" *Engineering & Mining Journal* April 10, 1915, p645-647.

Tinney, W.H. *Gold Mining Machinery: Its Selection, Arrangement, & Installation* D. Van Nostrand Company, New York, NY, 1906 p71.

Nearly all of the mines examined for this study in remote districts in the Great Basin exhibited evidence of having relied on cord wood fuel for boilers, or they lacked evidence indicative of burning coal. Only mines in well-developed districts exhibited evidence typical of firing boilers with coal.

64. International Textbook Company *International Library of Technology: Mine Surveying, Metal Mine Surveying, Mineral-Land Surveying, Steam and Steam Boilers, Steam Engines, Air Compression* International Textbook Company, Scranton, PA, 1924 A23 p53.

Peele, Robert *Mining Engineers' Handbook* John Wiley & Sons, New York, NY, 1918 p2086.

Tilden, E.G. "Burning Wood Under Boilers" *Engineering & Mining Journal* March 6, 1909, p499.

65. The numbers of bricks for boiler setting was extrapolated from tables given in:

Topeka Steam Boiler Works *Topeka Steam Boiler Works Catalog B Steam Boilers, Power Plant Equipment, Plumbing and Heating Supplies* R.R. Donnelly Sons Co. 1923.

66. Croft, Terrell *Steam Boilers* McGraw-Hill Book Co., New York, NY, 1921 p37.

Every boiler setting analyzed for this study that was either partially or completely intact exhibited evidence that the boiler shell had been supported by the two types of buckstaves described in the text. Further, many buckstaves were made of salvaged railroad rail or four-by-four timbers.

67. Tinney, W.H. *Gold Mining Machinery: Its Selection, Arrangement, & Installation* D. Van Nostrand Company, New York, NY, 1906 p61.

Many small and medium-sized mines in remote districts examined in conjunction with this work featured settings built with fieldstone masonry lined with firebrick and red brick. In many cases boiler settings at remote mines were mortared with sand, and boiler settings in well-developed districts tended to be assembled with cement mortar. The Banker Mine in Colorado's Collegiate Mountains featured a portland concrete setting for two return tube boilers erected during the 1890s.

68. Nearly all boiler settings inspected for this study featured half-facades, and only several of the largest mines possibly had boilers with full facades.

69. Ihlseng, Magnus *A Manual of Mining* John Wiley & Sons, New York, NY, 1892 p581.

International Textbook Company *International Library of Technology: Mine Surveying, Metal Mine Surveying, Mineral-Land Surveying, Steam and Steam Boilers, Steam Engines, Air Compression* International Textbook Company, Scranton, PA, 1924 A23 p53.

Keystone Consolidated Publishing Company Inc. *The Mining Catalog: 1925 Metal-Quarry Edition* Keystone Consolidated Publishing Company Inc., (no location given), 1925 p115.

Peele, Robert *Mining Engineers' Handbook* John Wiley & Sons, New York, NY, 1918 p2086.

Stein, Ernest "A Home-Made Boiler Feed Water Heater" *Engineering & Mining Journal* June 23, 1904, p1009.

70. Croft, Terrell *Steam Boilers* McGraw-Hill Book Co., New York, NY, 1921 p18, 53.

Greeley, Horace; Case, Leon; Howland, Edward; Gough, John B.; Ripley, Philip; Perkins, E.B.; Lyman, J.B.; Brisbane, Albert; Hall, E.E. "Babcock and Wilcox Boiler" *The Great Industries of the United States* J.B. Burr, Hartford, CT, 1872.

International Textbook Company *A Textbook on Metal Mining: Steam and Steam-Boilers, Steam Engines, Air and Air Compression, Hydromechanics and Pumping, Mine Haulage, Hoisting and Hoisting Appliances, Percussive and Rotary Boring* International Textbook Company, Scranton, PA, 1899 A18 p35.

Linstrom, C.B. & Clemens, A.B. *Steam Boilers and Equipment* International Textbook Co., Scranton, PA, 1928 p30.

Peele, Robert *Mining Engineers' Handbook* John Wiley & Sons, New York, NY, 1918 p2083.

Thurston, R.H. *A Manual of Steam Boilers: Their Design, Construction, and Operation* John Wiley & Sons, New York, NY, 1901 p34.

Tinney, W.H. *Gold Mining Machinery: Its Selection, Arrangement, & Installation* D. Van Nostrand Company, New York, NY, 1906 p63.

71. Several large, well-financed mines recorded for this study featured the remains of water tube boilers: the Locan Shaft in Nevada's Eureka Mining District, the Vindicator and Golden Cycle mines in Colorado's Cripple Creek District, and the Silver Lake Mine, Iowa Mine, and Unity Tunnel in Colorado's Silverton District. The boilers were erected after 1900.

72. Twitty, Eric "From Steam Engines to Electric Motors: Electrification in the Cripple Creek Mining District" *The Mining History Journal* Fifth Annual Journal, 1999.

73. Ihlseng, Magnus *A Manual of Mining* John Wiley & Sons, New York, NY, 1892 p80.

74. International Textbook Company *A Textbook on Metal Mining: Steam and Steam-Boilers, Steam Engines, Air and Air Compression, Hydromechanics and Pumping, Mine Haulage, Hoisting and Hoisting Appliances, Percussive and Rotary Boring* International Textbook Company, Scranton, PA, 1899 A23 p5.

Peele, Robert *Mining Engineers' Handbook* John Wiley & Sons, New York, NY, 1918 p1126.

Russell, H.A. "Electric Hoists for Mine Service" *Mining & Scientific Press* Feb. 8, 1913 p236-240.

Twitty, Eric "From Steam Engines to Electric Motors: Electrification in the Cripple Creek Mining District" *The Mining History Journal* Fifth Annual Journal, 1999.

75. Twitty, Eric "From Steam Engines to Electric Motors: Electrification in the Cripple Creek Mining District" *The Mining History Journal* Fifth Annual Journal, 1999.

76. Hoover, Herbert C. *Principles of Mining: Valuation, Organization, and Administration* McGraw-Hill Book Company, Inc., New York, NY, 1909. P130.

Rickard, T.A. "Electricity or Steam For Hoisting" *Engineering & Mining Journal* May 2, 1904 p881.

Walker, S.F. "Electric Hoisting in Mining Operations" *Engineering & Mining Journal* Nov. 19, 1910 p1014-1016.

77. "Electric Mine Hoists" *Mining & Scientific Press* Oct. 13, 1900 p433.

International Textbook Company *International Library of Technology: Hoisting, Haulage, Mine Drainage* International Textbook Company, Scranton, PA, 1906 A50 p36.

Keystone Consolidated Publishing Company Inc. *The Mining Catalog: 1925 Metal-Quarry Edition* Keystone Consolidated Publishing Company Inc., (no location given), 1925 p466, 473, 476.

Mine & Smelter Supply Co. *Catalog No. 92 Supplies & Equipment* R.R. Donnelly & Sons Co., 1937 p552-554.

Walker, S.F. "Electric Hoisting in Mining Operations" *Engineering & Mining Journal* Nov. 19, 1910 p1014-1016.

78. Bushnell, Andrew F. "Electrically Driven Air Compressors" *Engineering and Mining Journal* Nov. 2, 1907 p823.

79. Twitty, Eric "From Steam Engines to Electric Motors: Electrification in the Cripple Creek Mining District" *The Mining History Journal* Fifth Annual Journal, 1999.

80. Articles and advertisements announcing the arrival of practical single-drum electric hoists and other mine machines began appearing in substantial numbers in mining industry trade journals shortly after 1900. Articles and advertisements for double-drum hoists, however, began appearing during the mid-1910s. Analysis of historic mine sites in association with this study reflects the trend noted above.

81. "Electric Hoist at Calumet & Arizona" *Engineering & Mining Journal* Jan. 20, 1912 p162.

Hamilton, J.W.H. "Electric Mining Hoists" *Engineering & Mining Journal* Sept. 22, 1906 p537-540.

Lewis, Robert S. *Elements of Mining* John Wiley & Sons, Inc., New York, NY, 1946 p188.

Rushmore, D.B. "Electric Hoists for Mines" *Engineering & Mining Journal* Dec. 16, 1916 p1177.

Staley, William *Mine Plant Design* McGraw-Hill Book Co., New York, NY, 1936 p141.

Walker, S.F. "Electric Hoisting in Mining Operations" *Engineering & Mining Journal* Nov. 19, 1910 p1014-1016.

82. Botsford, H.L. "Types of Headframes" *Engineering & Mining Journal* Oct. 13, 1913 p690.

Burr, Floyd "Design of Headframes" *Engineering & Mining Journal* April 7, 1917 p611.

Fitch, Charles H. "Head Works Framing - Primary Notions" *Mining & Scientific Press* June 6, 1903 p335.

Fitch, Charles H. "Head Works Framing" *Mining & Scientific Press* June 13, 1903 p380.

Forsyth, Alexander "The Headframes of Shafts at Cripple Creek" *Engineering & Mining Journal* March 7, 1903 p366.

Ihlseng, Magnus *A Manual of Mining* John Wiley & Sons, New York, NY, 1892 p91.

International Textbook Company *A Textbook on Metal Mining: Steam and Steam-Boilers, Steam Engines, Air and Air Compression, Hydromechanics and Pumping, Mine Haulage, Hoisting and Hoisting Appliances, Percussive and Rotary Boring* International Textbook Company, Scranton, PA, 1899 A23 p105.

International Textbook Company *International Library of Technology: Hoisting, Haulage, Mine Drainage* International Textbook Company, Scranton, PA, 1906 A53 p31.

Ketchum, Milo S. C.E. *The Design of Mine Structures* McGraw-Hill Book Co., New York, NY, 1912 p41.

Peele, Robert *Mining Engineers' Handbook* John Wiley & Sons, New York, NY, 1918 p926.

"Something More of Headframes" *Mining & Scientific Press* Dec. 30, 1905.

"Variations in [headframe] Construction" *Mining & Scientific Press* Aug. 22, 1903.

83. Forsyth, Alexander "The Headframes of Shafts at Cripple Creek" *Engineering & Mining Journal* March 7, 1903 p366.

Ihlseng, Magnus *A Manual of Mining* John Wiley & Sons, New York, NY, 1892 p88.

Ketchum, Milo S. C.E. *The Design of Mine Structures* McGraw-Hill Book Co., New York, NY, 1912 p41.

Peele, Robert *Mining Engineers' Handbook* John Wiley & Sons, New York, NY, 1918 p926.

Staley, William *Mine Plant Design* McGraw-Hill Book Co., New York, NY, 1936 p73.

Young, George *Elements of Mining* John Wiley & Sons, New York, NY, 1923 192.

The Author determined the multiple landings associated with the Joe Dandy, Nichol, and Vindicator mines from field inspection.

84. Botsford, H.L. "Types of Headframes" *Engineering & Mining Journal* Oct. 13, 1913 p690.

Denny, G.A. "A-Type Timber Headframe" *Engineering & Mining Journal* March 30, 1914 p1100.

Forsyth, Alexander "The Headframes of Shafts at Cripple Creek" *Engineering & Mining Journal* March 7, 1903 p366.

Peele, Robert *Mining Engineers' Handbook* John Wiley & Sons, New York, NY, 1918 p926.

85. International Textbook Company *International Library of Technology: Hoisting, Haulage, Mine Drainage* International Textbook Company, Scranton, PA, 1906 A53 p35.

Ketchum, Milo S. C.E. *The Design of Mine Structures* McGraw-Hill Book Co., New York, NY, 1912 p7.

Peele, Robert *Mining Engineers' Handbook* John Wiley & Sons, New York, NY, 1918 p935.

Nearly all of the headframes examined by the author for this study were constructed with timbers of the dimensions discussed.

86. "Concentrates [headframe notes]" *Mining & Scientific Press* Oct.1, 1904 p220.

Mentzel, Charles "Prospecting Headframe" *Engineering & Mining Journal* Oct. 5, 1912.

"Simple Head Frame Construction" *Mining & Scientific Press* Feb. 6, 1904 p94.

The author analyzed the last two types of headframe foundations exposed in areas of subsidence around collapsed shaft collars, and in waste rock dumps partially removed.

87. Botsford, H.L. "Types of Headframes" *Engineering & Mining Journal* Oct. 13, 1913 p690.

Colliery Engineer Company *Coal & Metal Miners' Pocketbook* Colliery Engineer Company, Scranton, PA, 1905 [1893] p120.

88. International Textbook Company *International Library of Technology: Hoisting, Haulage, Mine Drainage* International Textbook Company, Scranton, PA, 1906 A53 p35.

89. Burr, Floyd "Design of Headframes" *Engineering & Mining Journal* April 7, 1917 p611.

Denny, G.A. "A-Type Timber Headframe" *Engineering & Mining Journal* March 30, 1914 p1100.

Forsyth, Alexander "The Headframes of Shafts at Cripple Creek" *Engineering & Mining Journal* March 7, 1903 p366.

International Textbook Company *International Library of Technology: Hoisting, Haulage, Mine Drainage* International Textbook Company, Scranton, PA, 1906 A53 p35.

Ketchum, Milo S. C.E. *The Design of Mine Structures* McGraw-Hill Book Co., New York, NY, 1912 p7, 376.

Peele, Robert *Mining Engineers' Handbook* John Wiley & Sons, New York, NY, 1918 p935.

"Some Types of Headframes" *Mining & Scientific Press* Dec. 16, 1905 p410.

"Steel Headframe at the Gwin Mine [California]" *Mining & Scientific Press* Jan. 2, 1904 p5.

"Steel vs. Timber Headframes" *Mining & Scientific Press* July 11, 1903 p18.

Young, George *Elements of Mining* John Wiley & Sons, New York, NY, 1923 192.

90. Binckley, George S. "The Rational Design of Headframes for Mines" *Mining & Scientific Press* June 10, 1905 p374.

Forsyth, Alexander "The Headframes of Shafts at Cripple Creek" *Engineering & Mining Journal* March 7, 1903 p366.

Halloran, A.H. "The Leonard Headframe" *Mining & Scientific Press* June 30, 1906 p432.

"Steel Headframe of the Parrot Mine, Butte, Montana" *Engineering & Mining Journal* June 21, 1902 p862.

"Steel Headgear at Cripple Creek, Colo." *Mining & Scientific Press* Dec. 8, 1900 p573.

91. The foundations for steam hoists were characterized from field analyses of numerous historic mines in California, Colorado, Montana, Nevada, New Mexico, and Utah.

92. Stromberg-Carlson Telephone Manufacturing Co. *Stromberg-Carlson: Telephones, Switchboards, Radio Apparatus, Supplies* Stromberg-Carlson Telephone Mfg Co., Rochester, NY, 1925 p179.

The characteristic foundations and artifacts for electric hoists were determined from analysis of field data collected by the author in Colorado, Nevada, and Utah.

93. Descriptions of the remains of boiler settings were synthesized from the author's field work at historic mine sites in Colorado, Nevada, New Mexico, and Utah.

94. Glover, Margaret L. *Log Structures: Criteria for Their Description, Evaluation, and Management as Cultural Resources* Thesis Manuscript, Department of Anthropology, Portland State University, Portland, OR 1982 p25.

Ketchum, Milo S. C.E. *The Design of Mine Structures* McGraw-Hill Book Co., New York, NY, 1912 p359.

The author observed industrial log mine buildings in some of Colorado's remote mining districts.

95. "Iron and Steel Buildings for Mining Plants" *Engineering & Mining Journal* March 13, 1897 p256.

Most of the mine buildings recorded and observed by the author that pre-dated the 1890s were sided with wood, while the structures post-dating around 1900 were sided with corrugated sheet iron.

96. "Iron and Steel Buildings for Mining Plants" *Engineering & Mining Journal* March 13, 1897 p256.

Ketchum, Milo S. C.E. *The Design of Mine Structures* McGraw-Hill Book Co., New York, NY, 1912 p236, 383.

Roth, Leland M. *A Concise History of American Architecture* Harper & Row Publishers, New York, NY, 1979 p173.

97. "Cripple Creek Mines in 1899" [Hull City] *Engineering & Mining Journal* Jan. 30, 1900 p50.

"Cripple Creek in 1908" [Vindicator] *Mining & Scientific Press* Jan. 2, 1909 p41.

Division of Mines & Minerals Colorado State Archives, Denver, CO

Mine Inspectors' Reports: Teller County File # 48440 Gold Coin Mine

Mine Inspectors' Reports: Teller County File # 48441 Hull City Mine

Mine Inspectors' Reports: Teller County File # 48516 and 48441 Vindicator Mine

Forsyth, Alexander "The Headframes of Shafts at Cripple Creek" *Engineering & Mining Journal* March 7, 1903 p366.

98. Peele, Robert *Mining Engineers' Handbook* John Wiley & Sons, New York, NY, 1918 p2095.

The author characterized shaft houses from the examination of numerous historic photographs, and from the analysis of the remains at historic mine sites in Colorado, Nevada, and Montana.

99. "General Mining News: Colorado, El Paso County" *Engineering and Mining Journal* April 1898.

100. The author characterized the remains of mining buildings from an analysis of historic mine sites in Colorado, Montana, Nevada, New Mexico, and Utah.

101. E.I. DuPont de Nemours & Company *DuPont Blasting Powder* E.I. DuPont de Nemours & Co., Wilmington, DE 1917 p15.

Institute of Makers of Explosives *Safety in the Handling & Use of Explosives* Institute of Makers of Explosives, Washington DC 1934 p11.

Peele, Robert *Mining Engineers' Handbook* John Wiley & Sons, New York, NY, 1918 p155.

Young, George *Elements of Mining* John Wiley & Sons, New York, NY, 1923 p107.

102. Crampton, Frank A. *Deep Enough: A Working Stiff in the Western Mine Camps* University of Oklahoma Press, [1956] 1982 p30.

103. Schreier, Nancy B. *Highgrade: The Story of National, Nevada* Authur H. Clark Co., Glendale, CA 1981 p81.

Wyman, Mark *Hard Rock Epic: Western Mining and the Industrial Revolution, 1860-1910* University of California Press, Berkeley, CA 1989 [1979] p81.

Young, Otis E. *Western Mining* University of Oklahoma Press, 1987 [1970] p223.

Chapter 5: In the Shadow of the Fortune Seekers: Mining During the Great Depression

1. Abbott, Carl; Leonard, Stephen; McComb, David *Colorado: A History of the Centennial State* University Press of Colorado, Niwot, CO, [1982] 1994 p276.

Elliot, Russell R. *Nevada's Twentieth Century Mining Boom: Tonopah, Goldfield, Ely* University of Nevada Press, Reno, NV, 1988 [1966] p158-164.

Smith, Duane *Rocky Mountain West: Colorado, Wyoming, & Montana 1859-1915* University of New Mexico Press, Albuquerque, NM, 1992 p226.

Watkins, T.H. *Gold and Silver in the West: the Illustrated History of an American Dream* Bonanza Books, New York, NY, 1971 p161.

2. McElvaine, Robert *The Great Depression: America, 1929-1941* Times Books, New York, NY, [1984] 1993 p164.

3. Altenbach, J. Scott, Mining Historian Personal Interview: Albuquerque, New Mexico Sept. 27, 1998. Altenbach interviewed several miners who worked in Leadville, Colorado during the Depression, and they discussed the regular use of hand-steels for drilling and blasting low-grade ore in abandoned stopes.

Sporr, Ray, Mining Engineer & Gold Miner 1930s - 1950s Personal Interview July 20, 1993, Delta UT 1993. Sporr recounted that he and other miners drilled with hand-steels instead of machine drills during the Great Depression.

The author observed hand-steels and hand-drilled blast-holes at mines that had been worked and subsequently abandoned during the Great Depression in Colorado, Nevada, and California.

4. Eaton, Lucien *Practical Mine Development & Equipment* McGraw-Hill Book Company, New York, NY, 1934 p129.

Tillson, Benjamin Franklin *Mine Plant* American Institute of Mining and Metallurgical Engineers, New York, NY, 1938 p17-19.

5. The popular types of Depression-era ventilation fans were determined from field analyses of historic mine sites in Colorado, Nevada, New Mexico, and Utah.

6. Ingersoll-Rand Drill Company *Imperial Type 11 Air Compressor: Instructions for Installing and Operating* Rand Drill Co., New York, NY, 1915.

Rehfuss, Louis A. and Rehfuss, W. Clifford "Portable Mining Equipment for Prospectors" *Engineering & Mining Journal* June 10, 1916 p1025.

The author observed and recorded upright two-cylinder compressors mounted onto both factory-made trailers and stationary foundations.

7. "Economical Air Compressor Drive" *Engineering & Mining Journal* April 13, 1918 p685.

Nelson, S.T. "Development of Reciprocating Air Compressors" *Engineering and Mining Journal* Sept. 27, 1919 p533-536.

"Two-Stage Power-Driven Angle Compressor" *Engineering & Mining Journal* March 28, 1914 p667.

A review of mining trade journals and machinery catalogs determined that angle-compound compressors were introduced to the market during the 1910s and they experienced slight popularity during the 1930s.

8. Most of the Depression-era mines inspected by the author featured either belt-driven duplex or small straight-line air compressors.

9. Peele, Robert *Mining Engineers' Handbook* John Wiley & Sons, New York, NY, 1941 [1918] A15 p16.

The author analyzed both intact V-cylinder compressors and foundations at mines in Colorado.

10. The author characterized the wide variety of foundations constructed for Depression-era machinery at historic mine sites in Colorado, Montana, Nevada, and Utah.

11. Eaton, Lucien *Practical Mine Development & Equipment* McGraw-Hill Book Company, New York, NY, 1934 p86, 295.

Lewis, Robert S. *Elements of Mining* John Wiley & Sons, Inc., New York, NY, 1946 [1933] p187.

Staley, William *Mine Plant Design* McGraw-Hill Book Co., New York, NY, 1936 p137.

Young, George *Elements of Mining* John Wiley & Sons, New York, NY, 1946 p203.

Zurn, E.N. *Coal Miners' Pocketbook* McGraw-Hill Book Co., New York, NY, [1890 Colliery Engineering Co.] 1928 p760.

In keeping with the above references, most of the large mining operations active during the 1930s analyzed by the author used electric hoists and electric compressors, instead of machinery run by other sources of power.

12. Lewis, Robert S. *Elements of Mining* John Wiley & Sons, Inc., New York, NY, 1946 [1933] p187.

Staley, William *Mine Plant Design* McGraw-Hill Book Co., New York, NY, 1936 p139, 141.

Young, George *Elements of Mining* John Wiley & Sons, New York, NY, 1946 p205.

13. The performance of electric hoists has been synthesized from data in mining trade catalogs.

14. The Cripple Creek Mining District contains numerous mines in which small mining companies retrofitted extant steam hoists with electric motors.

15. The author examined several mines which used compressed air to power hoists in Colorado, Montana, and Nevada.

16. The author analyzed hoisting systems that fall into the general category described in the text in mining districts in Colorado, California, and Nevada.

17. Cripple Creek abounds with examples in which mining companies retrofitted steam hoists with motors.

Sara, Tim; Twitty, Eric; Grant, Marcus *Mining and Settlement at Timberline* Paragon Archaeological Consultants, Inc., Denver, CO, 1998 8-126.

18. Most headframes erected during the 1930s examined by the author were constructed of timbers either butted together or set in shallow square-notch joints. The most common styles for well-built headframes were four and six post derricks for vertical shafts, and squat A-frames for inclined shafts.

Sara, Tim; Twitty, Eric; Grant, Marcus *Mining and Settlement at Timberline* Paragon Archaeological Consultants, Inc., Denver, CO, 1998 8-126.

This reference provides examples of reused headframes in the Cripple Creek District.

19. Hoover, Herbert C. *Principles of Mining: Valuation, Organization, and Administration* McGraw-Hill Book Company, Inc., New York, NY, 1909 p123.

Ketchum, Milo S. C.E. *The Design of Mine Structures* McGraw-Hill Book Co., New York, NY, 1912 p34.

Lewis, Robert S. *Elements of Mining* John Wiley & Sons, Inc., New York, NY, 1946 [1933] p205.

Young, George *Elements of Mining* John Wiley & Sons, New York, NY, 1946 p171.

20. Summarization of the buildings erected by well-capitalized Depression-era mining companies was synthesized from an analysis of historic mine sites in California, Colorado, Montana, Nevada, and Utah, as well as from historic photographs.

Chapter 6: Riches to Rust:
Interpreting the Remains of Historic Mines

1. Firebaugh, Gail S. "An Archaeologists Guide to the Historical Evolution of Glass Bottle Technology" *Southwestern Lore* Sept. 1989, p49.

Jones, Olive R. and Sullivan, Catherine *The Parks Canada Glass Glossary* National Historic Parks and Sites Branch, Parks Canada, Ottawa, 1985 p13.

Rock, James T. "Basic Bottle Identification" Unpublished Manuscript Yreka, CA, 1990 p25.

The author documented aqua and amethyst bottle and window glass at historic mines that closed prior to 1920. Mines that operated after 1920 featured selenium glass.

2. Busch, Jane "An Introduction to the Tin Can" *Historical Archaeology* Vol. 15 No.1 1981.

Rock, James T. *Tin Canisters: Their Identification* Jim Rock, Yreka, CA, 1989 p50.

3. Busch, Jane "An Introduction to the Tin Can" *Historical Archaeology* Vol. 15 No.1 1981.

Rock, James T. *Tin Canisters: Their Identification* Jim Rock, Yreka, CA, 1989 p60, 65.

4. Wyman, Mark *Hard Rock Epic: Western Mining and the Industrial Revolution, 1860-1910* University of California Press, Berkeley, CA 1989 [1979] p127.

Young, Otis E. *Western Mining* University of Oklahoma Press, 1987 [1970] p221.

Underground mine workings of company-operated mines accessible today exhibit characteristics that are distinct from the mines run by groups of lessees. The workings of company-run mines tend to be clean, well-timbered, well-structured, and debris was disposed of in an orderly manner. The workings of leased mines are poorly structured and exhibit signs of an excessive use of explosives, the timbering is substandard, and trash poorly disposed of.

5. "General Mining News: Colorado, El Paso County" *Engineering and Mining Journal* May 30, 1896.

INDEX